D1247023

Project Management in New Product Development

ABOUT THE AUTHOR

Bruce T. Barkley is the author of *Integrated Project Management,* and *Project Risk Management*, and is co-author with James Saylor of *Customer-Driven Project Management: Building Quality into Project Processes* (all from McGraw-Hill). *Customer-Driven Project Management* has been translated into Chinese, and is on the Project Management Institute Best Seller List.

Mr. Barkley is a senior faculty member and project management curriculum manager with DeVry University, Keller Graduate School of Management in Atlanta. He teaches management courses and chairs the Keller Project Management Faculty Forum in the Atlanta metropolitan region. DeVry/Keller is one of the largest producers of quality graduate project management (MBA and MPM) degrees in the world in a unique online and onsite learning format. Mr. Barkley has managed the Project Management Office (PMO) with Universal Avionics, Inc., Atlanta Office, and served as Vice President of The Learning Group Corporation of Rockville, MD, a project management consulting company. Mr. Barkley was a member of the Senior Executive Service in the federal government in Washington, DC, and served four cabinet secretaries—Transportation, Environment (EPA), Office of Management and Budget, and Heath and Welfare—in a variety of career management positions.

He has a bachelor's degree from Wittenberg University and master's degrees from the University of Cincinnati and The University of Southern California. He has designed and delivered a wide range of online and classroom project management courses for DeVry/Keller, for the University College, University of Maryland, and for various business enterprises. He was awarded the Excellence in Teaching award by the University of Maryland.

Project Management in New Product Development

Bruce T. Barkley, Sr.

New York Chicago San Francisco Lisbon London Madrid
Mexico City Milan New Delhi San Juan Seoul
Singapore Sydney Toronto

The McGraw·Hill Companies

Library of Congress Cataloging-in-Publication Data

Barkley, Bruce, date.
Project management in new product development / Bruce T. Barkley.
p. cm.
Includes bibliographical references.
ISBN 978-0-07-149672-8 (alk. paper)
1. New products. 2. Product management. I. Title.
TS170.B37 2007
658.5'75—dc22

2007021840

McGraw-Hill books are available at special quantity discounts to use as premiums and sales promotions, or for use in corporate training programs. For more information, please write to the Director of Special Sales, McGraw-Hill Professional, Two Penn Plaza, New York, NY 10121-2298. Or contact your local bookstore.

Copyright ©2008 by The McGraw-Hill Companies, Inc. All rights reserved. Printed in the United States of America. Except as permitted under the United States Copyright Act of 1976, no part of this publication may be reproduced or distributed in any form or by any means, or stored in a data base or retrieval system, without the prior written permission of the publisher.

1 2 3 4 5 6 7 8 9 0 DOC/DOC 0 1 2 1 0 9 8 7

ISBN 978-0-07-149672-8
MHID 0-07-149672-6

This book is printed on acid-free paper.

Sponsoring Editor
Larry S. Hager

Acquisitions Coordinator
Alexis Richard

Editorial Supervisor
David E. Fogarty

Project Manager
Rasika Mathur

Copy Editor
Susan Hobbs

Proofreader
Elise Oranges

Indexer
Steve Ingle

Production Supervisor
Richard C. Ruzycka

Composition
International Typesetting and Composition

Art Director, Cover
Jeff Weeks

Information contained in this work has been obtained by The McGraw-Hill Companies, Inc. ("McGraw-Hill") from sources believed to be reliable. However, neither McGraw-Hill nor its authors guarantee the accuracy or completeness of any information published herein, and neither McGraw-Hill nor its authors shall be responsible for any errors, omissions, or damages arising out of use of this information. This work is published with the understanding that McGraw-Hill and its authors are supplying information but are not attempting to render engineering or other professional services. If such services are required, the assistance of an appropriate professional should be sought.

To the thousands of hard working, adult graduate and undergraduate students at DeVry University/ Keller Graduate School of Management, Atlanta, and at The University College, University of Maryland, who have provided me over the past 35 years with wonderful opportunities to learn from them—undoubtedly more than they learned from me. I am inspired by their dedication to improving themselves through part-time college and graduate work despite the challenges of everyday living.

Contents

Acknowledgments

The author would like to acknowledge the following sources for this book:

- The Universal Avionics Systems Corporation, Instrument Division, for valuable experience in supporting and managing new, integrated product development projects and processes, and writing program manuals and policy documents and conducting analyses in the program management office.
- The Alumax Aluminum Company (now Alcoa, Inc.), where the author was a project management and organizational development consultant, for valuable experience and case material in integrated strategic and portfolio planning and SWOT analysis in a manufacturing work setting.
- Students and faculty at DeVry University and Keller Graduate School of Management, Atlanta, where the author serves as a senior faculty member and curriculum manager for project management, for valuable stories, cases, and exercises in integrated project and new product risk and cost management, which serve as the basis for material in the book. Special thanks to three excellent MBA students who contributed to this book: Maria Thompson, Eliska Johnson, and Francina Price.
- Marketing consultant Sue Harris, who has had extensive experience with Sonoco Products Co. in the marketing process introducing a new product to the world of retail business, and who freely shared her insights with the author.
- NPLearning and Ken Westray, its energetic and innovative leader. NPLearning is a progressive new product development firm, that has pushed the envelope in this field for years.
- The Project Management Institute, Project Management Body of Knowledge (PMBOK), 2004.

Introduction

This book is about managing new product development using project management concepts and tools. The author has been guided by seven key principles in addressing how to manage new product development projects. These principles are addressed here briefly and then wrapped up again in Chapter 10. These principles are as follows:

Principle #1. Develop project management and new product development processes, and integrate the two. The theme of the book is that new product development is a largely technical and developmental process that must be carefully managed to control time, resources, and quality. The fundamental contribution of project management tools and techniques is to enable project managers to make the business case for a new product. Both the project management process and the new product development process must be defined in order for this integration to work.

Principle #2. Open the company to new ideas and new partners. The organization has to be open to new concepts and ideas, and sometimes has to take a proactive approach to find new ideas and partners on a global scale. New product development is no longer an "internal" process.

Principle #3. Define measures for choosing new product projects. The development of a new product portfolio requires analysis and evaluation using at least three measures: financial performance for the business, alignment with business planning and strategy, and identification of risks and contingencies.

Principle #4. Create a way through project reviews to stop bad products. New products have a way of moving through the system despite overwhelming evidence of potential failure in the marketplace. One of the key project concepts we explore is the *project review*, which enables management to make go or no-go decisions at each product development phase.

Principle #5. Choose project managers who understand technology. Project managers must be trained and developed to manage time, resources, and quality, but they also must have a working knowledge of the technology inherent in any new product development. Project management is more than an administrative function; it implies technical judgments as well.

Principle #6. Build cross-functional teamwork and accountability. The new product team is important. If the team is narrowly structured to reflect only design and development issues, and not production, marketing, and sales concerns, the team is likely to produce prototypes that work, but that cannot be produced and marketed at a profit.

Principle #7. *Ad hoc* it when necessary to succeed. Sometimes you simply dispose of process and procedure and jump into the market with a new product because you have to. Success is often achieved in the field through trial and error and pure determination.

The essence of these principles can be wrapped up in single sentence:

> Create a loose-tight process that loosens up the company and opens it to new ideas and concepts, but once you decide to go with a new product and fund and develop it, manage and control the process using disciplined process definitions and proven project management tools.

The reader will be introduced in this book reflecting a series of processes and phases, some of them technical and others simply common sense, for managing new product development. Some of these processes, may not provide the kind of detail an engineer might be looking for. The author preferred to stay on the high ground in these discussions and let the reader fill in the gaps with technical details tailored to specific organizations. For instance, we do not go into detail on various configuration management packages available to preserve the structure and components of new products in development, depending on the reader to pursue that level of inquiry.

In some cases we provide perhaps more detail and structure than the typical reader will want. For instance, we provide many illustrations of a key project planning tool, the Gantt chart, with detailed tasks, linkages, and assigned resources. Some readers may be bored by this focus on administrative tools, but our purpose here is to offset the common bias against this kind of tool among new product and marketing professionals.

The author is a student of organizational behavior and leadership, as many of our readers will be. We mention this because we try to place new product development and project management in the context of a company culture. Culture sets the boundaries for management and employee behavior, sometimes unconsciously, and thus deserves some attention. What Pepsi Cola does with a given project in its new product development process may not align with what Coca Cola would do with the same project. Thus "one size fits all" does not work very well when applying standard models of work to new product development. The reader is advised to grasp the conceptual material here and to then work to integrate the concepts into the target organization, whether it be a given company or a case study for training or education purposes.

The reader will also see a bias in the book toward application of management and administrative tools that might be inconsistent with the thinking and instincts of marketing and sales professionals. We see in marketing and sales *circles* an increasing tendency to depend on product branding and pricing to sell products in the marketplace. With this focus comes an underestimation of the design and development challenges in getting today's new products to market tomorrow. This is why we spend some time on cross-functional teams that reflect both the early product development and later marketing phases in their composition. The truth lies somewhere in the middle of this wide-ranging spectrum of activity, but the smart program manager accountable for the *whole* process will build those interests into the team early in the process.

The reader will also see in this book considerable ambivalence in applying some of the newer project management tools that focus on the big picture and leave the details to team members. For instance, the so-called critical chain concept aims for single bottlenecks and advocates the phasing of projects into the pipeline one at a time instead of multi-tasking the workforce. The results are not in yet on whether critical chain theory really works.

Another debate in project management circles has to do with the strategic value of projects, for example, project portfolio management. New thought about the strategic application of projects focuses more on alignment of projects and project selection. This focus on project selection means that projects need to be planned early and fleshed out in order to decide whether to undertake them in the first place.

The author sees both sides of these kinds of healthy indicators of change in the project management field and tries to portray those sides in the discussions when possible.

This book is different in many ways from past treatments of new product development and project management. This is a *managerial* view of the new product development process, the perspective of company management in the new product business. The view is driven by the need to grow the business, control resources and time to market, monitoring how the process is going against business strategy, and when to proceed and when to "pull the plug" on a bad product. The book takes an integrative perspective on the process, not bound by narrow views of consumer products or services or traditional marketing concepts that sell whatever product is produced. Emphasis is given to what can and does go wrong with many new product development and marketing initiatives because *things weren't planned very well.*

While the book provides a reasonably detailed discussion of new product development from a technical view, its purpose is to put that process in a management framework, to embody technical process in managerial context. New *products* are not *projects* until they evaluated, planned, scoped-out, scheduled, budgeted, managed, and monitored by managers. It is not technical process failure that inhibits the successful introduction of new products into the economy—it is typically managerial and marketing failure.

A short journey through the history of these two fields—product development and project management—may be helpful here.

What is new product development? There is a presumption in many quarters that new product development refers essentially to consumer products—not system or process products. That is, the concept of *product* is confined in this *marketing* sense to products for consumption, and marketing is the process of ensuring that customers are attracted to and buy the product. However, the other view is that new products can be system or process products as well, that the development of new ideas and concepts for new products as a part of a system, e.g., an electronic instrument for a business aircraft that will enhance pilot performance, or a new service concept that provides a new broadband platform for public safety communication, are new "products/processes/services."

So it turns out that the field of new product development has been driven largely by *marketing and market launch* views of the world, and more recently by a focus on the new product stage-gate process articulated by Robert Cooper. In the current literature, the process is seen as a logical sequence of generating new products, making the business case, testing and prototyping, marketing, "launching," and distributing the new product. New product activity is often viewed as "separate" from the rest of the company's product design and production processes, somehow placed in a distinct category and treated as such. This reflects the propensity of new product developers to think of themselves as "non-operational" and non-routine in their perspective. The process is rarely seen from a managerial view, e.g., what does the process cost, how long does it take, who is doing the work and how well are they doing, when should we stop it, and will it help the business grow.

On the other hand, the field of project management has been driven by narrow views of a project as a schedule and a budget. More recently we see a focus on portfolio management and how projects are selected and critical chain management focusing on resource bottlenecks and shortening task durations. Project management has been articulated traditionally as a set of planning and controlling tools to get products done on time and on budget. Only recently has the field begun to look at the broader perspective on projects, e.g., that projects are rarely managed as distinct pieces of work, that projects are all creative and innovative because of the changes that take place during their life cycle, that it doesn't matter that a project is on time and within budget if it does not hold promise to grow the business, and that all product development efforts must be *customer-driven*. A relatively new concept, critical chain theory, now has nudged the field to look at its own propensity for micro managing schedules, and changes the focus to bottlenecks, slimmed down task estimates, and monitoring the big picture. But still, project management and administrative concerns are seen by many professional engineers and marketing people in new product development as not worth their time.

I see these historic and narrow conceptual boundaries in these two fields as inhibitors of imagination and understanding, as evidences of Thomas Kuhn's (*The Structure of Scientific Revolution*) paradigms that restrict a full view of what is really happening out there and what should happen out there. Paradigmatic change occurs only when someone can overcome the blinders of constrained thought processes to question conventional wisdom and to get at reality for those in a real work setting and who muddle through the messy world of the competitive global marketplace.

The fact is that most management thought suffers from the lack of a "full and realistic screen." My experience is that success in business is a function of a wide variety of factors unrelated to project management, some driven by outside economic and social factors and forces which create the conditions for business growth. This is not to say that leaders and managers cannot steer companies in the right direction, given their reading of the outside world and their own organization's capacity, or that the generation of new ideas and new products cannot influence the market and create business growth and profitability if they are properly nurtured, managed and delivered. But there is no substitute for planning. We often see that our luck seems to get better with good project planning.

This practical view guides my treatment of this important subject and helps me view new product development with new eyes, with more focus on what actually happens when business succeeds or fails in this process of innovation and performance, and how managers actually *manage* under the daily stress of the new product work setting in the real world—where many factors are actually *unmanageable*.

I am driven by the need to see things as they are instead of how they should be. Then when I see those things clearly, I begin to see how they could be. It is the backward mapping process that starts with what happened and then backs up to how it happened and how it might have happened differently or "better."

To illustrate my point, let's take my friend the engineer. We'll call him Bill Close. Bill is an electrical engineer in an electronic instrumentation company producing avionics products. Bill is a functional manager, not a project manager; he manages electrical engineers who serve on projects to produce new avionics instruments for mid-sized aircraft. He could very well be an engineer for a toy manufacturer or a home improvement tool producer.

Bill works in an environment in which the production of new equipment is driven by factors largely out of his control. Ideas for new products come from management, customers, company ownership, competitors, and suppliers. New ways of meeting future customer needs coagulate on the fringe, during the production, distribution, and consumption of conventional products, not from starry-eyed engineers who see visions of new products in their sleep. New innovations are embedded in the current process of a company, not in some collection of autonomous, and off-center, "integrated ideation and product teams." Without grounding, new products do not survive.

Bill's people are managed to a certain extent in a typical new product environment by project managers who share responsibility with him for the performance of the product. He can be very skeptical of a new product process run by a project manager whose major focus is on time and budget and not on the performance and quality of the product. And marketing people continue to press for getting the product out before it is ready. To Bill, these company forces threaten good engineering.

Further, it is my observation from years of watching and working with companies and public agencies, that new products and services induce rather than follow customer need. "Needs" are not discovered; they are created and induce economic demand in the marketplace because they create value as incremental improvements in the current material and service world. Therefore, managing new products and services is most effective when positioned on the *edge* of the market, on the incremental periphery of the way people and companies live and breath each day. Evolutionary change prevails in the process of successful new product development because fundamental change does not occur in major *breakthrough* steps as much as in controlled evolution of ideas and collaboration. We saw in Universal Avionics, Inc. that aviation instrumentation and systems improvements were not invented over night, but came from many years of trial and error, incrementally. And leaders do not generate change in their companies over night simply because of their charm and charismatic disposition. This uniquely American view of invention and innovation suffers from the general naiveté of our Hollywood view of how "great" things happen and how leaders lead. It is typically professional people working each day together who create market and business value in new products to serve customers.

My point with my friend Bill is that he is not a project manager and does not belong to any project team, but he is vital to the success of any product development effort in his company because his responsibility is technical and engineering process, quality, professional development of his staff, and keeping his eyes on technical quality. He casts an uneasy eye on managerial decisions that short cut good engineering.

In today's global economy, businesses are increasingly challenged to generate and translate new, innovative ideas into new products and services, and then to get those products and services to market—cheaper, better, and faster. At the same time, project and product managers face the difficulty of stopping bad ideas at critical points or "gateways" in the process before they become expensive project and product—and company—failures. These "go and no-go" decision points occur in what is termed *project review stage*.

This book will integrate the two fields—new product development and project management—into one practical treatment of how to *drive* new concepts and products to market faster, better, and cheaper. The basic issue is how to move new products and services quickly from concept to product to market as a *managed and seamless* process, free of technical and handoff bottlenecks, technical problems, and delays, and how to ensure that bad products are stopped at key *review* points, before they become product and project failures.

New product development, as defined and supported by the Product Development Management Association (PDMA), centers on customers, product designs, marketing, launch plans, and the business case for a product. The PDMA community is dominated by marketing people and marketing ideas, not project management ideas. However, PDMA-oriented guidance is beginning to reflect the managerial and organizational issues in producing successful new products.

On the other hand, resources, schedules, Gantt charts, budgets and costs, and teams drive the project management field. PMI (Project Management Institute)-oriented books typically address the managerial and resource issues in producing deliverables in an organizational environment. Project management and the traditional PMI process and certification process focuse on resources, scopes of work, contracts and proposals, schedules, budgets, and teams, providing the essential administrative, organizational, and managerial framework for designing and producing a deliverable. But classic project management does not push the frontier of new product development or marketing. Project management has a process and managerial focus.

Many good authors have addressed product development. Robert Cooper in *Winning at New Products and Portfolio Management for New Products*, Michael McGrath in *Next Generation Product Development,* and Bean and Radford in *The Business of Innovation* are good examples of excellent references. But there is little out there on managing new product development. See our bibliography for other good sources, especially on new product development from a marketing aspect.

Classic project management texts address the PMI PMBOK process but rarely get into hard product development and engineering cases; Kerzner, Lewis, Larson, and others have produced books and texts in project management, and some include good case studies. But project management books rarely see new product development as the key, generic process structure for the deliverable.

A common process for both fields is portfolio development, but new product people see this as a business case and marketing process, where project managers tend to think in terms of feasibility, financial performance, strategic alignment, and the right number of projects in the pipeline.

The managerial implications of new product development come especially to light when considered in the context of multiple projects and resource management challenges. How does a program (multiproject) manager keep track of several new product projects in a portfolio and ensure a good balance and mix, consistent management, and good decisions on progressing from one phase to another?

Internet Impact on New Product Development

The Internet is changing the new product development process such as Internet sources as through Wikipedia, YouTube, and Innocentive. Models, simulations, and basic performance information on new products are now instantaneously available through the Internet, freeing new product developers to generate new ideas and demand for new information instead of simply researching for current product information. *Virtual teams* now dominate the process. This means that innovation and creativity will be increasingly focused on truly incremental product and service concepts, not reinventing the wheel. Engineers can find out easily what now works, what is in the pipeline for development, and what is *fair game* for new development. Customers and client groups now can participate in new product development through Internet market research and direct involvement, thus the design and testing process cannot be seen as more concurrent and integrative, and less sequential.

Risk and New Product Development

As evidenced in Figure I-1, there is a change going on in new product development process. This change indicates movement from pure marketing to control of resources and time, to integration of new product development into the mainstream of company activities, to move emphasis on go or no-go decisions, and to ensuring that new product managers have adequate technical background. These changes portend general transaction to a more controlled, managed process.

Business risk is embedded in managing new product development, from beginning to end. To a very real extent, new product development is business risk management, the design and testing of a company product and/or service that carries with it risks, opportunities, and contingencies. Project risk management is an art, not a science. I have always been skeptical of scientific and overly quantitative answers to complex social, organizational, and project outcomes, especially when products, customers, products, and markets are involved. I think risk can be *stewarded* and managed by good planning and analysis, but in the end it is often the *gut feel* of a project manager that turns a project in the right direction and overcomes risk.

We tend to look for ways to control business and new product outcomes that sometimes simply cannot be controlled. It is as if there is some underlying need

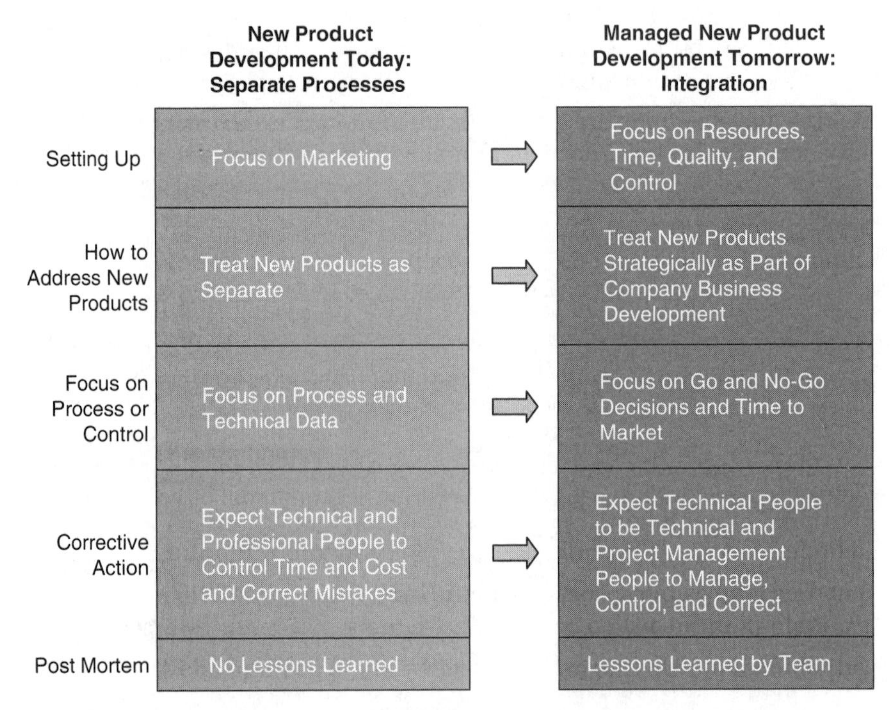

Figure I-1 Changes in new product developments.

to explain why things go "south" in a complex endeavor or new product project in systematic terms, and as if the world of human systems operated in a predictable and controllable way. We seek answers for all failures to *fix* them, yet we often do not know what factors were important. We assume that a *system* is in place and if the system fails we want to find out why it failed. When a business or new product project fails, we conclude that "somehow this failure could have been avoided if we had just studied and analyzed the risks a bit more, perhaps drilled a little deeper into the inherent impacts and probabilities."

Such an approach assumes that risks and failures operate in a predictable way; that the factors that lead to risk events and failure can always be identified, catalogued, and controlled; and that more analysis will uncover the *secret to the mystery*. The principle is that we should be able to identify what might happen, what the probabilities are, what the impacts are, and how to respond. It assumes we can find attribution, that we can attribute failure to key events or circumstances. It is true that the root causes of new product project failure are rarely a mystery—they often have to do with business performance, market conditions, leadership bias, and lack of support. They rarely have to do with technological failure—the engineers will usually find a way—it's the organization that cannot produce success.

The problem is complicated by the variety of definitions among stakeholders of *failure and success*. One person's failure is another person's success. A project that overcomes technology risk can deliver within budget and schedule and be termed a success by the project team, but it is possible that this same deliverable cannot be manufactured, or that the customer is not happy with the outcome, or that the business itself fails for reasons that have nothing to do with the project.

The drive to *mystify* risk assumes that there is always one *true* risk involved in every factor, task, or project, and that to solve the risk mystery we have to go to extreme limits to identify and quantify that risk. This makes the subject more complicated than it needs to be—and assumes that it is within our grasp to capture all the root causes of risk. Somehow, if we can establish that the risk of failure of a team task to integrate an information system is 66 percent rather than 24 percent, we make decisions based on an unreal confidence in science to predict things like the economy.

What is missing here is the fact that businesses and projects are human, not mechanical systems. Despite our increasing propensity to consider the study of organizational and new product development efforts in business to be a science rather than an art, human behavior is often unpredictable and counterintuitive. Despite our understanding of complex systems, we cannot identify all of the factors that contribute to risk and success even if we all agree on the definitions of those terms.

In addition, technical professionals and engineers—especially those in new product ventures—have developed their own language and values, which sometimes complement but often conflict with good project risk management. Theirs is a focus on the product and sales, not on business alignment, growth, and financial performance. The people and communication issues in engineering and product development are not unique, but they are accentuated by a working "axiom" of engineering project management—engineers communicate through channels and thought processes sometimes at odds with *cheaper, faster, better*. But they are inherently good risk managers. Engineers and technicians are often conflicted in a project management setting by time, cost, and organizational constraints that require they take short cuts to good engineering and risk management. They are challenged by risk and typically want to get it right, rather than getting it on time and at lowest cost. For instance, the measure of "mean time between failures" (MTBF) is often applied to new products, e.g., electronic and technical equipment, and tests are designed to ensure that products perform under stress at the intended MTBF. MTBF is a risk indicator; the risk of failure is quantified by repeated tests and documentation, thus a quantitative probability can be applied to its future performance. But in most circumstances MTBF is not applicable or suitable because user settings and environments cannot be controlled to really predict all the circumstances a product will experience. And the customer may not be interested—or will not pay for—a certain level of MTBF. But engineers typically would like to get MTBF down to zero if they can—an application of Six Sigma thinking—even at the cost of on-time delivery and budget.

Another complication in project risk management is the resistance to change in the project management or supplier team, as well as in the customer's organization. Typically a complex project and its outcomes trigger the need for organizational change, thus surfacing the resistance of those who do not see the value of change. For instance, a new electronic product produced through a product development project can alter the priorities of the customer's organization as the new product is phased into marketing and sales. The priority on this new product can upset an ongoing dynamic in the organization long supported by the *old, replaced* product. The risk here is that employees will resist change and undermine new product delivery unless the following factors are in place:

1. Top management support
2. Clear vision
3. Incentives to accept change
4. Incentives to take risks
5. Clear communication
6. *Walk the walk,* the day-to-day process through which management produces on its promises

I am optimistic that there are useful tools to manage new product risk, and that these tools lie in core business and project planning and management processes. I believe that project risk can be *stewarded,* but not always controlled through good planning and scheduling—and critical thinking. Through the application of risk management tools outlined and illustrated in this book as part of the planning and control process—and separate from it—risk can be managed.

Our approach is to broaden and simplify the new product risk concept at the same time, offer useful tools and best practices, integrate risk into the strategic, business, and project planning and control processes, and to offer exercises and cases for learning purposes. This book shows how to apply the Project Management Body of Knowledge (PMBOK) on project management and risk to new product development, but goes beyond it in many respects to address new product applications.

While we treat the *team* and *organizational behavior* issues inherent in new product development more extensively in Chapter 9, there is a thread running through this entire book on the importance of people in the process and integrating the new product team into broad business processes so that they don't get too far from the center line and core competency. This ensures that the tendency of new product teams to slip toward undisciplined, *skunk work teams* is offset with the realities of business discipline. These teams can become isolated from the mainstream so that it is difficult to manage and control them unless there is a program-level management presence and solid business processes in place to bring them occasionally in line.

Flow of the Book

The flow of the book is shown in Figures I-2 and I-3.

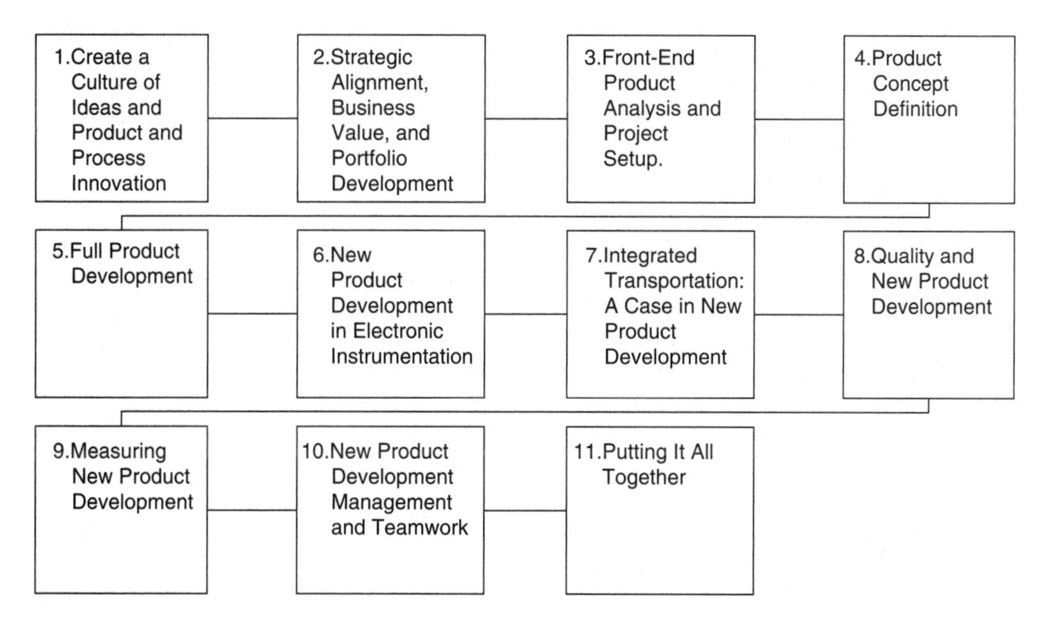

Figure I-2 Flow of the book.

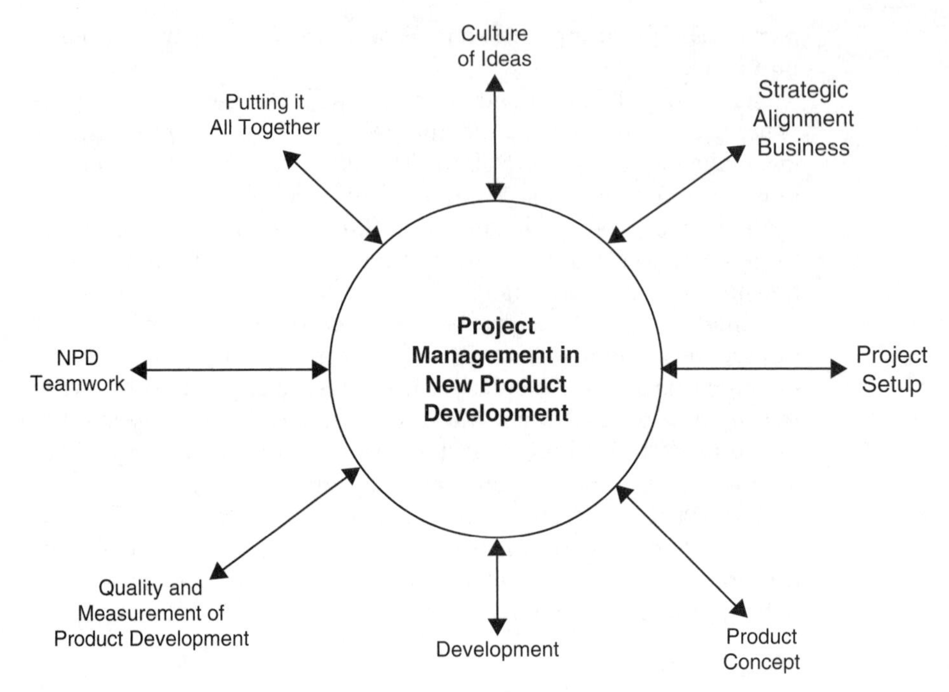

Figure I-3 Book flow.

Chapter 1 begins with a discussion of how to create a culture of ideas and innovation in the company or agency. Organizations need to be freed up to empower people to think ahead and to conceptualize new customer needs and new products and services to serve them.

Chapter 2 addresses strategic alignment, business value, and portfolio development. This discussion centers on how to articulate business plans and strategies, how to align ideas and product concepts with those plans, and how to develop a new product and service portfolio.

Chapter 3 goes into how to establish a project management system to support new product development, and how to do project setup. This is where we identify what management tools are necessary to control new product development without suffocating it.

Chapter 4 examines product concept definition and how the product is created out of a concept proposal. This is where we first make the point that each phase of new product development should be aimed at the final project review for that phase. This is the point at which the go or no-go decision is made. Project reviews make decisions either to pass the product to the next phase, or to terminate it. Project review is a key management control point.

Chapters 5 and 6 delve into full development of the product with all the marketing and product development activity necessary to prepare for market launch, and offer an example of full development for a representative new product, a new electronic instrument.

Chapters 7 and 8 discuss quality, Six Sigma, and other measurement tools and techniques for ensuring process improvement and full empowerment of the team.

Chapter 9 goes into project management and teamwork, and how to establish a new product team that effectively delivers.

Chapter 10 puts everything into a set of seven principles, as we try to "put it all together."

Appendix A provides a generic work breakdown structure for new product development with task descriptions.

Appendix B outlines a new product training or college course, with a weekly schedule and course objectives.

Appendix C includes agenda items to help new product development teams to organize and deliver new products more efficiently.

In this book, we see the whole new product development process broadly, including integrating organizational preparation, new product portfolio, finance, new product development, production, marketing, outside partners, and distribution—and ultimately the rollout to the marketplace. A challenge from this perspective is how to establish accountability schemes in the organization so that each of these processes works seamlessly with the others. Deciding who is in charge of what in this process is part of the leadership role in new product development.

Chapter

1

Create a Culture of Ideas

The Soul of Innovation and Creativity

The essential ingredients in delivering successful new products and services are the key people involved and their creative ideas. It is not process, it is people. Success comes from a vibrant and energetic organization that encourages its creative members and partners to think innovatively about what they are doing. They think about where the company is going, and what it can do in the future. They should feel free to generate and integrate new ideas, products, service concepts, and processes into the system. In the best companies, new product development is a mainstream activity driven by vision, organizational energy, top leadership, and lastly, process.

The Story of *Quikmate*®: Introduction of Sonoco Products Co. Plastic Grocery Sacks

Today's literature on new product development suggests an overwhelmingly rational, systematic process, controlling decisions at each phase. Rational decisions and process seem govern organizational behavior. But nothing could be farther from the truth in looking at the dynamic and unsystematic nature of how new products are really managed, introduced, and delivered. Urged by the inevitable tendency for academics to idealize and simplify this world of new products, new product development is seen filtering through a managed decision process from the bottom-up, going into a scheduled and predictable process of product development, testing, and marketing, and finally appearing at the front door of excited customers willing to stop what they are doing to use the new product and pay for it.

While process is important, what makes new product development succeed in companies like Sonoco Products Co. is strong, informed leadership at the top, and the collective dedication and discipline of professionals and support people trying and trying again and again to get it right. Notions of new features occur during the process, and some are integrated into product design; others are rejected because they don't sound right. Schedules are regularly violated and budgets ignored while teams work to get the product right for market testing. Marketing people over-commit and over-promise and engineering and product development people understate and protect their time. Program and project managers try to control and guide, sometimes successfully but often getting in the way of real work. Reporting is spotty and full of half truths hanging on the hope and trust that things will come out OK in the end. Project management offices are seen by project teams as administrative and bureaucratic, often feeding the flames of distrust and management paranoia.

To place a framework of project management systems on this muddling through process is dangerous but necessary because it turns out that some project management tools, e.g., work breakdown structures, schedules, budgets, utilization rate reports, and risk analyses *are* helpful to project teams. They help to guide day-to-day activity, much like job descriptions help to clarify interdependencies and interrelations of work and product.

New product development is not a science, it is an art. The process in reality does not flow consistently and predictably and does not always respond to traditional management actions such as directing, controlling, and reviewing. The process is driven not by schedules and milestones but by the sheer energy and commitment of people dedicated to change and opportunity. That does not mean that traditional project management tools are not useful to guide the process, just that one cannot fully understand how a successful new product is designed and introduced simply by looking at product development processes and information. This book is grounded on the proposition that while project management can help to guide new product development and deployment, the key drivers of success are leadership and commitment.

A good example of this is the introduction of the plastic sack to grocery stores in the 1980s by Sonoco Products Co. Driven by the vision and energy of a product leader, H. Gordon and the corporate leadership of F. Bennett (Ben) Williams, the Senior Vice President at Sonoco Products Co., plastic sacks were brought to market with little formal planning and scheduling. The initial focus of the company on what it thought would be the major bottleneck, the customer, turned out to be wrong as they began to see that the stores themselves were the biggest obstacle to change. Yet the introduction of this new product was successful against almost any criteria of performance.

Lets look at what happened. Sonoco introduced a new product, Quikmate, into grocery stores in the 1980s. Quikmate was a new plastic sack produced and manufactured to replace paper sacks. Sonoco's experience demonstrates the importance of energetic top entrepreneurial leadership working with a flexible sales force with "sleeves rolled up" to do whatever was needed to prove the benefits of the new product to client stores. The story also illustrates the need to support product development and the drive to marketing a new product with strong company project planning processes when it came to delivering on promises.

Sonoco's plastic (high density polyethylene) grocery sacks were designed and developed to replace the paper sacks which dominated the market up to the 1980s in the United States. The concept originated in Europe; an American businessman acquired a sample in the late 1970s, developed the concept for the US market, and sold it to Sonoco. He patented the product, stayed on with the company to provide the vision for success of the product, drove product introduction and delivery, and grew the business to profitability through the product. Some of those who were involved with the project, such as Sue Harris who was key to marketing the product in the field, testify that without his leadership and inspiration, the product would not have survived the obstacles to success in the field.

Despite the fact that the move to plastic sacks had occurred in many other countries, the US was still behind on this technology at that time. Market surveys had shown that the sacks were clearly better in many ways, easier for customers to carry, more flexible, less costly, easy to store, and reusable. There were environmental issues with the sacks, thus the marketing approach had to deal with not only government regulations but a groundswell of opinion from environmentalists against plastic products in the 1980s.

Confirming the importance of organizational learning, Sonoco *learned as it went*. Sonoco had not anticipated the major problem in market launch, which turned out to be the most difficult hurdle in the marketing channel—the stores themselves. Sonoco had assumed that the stores and their employees would embrace the change and concentrated their initial market planning on customers. But it worked out that the customers were not the problem. Not only did many store employees have difficulty with the change, some of them actually refused to use the new sacks because early versions were difficult to open and pack. As with most new products, there was tremendous resistance to change.

The market launch was messy and full of disconnects and trial and error. Because the first versions of the sacks were difficult to work with, company sales people including Sue Harris and H. Gordon Dancy, met regularly with development and manufacturing personnel to redesign the sacks and more importantly the dispensing system. Sonoco encouraged communication between field sales, marketing, and manufacturing in order to assure that the product and dispensing system design could change as marketing created useful field data on users and problems.

Sonoco teams attended trade shows, held customer focus groups, conducted surveys, and promoted the product during the formative years. It took many months of work and promotional activity to get the first *breakthrough*, with a grocery chain called Luckys. Luckys committed to the product in its stores on a handshake and despite the urge of company managers and lawyers to write a contract to solidify the sale, the company kept the commitment informal.

Later, more success with the Kroger chain management and others brought product sales up to a breakeven point in the mid-1980s, a point at which the company could begin to see return on their investment. In the end the product was successfully introduced and became part of the mainstream grocery system in the late 1980s and beyond.

As with any new product, success led to competition. Sonoco's experience with the growth cycle of the product and dispensing system is also useful to understand as it relates to how marketing and product support work. While their product was patented, Sonoco eventually faced fierce competition from domestic and foreign manufacturers in quality and price. So as the product matured, Sonoco lost some of its market share.

It is also interesting that Sonoco financed the process despite the drawbacks and problems because the company was committed to gaining market share. With a less energetic leadership team behind the product, Sonoco could well have dropped the product because of the unanticipated delays in selling its early production lots and gaining longer-term contracts. The key was an energetic and charismatic leader and sales force linked into company management at any time they needed support.

Production was a challenge in the early years until the German equipment and manufacturing process the company used could be tailored to the US product and marketing process. Marketing had to communicate back directly to engineering and manufacturing to adjust to customer feedback and needs in the field.

Those who worked in this environment admit that they often "winged it" in the early years. They hired a bunch of new college grads to test and train stores in the field. They came up against packing problems and had to come up with designs in product racks for delivering the product that worked for store employees. They were not driven by formal marketing plans and documents; the key here is that once a product is introduced into the chaotic and unpredictable marketplace, it is the personal relationships and capacity to adapt to quickly changing market factors that lead to success. Sonoco also provided incentives to the sales team so that they knew they would share in the benefits of successful introduction of the product. But in the end it was not incentives; it was the excitement of success!

As we will see in similar experiences at Universal Avionics, Inc., covered later, the Sonoco Products story confirms that to be successful in new product market launch initiatives, a company must have a value-added product that meets customer needs, a dedicated and hard working sales force with total access to top management when necessary, and most importantly, an energetic leader who knows the business and is willing to sacrifice to achieve success.

In successful new product programs and companies, business associates are not risk averse. They know *and feel* that their professional careers are enhanced by generating new ideas and opportunities in their domain, but they also know that their advancement is not solely determined by the success of a new product. This frees them from the fear of communicating bad news, if necessary, or terminating a project midstream because it makes no sense in the market. The best ideas for new products come from leaders and those who do the day-to-day work of the company, from market research, and from new product teams that are inspired by visionary leadership.

This target organizational environment for this successful process is a *culture of new ideas*, the organizational tone generated at the top that encourages a free flow of feedback *into and out of the company* on current products, opportunities, services, the competition, customers, and market operations. This does not happen naturally or easily because new ideas are not always welcomed in companies that focus on developing and marketing current products and services. New ideas tend to challenge the current way of doing business—they change processes and strongly held paradigms.

New Organizational Structure for New Products

We live in a global economy in which the turnaround of structural and technological change is rapid and sometimes unpredictable. Business processes are changing even as we speak, and traditional jobs and work settings are becoming virtual, *e-driven*, and mobile. This atmosphere of change creates the need to structure the organization so that (1) people can create new products *before* they are anticipated by the customer, marketplace, or competition, and (2) the company can design, develop, and produce new products better, faster, and cheaper, and get them to market. Development life cycles have gone from years to months to days.

Traditional ways of organizing around product lines, markets, or geographic locations do not seem to encourage new thinking about products in a broader context. How does a company re-organize to avoid being captured by *narrow* views of the customer determined by traditional organizational structure? The answer is that organizing around processes appears to be helping successful twenty-first century businesses to generate creative thinking. It is in a process framework that companies can best *see* their customers in a system. Processes can describe various systems and interactions that lead to product success, thus encouraging people to think in terms of *process domain* and giving them a better chance of coming up with new processes and products to serve customers.

New Products and Outsourcing

To get new ideas into the process, sometimes it pays to outsource creativity and front-end new product development to capture *domains* that the company itself cannot create. Outsourcing can increase the capacity of the company to see new processes and products differently; outsourcing increases the chances of successful new product introductions. This conclusion must be conditioned, however, on the presumption that the relationship with outsourced partners is more than a simple contract; that indeed the partner is a committed, long-term associate in a given market, and therefore can be trusted to act in the interest of the partnership.

Organizational Learning

We have also seen that successful new product rollouts occur in organizations that learn and document their insights and processes. These organizations grow and mature. Organizational learning refers to a measurable process within a company or agency that captures insights, experiences, and lessons learned, and has an understanding of their customer domains. These organizations use this learning to get a leg up on the competition. New product development effectiveness seems to improve when the company brings experience and insight to creative people. This results in a powerful combination of individual and team creativity joined with company insights into particular development and market domains. As an example, Microsoft, Inc. seems to have developed that synergy between the organization and the professional that creates the conditions for new product success in their particular domain.

Seven Key Strategies

There are seven key strategies in assuring that the company or agency creates a culture of ideas. Company leadership must:

1. State that new product development and *opportunity generation* **are** the business.
2. Remove barriers to the generation of new ideas.
3. Provide a system of information and feedback on current products and services.
4. Create a positive, perhaps virtual *place* for new ideas to incubate.
5. Generate a flexible process of filtering, evaluating and transitioning new ideas into a portfolio and new product development process.
6. Manage and control new product development and marketing.
7. Create real success stories to demonstrate that new ideas can produce products, that grow the business.

State that new product development is the business

Corporate leadership must articulate the importance of new product development as the core of the business,that makes it grow. This priority must be seen in every decision and action taken by the management

Remove barriers

Removing barriers to the generation of new ideas involves eliminating the fear of failure and providing inspirational leadership, incentives and organizational platforms for good ideas on new products and services. Employees, associates, and stakeholders see opportunity in their day-to-day work settings simply because they see what works, what doesn't, and what would work in specific situations to produce new product or service opportunities. They see opportunity and risk on one screen. These opportunities include process improvements that could substantially improve company efficiency and effectiveness, product improvements that could enhance marketing and sales, and service improvements that could help sustain product life and longer-term customer relationships. But these opportunities will not surface to management and company leadership unless employees are encouraged to generate them as an integral part of their jobs and without fear of rejection.

Promote return on creativity

Company people need to know and feel that there is a *return on creative solutions* to market and customer needs, and that the business places a high priority on new ideas that can return profits and margins. They should know that assessing the return on creativity is the objective of a new product development process, and that employees will be rewarded for coming up with new concepts that create business value. On the other hand, employees should also know that if the return is judged not to be worth the cost of development and marketing, the decision to terminate a product midstream will not reflect on those who came up with the idea.

Management should encourage thinking "out of the box." Thomas Kuhn referred to paradigms in his classic, *The Structure of Scientific Revolutions.* He said that *paradigms* are traditional concepts or structured views of the world that restrict scientists (and people in general) from seeing new data that does not agree with the prevailing mindset. Paradigms operate to filter out information so that strongly held concepts are not challenged, but rather are continuously confirmed despite data and information to the contrary. New product ideas test and challenge paradigmatic, locked-in views of a product, service, or market in a company or agency, and thus often face difficult challenges. Management cannot see the potential success of a product because it is inconsistent with their *paradigm* for the company. To avoid paradigmatic barriers, management must allow new concepts to see the light of day outside the normal chain of command, and outside the operating systems of the company.

Providing information and feedback

How do leaders and their employees generate new ideas for products and services, and what does the system of feedback look like? Sometimes, as in the Sonoco case, top management becomes the driver of a new product. In other cases, the energy will come from the middle of the organization. The system of feedback is a visible and working channel for employee ideas through email, suggestion forms, project reporting, and hard copy recommendations from every level of the organization. If the system is not visible and working—and taken seriously by management—it will quickly die of its own weight.

Barriers are removed by enabling fresh views of the marketplace. The process begins with empowering people to be *themselves,* and to openly communicate with each other on improvements and opportunities they see. Further, employees are invested and engaged in the company's success because they own it, either literately or figuratively.

Creating a virtual place for new ideas

New product and service ideas often incubate in the organization before they can be conceptualized and defined. This incubation process requires a place—a Website or physical office—that represents the company's incubation laboratory. This is where ideas are placed for initial review and development by a staff of associates whose primary responsibility is to incubate new ideas from within the company.

Generating a filtering process

These ideas are filtered, evaluated, and transitioned by marketing staff into potential product characteristics and features and new project definitions. This would include risk assessment and response, using a risk matrix to categorize new concepts and identify their risks, impacts, probabilities, severity, and contingencies. We provide more on the transition from idea to product in Chapter 4.

Demonstrating successful ideas

There is nothing like making success visible to all. In other words, good companies and agencies publicize new product successes as stories and reports in company newsletters and other media, making sure that all company people see that the process of generating, developing, and marketing new products works to grow the business. Work hard to create those stories if you don't have any. The Sonoco success story was visible to all the employees early; they saw and felt the winning attitude that generated success.

Organizational Agility

Organizational agility is the metaphor for responsiveness and energy. Sonoco was flexible when it had to align itself with the issues and values of store employees who initially rejected the plastic sack concept. Leadership creates this attribute by injecting excitement and purpose into the company, and empowering its employees to take action on the spot. This process begins with ownership, leadership, and management, which must be trusted to articulate a new product mission and integrate the mission into the fabric of the company. Management walks the talk by reinforcing the value of innovation and creativity.

Creative intelligence and new products

What kinds of people generate new product ideas, and how are these people encouraged by their organizational environments? If we could answer this question, we would be closer to knowing how to increase the probability that a certain company, group, or individual is more capable of producing new products than others. Howard Gardner has researched and written about multiple intelligences, including the intelligence associated with creativity and new products.

In his book, *Intelligence Reframed: Multiple Intelligences for the 21st Century,* Howard Gardner (Basic Books, 1999) states on page 116:

> My definition of creativity has revealing parallels with, and differences from, my definition of intelligence. People are creative when they can solve problems, create products, or raise issues *in a domain* in a way that is initially novel but is eventually accepted in one or more cultural settings. Similarly, a work is creative if it stands out at first in terms of its novelty but ultimately comes to be accepted with a domain. The acid test of creativity is simple: In the wake of putatively creative work, has the domain subsequently been changed?

The essential point is that creative people seem to be able to be creative in certain domains, those they know something about, certain frameworks of thinking in which they intellectually reside. Further, Gardner asks not who or what is creative, but where creativity is. What kinds of environments encourage creativity and allow those who are creative to "dwell in their domains?" Three elements seem to join to produce an environment for creative thought:

- An individual creator with his or her talents, ambitions, and personal foibles
- A domain of accomplishment that exists in the culture
- The field, a set of individuals or institutions that judge the quality of works produced in that culture

Companies that identify creative people in their hiring policy and create organizational cultures that enable creative people to work in their domains have the best chance of generating new product ideas. Sometime leaders must combine with creators to assure that new ideas survive and then move to product concepts and prototypes. This is the reason that project *managers* must lead processes, while *creators* must think and act in their domains. A software engineer with in-depth experience in designing embedded software in electronic instrumentation is in a domain few others populate. But new ideas about embedded software are going to come from that engineer, not his or her project manager. It is the process of being able to spot good ideas and turn them into new products that characterizes good project management. After ideas are generated and preserved through aggressive leadership and management systems, then we have the powerful integration of creativity and leadership to introduce new products in the market.

Risk and New Product Development

Creating a culture of ideas involves not only surfacing new ways of inducing demand, e.g., creating new needs through new products, but also encouraging innovative ways to overcome risks. Overcoming risks creates product success. The process of risk management is critical in creating culture of ideas because as new ideas surface, so must potential risks and opportunities.

Educating associates on the relationship between risk, opportunity, and new product development helps to open the process of generating ideas. Ideas for new products automatically create risks because the process of taking a new product or service through concept, design, development, market launch, and support is expensive and time consuming. Yet opportunity is linked with new products because these products can generate new markets, and because a company's success in marketing a new product before the competition creates opportunity.

Risk: The organizational culture issue

Successful new product development requires the individual management of risk and business opportunity. Each participant has to understand that business is grown on new products with inherent risk. The offset to risk is hard work, informed decisions, flexibility, and the capacity to stop a new product process when appropriate.

Although risk is traditionally seen as an analytic activity (identifying and assessing risks in the project task structure, and applying decision trees, sensitivity analysis, and probabilities), the essence of risk management is the way the organization treats risk and the way you and your team think about a new product project. The challenge for the organization is to teach and train project leaders and team members to think in terms of risk, and to internalize the risk management and opportunity generation process into their daily routines. The assumption behind this approach is that risk management is "something I want my people to do in the normal course of their work," not "something I want a specialist to do later in the project as a separate exercise." Risk is a way of visualizing the project and its successful outcomes, and *seeing* potential pitfalls. You can't see risks if you are not looking for them, and you can't see opportunity without addressing risk and contingency.

So the successful management of risk is usually the product of a successful organization that has instilled into its people the importance of careful planning. Careful planning involves a core competence—the capacity to define uncertainty and risk, integrate risk identification and assessment into program and project planning, and build and sustain a support system for risk management that provides essential information when it is needed. But how does an organization build risk into its daily work, and how do executives use their leadership and institutional leverage to further good risk management as a part of creating a culture of ideas?

A culture of risk management competence

The successful risk management organization has five basic competencies:

- Active training and development in risk planning and management and how risk relates to opportunity
- Strong linkage between corporate planning and project planning, particularly between business analysis of threats and opportunities, and analysis of project risk
- Deep project experience in its industry
- Capacity to document project experience and "learn" as an organization
- Strong functional managers who address product quality as a risk reduction issue

Link corporate and new product planning

Strong ties between corporate strategic planning, including market analysis, and project planning and production ensure that the business "sees" its new product risks early in its planning. The business is then able to anticipate and dimension the risks it will face in designing and implementing projects that carry out its business plans. For instance, a telecommunications firm that performs SWOT analysis in its field may uncover potential threats to marketing new, breakthrough products in telecommunications cable technology. Planning broad marketing strategies and contingencies at the corporate level to address these potential breakthroughs (opportunities created from analysis of threats) helps the business support its selected projects that involve such new cable systems.

When risks and opportunities are fully aired and a project management system can control specific project costs, it is often possible to interest venture capitalists in financing specific new products, earmarking funds for particular new product enterprises.

Training and development in risk

Training and development programs that address risk identification, assessment, and response can help build professional competence in handling risk issues in real projects. Such training would include a curriculum in:

- Building a work breakdown structure (WBS)
- Identifying risks in the WBS
- Producing a risk matrix

Project experience

A company that "sticks to the knitting," as Tom Peters and Robert Waterman called it in *In Search of Excellence* (Harper and Row Publishers, 1983) is in a better position to recognize and offset risk simply because its workforce is likely to have a better handle on the technology and process risks inherent in its core business. Whenever a business departs fundamentally from its core competency areas, it stands to experience unanticipated problems, which develop into high-impact and high-severity risks. Successful new products evolve typically from the company's basic "knitting," and not from product and market areas outside company competency.

Learning organization

A learning organization as Peter Senge describes it, is an organization that learns from its experiences by documenting risk impacts and contingencies, and does not reinvent the wheel each time it plans and implements a project. This means that lessons learned from past project experiences are incorporated in documentation and then embedded in training programs so that project managers learn from past experiences. Communication is open in such organizations, leading to a process by which project experiences are "handed down" to next generation project teams.

Functional managers

The existence of strong functional management ensures that the basic functional competency of the company in areas such as engineering or system development is backed up by technology leaders in the field. Key processes, e.g., product development, are documented, and product components are controlled through disciplined configuration systems. This means the risks of product quality failures that result from product component variation are minimized, simply because the company can replicate products and prototypes repeatedly for manufacturing and production without variation.

Building the Culture

Organization culture can be defined as the prevailing standard for what is acceptable in work systems, work performance, and the work setting. A *risk management culture* can be defined as the prevailing standard for how risk is handled. Organizations with a strong risk management culture have policies and procedures that *require* its workforce to go through disciplined risk planning, identification, assessment, and risk response project phasing. But employees are encouraged to take on risks with a positive attitude.

A mature organization does not treat risk management as a separate process, but rather embeds the risk process into the whole project planning and control process. Risk is an integral part of the thinking of its key people. In the same way that the quality movement matures to the point that quality assurance and statistical process control processes become institutionalized into the company rubric, risk assessment tools and response mechanisms become an indistinguishable part of a company mosaic in a mature organization.

Sustaining the culture of risk management is considered a major function of corporate leadership in the risk-planning phase. Although most organizations do not enter the risk-planning phase as a distinct step in the project planning process, best practice addresses potentially high-risk tasks, assigns probability implicitly to the process, and develops optional contingencies that may or may not be documented in a formal risk matrix. This is typically not a mysterious, mathematical process, but rather an open, communicative process in which key project stakeholders, team members, and the customer talk about uncertainty and identify key "go or no go" decision points. They often know where the key risks are in the project process because the project itself is grounded in addressing a risk that the customer is facing.

A good example of a strong risk management culture is found at the Keane Company.

Keane's risk process

Keane, Inc. a leading information technology firm, connects and integrates risk with cost and schedule estimating, e.g., identifying project risks and determining actions to minimize the impact on the project and to improve project estimates. In other words, Keane thinks in terms of risk as a guide for cost estimating, scheduling, and defining mitigation actions. There is an initial process that takes much of the guesswork out of estimating. Keane established a set of guidelines, techniques, and practices to pin down estimates and ensure that customers and stakeholders clearly understand associated risks. Keane emphasizes communication on the relationship between a given project estimate (project schedule and cost estimate) and how the estimate has handled risks and risk mitigation. Their experience is that project success does not depend as much on completely mitigating risk as it does on communicating risk up front so that stakeholders can make judgments and decisions along with the project team as things happen.

In building the culture for risk management, Keane suggests that its project cost estimators:

- Make sure they know the difference between negotiating and estimating. Estimating is the calculation of schedule and cost given the tasks at hand; negotiating is working out the differences between the estimate and a customer or client schedule and cost.
- Understand the variations in technical skill and how those variations can impact estimates.
- Be objective about your own work.
- Anticipate the need for precision, understand the timing for "order of magnitude," ballpark estimates, versus the need for more detailed budget and definitive estimates.
- Look at untracked overtime in building estimates from past work.

Keane advises its people to ask "who is at risk?" before you ask "what is at risk?" because the issue of risk is framed by those who are affected by it, not by some arbitrary quantitative formula. Different project stakeholders have different perspectives on risk and estimates, and indeed their perspectives change during the life of the project. It is best that risk assessment be guided by those who will suffer the consequences of risk and who will bear some or all of the cost of risk mitigation.

The role of the project planner/manager is to provide the process with parametric data to measure project and product performance. This is because the person asking for such data is going to use the estimate to make critical business decisions. For instance, if a client for a new information system is facing the possibility that the new system cannot handle the estimated user load during peak periods, that client must make a decision to either limit the user universe or upgrade the system. So the estimate of risk is key to the client decision process, and will affect client success.

Keane integrates cost and risk to better understand how risk affects project schedules. By training its people to identify risks from broader business and industry data, and to schedule risk planning and management activity into the project baseline schedule, the company delivers an important message to its people.

Risk analysis and mitigation

Keane sets the organizational tone for risk mitigation in its book entitled, *Productivity management: Keane's project management approach for systems development:*

> One way to approach a project realistically is to consider risks. Evaluating the risks should be an inherent part of estimating—the risk that the project will take longer, cost more, or require more effort than planned. The existence of risk is the reason an estimate is necessary in the first place. But every project possesses multiple variables, and when supplying an estimate, you should also apply a risk factor.

The Keane book involves a story about risk:

> One fall morning, a facility of ours shut down as a result of snakes moving into an underground electrical junction box, thinking they had found a comfortable winter hybernation spot. Their squirming caused an electrical short and small explosion that brought down the entire communication linkage.
>
> There is no way that we could have been aware of the potential of this snake situation, and it therefore falls in the category of uncertainty. Snake interference is not a variable considered in the management of a project. If, however, we view the snakes from the standpoint of "might our communications facilities go down during the course of this project?" then we have risk. We can make a statistical evaluation of how likely some unforeseen event is and estimate the dollar and time impact of that occurrence.

Addressing risk with scenarios

Keane, Inc. is a good example of a *projectized* company that uses risk scenarios to get its project teams to anticipate risks in the planning process. The company encourages the development of issue or scenario statements that pose potential problems—variations from the plan—in a project, and generate queries about the issue. For instance, Keane might encourage a project manager developing a new project information system to build the following question into an early project review session:

> After we develop and build this new project information system, what interface, platform, and network compatibility challenges is the customer going to experience with the system? What options will we have, and what is the likelihood of these challenges becoming insurmountable?

Performance incentives

Any organization building a risk-based culture must provide incentives for integrating risk into the project planning and control process. The incentive for handling risk is top management support and resources. Top management support comes when project management identifies and anticipates business risks that save the company time and money. Project managers who manage risks effectively are likely to be more successful in acquiring additional resources because they tend to have backup and contingency plans ready when risks occur.

The Johari Window

One of the major risks in any project is the tendency of its key project decision makers—especially the project manager—to overestimate what they know and underestimate what they don't know. The risk is that key people will take risks, but not manage risk. This means that the beginning of good risk management is the ability to know what the organization and its people can do, and what they cannot do.

The field of organizational behavior contributes a tool called the Johari Window (Figure 1-1) that is helpful when analyzing the personal tendencies of leaders and marketing project managers to "take risks."

The Johari Window, named after its inventors, Joseph Luft and Harry Ingham, is one of the most useful models that describe the process of human interaction and behavior. The four-paned "window" divides personal awareness into four different types, as represented by its four quadrants: open, hidden, blind, and unknown. The lines dividing the four panes are like window shades, which can move as an interaction progresses.

The following sections show the thinking process of a typical project manager when testing their grasp of what they know.

The open quadrant. *The "open" quadrant represents things that I know about myself, and that others know about me. For example, I know my name, and so do you, and you know some of my interests. The knowledge that the window represents can include not only factual information, but my feelings, motives, behaviors, wants, needs, and desires... indeed, any information describing who I am.*

The risk here is that what is open to some coworkers may not be open to the customer or a project sponsor. It is important that a project manager get to know the customer, key sponsors, or stakeholders as people. The focus is customer expectations; if there is an open process on expectations, the chances of managing risk are high.

The blind quadrant. *The "blind" quadrant represents things that you know about me, but of which I am unaware. For example, we could be eating at a restaurant, and I may have unknowingly gotten some food on my face. This information is in my blind quadrant because you can see it, but I cannot. If you tell me that I have something on my face, the window shade moves to the right, enlarging the open quadrant's area.*

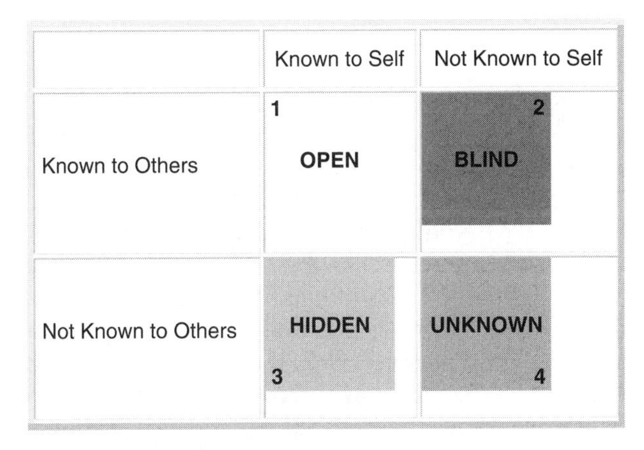

Figure 1-1 The Johari Window.

The risk factor here is that there may be variables that a competitor or customer knows about the organization that the project manager may not know. For instance, a current supplier to the project may have failed in the delivery of a similar component to a competitor, but the project manager is unaware of the situation.

The hidden quadrant. *The "hidden" quadrant represents things that I know about myself but that you do not know. For example, I have not told you, nor mentioned anywhere on my website, what one of my favorite ice cream flavors is. This information is in my "hidden" quadrant.*

The risk here is that a project team member may not be entirely open in divulging important information about their expertise and experience.

The unknown quadrant. *The "unknown" quadrant represents things that I don't know about myself, and that you don't know about me. This is the universe of new product development. In effect, this unknown quadrant represents the undiscovered market for new products, a realm of innovation and imagination that is created in the mind of a company associate. It is this kind of new ground we want people to discover every day in the innovative, new product organization. Customers have hidden needs and expectations that are not expressed in the marketplace; new product developers do not know what these needs are, but can discover them through good market research.*

The risk here is that there are factors at work that are unanticipated both by the project manager and the customer. The point is that the early generation of new ideas and concepts that might lead to successful new products and services comes from the unknown quadrant, the result of inquiry, investigation, and imagination. It starts with an individual who envisions a new market opportunity that is beyond the current reality; the thought process creates a new scenario in the marketplace that is unknown, a subjective reality. This is the beginning of seeing new products and services that might meet a currently unknown demand.

Personal, Project, and Organizational Risks

There is something very personal about the issue of risk and new products. In many companies, taking risks in new product concepts is rewarded in principle, but failure to take risks has its implications despite the company rhetoric. What the company is really saying is, "Go ahead and take risks with new product ideas, but take them only if you think you can succeed and produce value for the customer and the company. We will support you with data and information. Don't take risks frivolously."

For the business and project professional, risk is a personal issue because project risk is directly associated with personal risk. If a project manager fails to see and control risk, that project manager faces the prospect of being associated with a failed project. So the way a project team faces risk has implications for each team member personally—and for the team dynamics involved in a given project.

The way the company protects its employees and officers from risk is key as well. If the company is positioned to absorb the cost of failure, the program or project manager is more likely to take the risk. Thus the propensity to accept risk and manage it successfully is partly a function of organizational support; if my company supports me, I will address risk and make the best decisions I can, but I will want to let my top management know the risks as I see them so that if the risk is not successfully controlled, it will have been a company-wide decision, not a personal one.

In sum, the organization must position itself for risk, and must empower and enable its business and project people to address and take risks; however, there must be an open, organization-wide process for addressing and absorbing risk. If these conditions don't exist, the project manager is not "incentivized" to address risk and will avoid it—often at the expense of opportunity.

The New Product Risk Framework

It is important to see new product development and risk as a business-wide challenge. After all, business enterprise itself is a risk, and that is what makes success and payoff satisfying to the business entrepreneur. Project risk is simply a microcosm of the overall business challenge, and the fate of every project lies first in the capacity of the parent company to create conditions for success. As we have said, project risk starts with the business itself, its market position and business viability, its partnerships and vendor relationships, and the economic risks the business itself faces, as well as customer and client risks.

The author learned a valuable lesson in project risk management working with a leading avionics product company, a product development, engineering, and manufacturing company that used project management systems at the division level, but had no project management systems in the corporate "head-shed."

As part of the support project management office, the author's role was to ensure that there was support to the project managers, and that there was a clear project management policy and process in place. I also ensured that standards were enforced, and served as an assistance project manager in several functional areas such as procurement and cost capture. The work was in a regional facility in the east, while corporate was in the west run by a single owner and a small corporate staff largely without corporate program management competence.

This major producer of avionics equipment had several new product development projects going on in the engineering plant involving mechanical, software, and electrical engineers, supported by procurement, acquisition, and accounting staff, and a human resources office. The product involved embedded software and regulatory requirements for avionics equipment, and much of the process was testing and retesting prototypes against standards. In addition to the work underway, the regional facility had been *awarded* (by the owner's decision) a system program from another region that had not been successful with a system upgrade. New staff was hired to handle the project. At the same time it was authorizing this hiring process at our facility, corporate was experiencing downturns in sales and marketing efforts, and consolidating facilities to reduce costs. But the owner made the decision and we implemented it.

Later the same year the company had to conduct a major downsizing, cutting many of those same staff members hired to run the new program, plus other valuable and high-performing engineers. The reason it had to downsize was that the corporate investment source was unwilling to forward additional funding until the company cut costs.

We learned that project risk cannot be separated from business risk in general, and that the effectiveness with which a company identifies broad threats and risks in its business planning will establish the conditions for successful management of project risks.

Project risk is inherently business risk, and cannot be disassociated from the overall risks and threats faced by the business as a whole. Thus project risk management must start at the perimeter of the business and its relationship to its market environment. As the business identifies its threats, competitors, and risks, it provides the basic wherewithal to identify project risks. There is an inextricable linkage between the threats a company faces in technology, labor availability, or product development, and the threats a project team faces in producing a deliverable designed to implement the business plan and strategy.

The lessons are these: Risk is a *vertical* process, not just *horizontal*; that is, risk happens up and down the organization at the same time it happens in project planning management processes over time. Risk is a multidimensional and multiscaled syndrome that can affect you without warning if you are not *in the inside*. And risk does not often come in recognizable clothes, but rather sneaks up on you through the side door. Very often the key determinant risk is out of your control as a program manager, or even as a general manager, because risk stems from central leadership more than from project processes. The "risk as part of the business framework" concept looks something similar to Figure 1-2.

Knowing the business you are in helps anticipate risks inherent in the business. For instance, if I am project manager of a software development project in a software firm, I know from past experience and good corporate strategic planning that one of my risks—as a business—is going to be the "integration and debug" process. This includes the time and effort to ensure that a new computer program or code works on the user's hardware platform. I can almost bet that this task will be one of my risks in any such project. I can differentiate that very distinct risk from a general uncertainty about whether there will be any demand for the product once it is ready. I can "work" the risk in my project; the uncertainty about market demand can also be worked, but I can't control it.

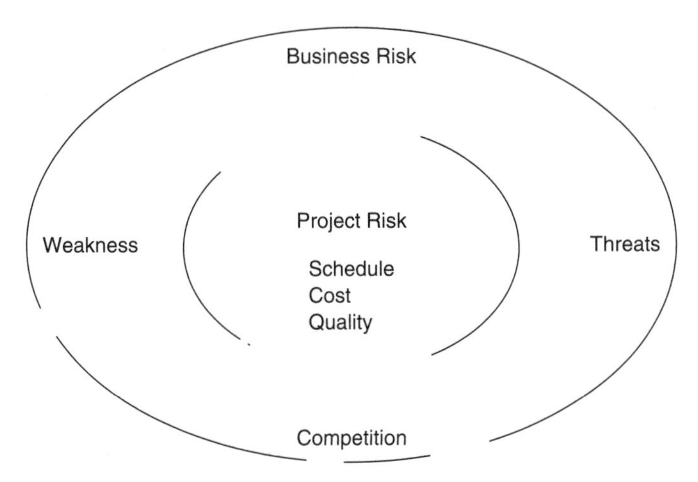

Figure 1-2 Business framework for risk.

The risk in this software development process can be identified, defined, and ranked as part of a process called qualitative risk assessment. If I wanted to quantify the probability that a debug problem will actually happen and create a major schedule slippage and/or quality issue, I would do a quantitative assessment of probability. That process might result in my estimating that there is an 80 percent probability of a debug problem not getting fixed within my estimated and scheduled "most likely" task duration of three weeks. I might then identify two alternatives—a worst case and a best case, based on various assumptions, and plug them into my schedule using the "PERT" tool discussed in Chapter 3.

Thus the reason that project risk starts with the business itself is that any project or portfolio of projects coming from the business pipeline is typically aligned with the business competencies and capacity. The business leadership has chosen a project because it believes the payoff of a project is worth the investment in overcoming the risks inherent in doing the project. The product or service to be produced by the project is key to the success of the business in its industry niche.

Figure 1-3 illustrates the linkage of customers and business risk planning. Customer requirements generate business plans which look at the strengths, weaknesses, opportunities and threat (risks). New products are evaluated for risks and opportunities and selected. Programs and projects for new products are planned that make sense financially and from a risk standpoint. The individual WBS, schedules, and contingencies are flushed out.

Another Case in New Product Development: The Schneider Program

"I got the customer's approval to do the Schneider program," Lakeisha told Bill.

Lakeisha and Bill were project managers at Project Associates, Inc., a software and information technology company. "I've got four months to deliver the program to Schneider, including a new hardware platform, software code and

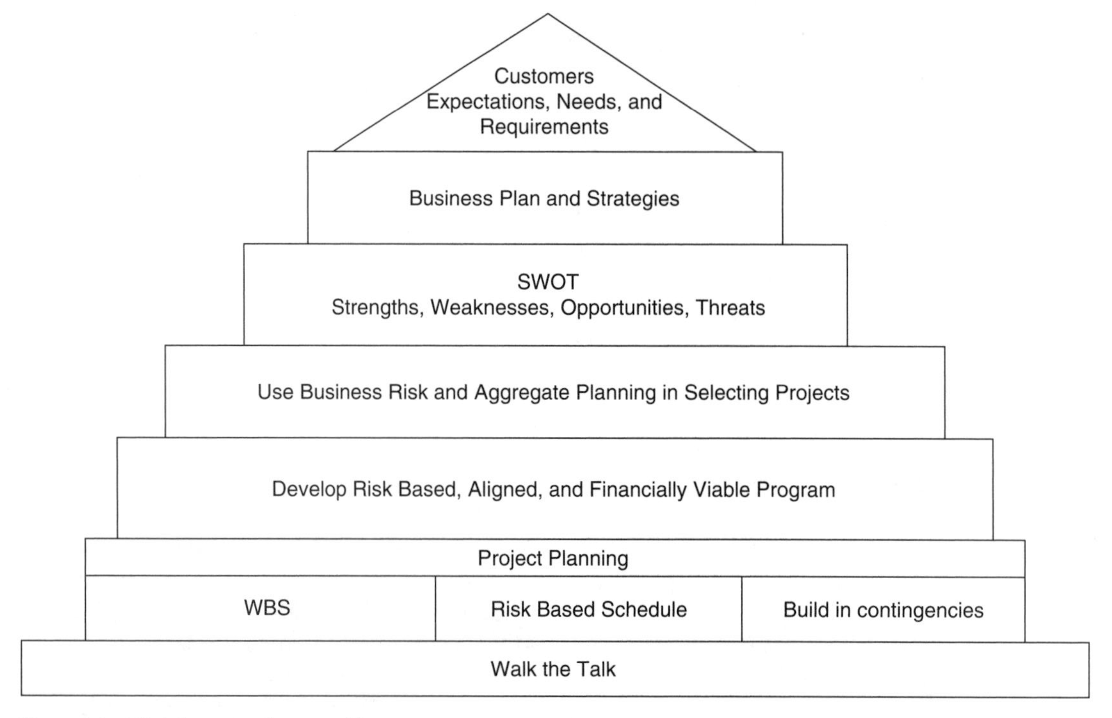

Figure 1-3 Risk framework pyramid.

documents, and a training manual." "I think it's going to be a blast—the biggest issue to me is the software. The hardware is a no-brainer."

Lakeisha had delivered her last project, the Mires program, well ahead of schedule. Bill, on the other hand, had not done well in his last project and was late and over budget. Lakeisha was eager to show that her last performance was not a fluke and that she knew what she was doing. She always suffered a little lack of confidence under pressure, but always came through. There was a subtle competition going on between Lakeisha and Bill, but they worked well together.

"I wouldn't get too excited just yet," Bill told her. "You know I was going to do that program, but I got sidetracked and the boss gave it to you. I've seen the specifications for the program, and there are a lot of risks. I think you've got at least six or seven months of work with the current team to produce the deliverable as I see it. And that is if you don't have any problems with platforms, software, people, our 'old' testing equipment, and good old unreliable Software Systems, our contractor. Are you going to go with the same team you used last time? Planning any risk assessment and contingency—you know, those scenario things?"

"Yes and no. I am going to use the same team—I think—but I don't have time for the risk stuff. Been there, done that. I will use a schedule from our last project for Smothers, which was a good run and we kicked butt. I have looked at the risks and the big ones are in the software graphics package and online training package. I've got a plan to shrink the schedule to four months by crashing some stuff and outsourcing. I read an article on outsourcing last week and a story included a great company that does just what I want them to do—or almost. That is going to cut my schedule by two months. This project is going to have to be on the fast track from the word go. I am going to write a cost-plus contract with them because procurement says that is the template they like to work with."

"Well, I hope you know what you're doing," Bill said to Lakeisha as they crossed the corridor. "I have seen a lot of people get burned by contractors, and you know that the hardware for this deliverable is new and will require some long lead times on parts."

"Cool it, Bill. I've picked a reputable contractor," Lakeisha said. "I checked his references and I am sure he'll do a great job—and he is going to accept most of the risk. I told him I would make progress payments on the cost-plus contract only if he was on schedule, and he agreed. He said he would put more people on the job if it turned out to be more complicated than he thought it was. I will just keep an eye on him. I understand that this is risky business and some risks are inescapable. But what an opportunity to show what we can do—think of the plus side of this thing if it goes! When I have this much to do in such a short time, I'm not going to waste the team's time on risk games and risk matrices—I know what I want and I know what the risks are."

Bill thought she should be more careful, but liked her spirit. He and Lakeisha had been over this ground before. He had learned some lessons on risk in his previous project, which failed because he was hit by an unanticipated shortage of key technical people and a surprise glitch in getting parts for the hardware platform. He had been reading about a new theory of "constraints" that focused on resource and equipment availability risks, not just critical path issues. But he had learned not to argue with Lakeisha when she had already decided what she was going to do.

Lakeisha met with the contractor and gave him the specification before her procurement office could get the scope out and signed, but they didn't have a problem with that and she didn't have time to go through formal signatures anyway. It turned out that the contractor, Mag Company, was headquartered in New Delhi, India, but had a local office. Their procurement people had said the specification made sense and that they would get on it right away—they too felt that could go ahead without a signed contract. They needed the work.

Six weeks later, Lakeisha called the local Mag Company project manager, Abdur Manat, to check on progress. "Everything is going great," he said. "But we have been working on a high-priority project for another company that came in just last week with a heavy job due yesterday, of course. So I have not made as much progress as I had hoped." Lakeisha responded, "Maybe I ought to have a schedule from you, particularly on the tough pieces of the work you see as potential problems." Abdur replied, "But I still have three and a half months to do two months of work, so I don't see any problems. Did you send me those specs on the hardware? By the way, did you say that the customer wants online training—what's that?"

Lakeisha hesitated, but disguised her concern in a positive expression. "That sounds fine," she responded. "Let me know if you need anything. I'll be back in touch in another six weeks and then we can talk about integration."

For the next several weeks, Lakeisha spent most of her time trying to put together the specifications for an online training program for the contractor, a task she hadn't anticipated. She also inquired on the lead times for the new hardware, but the design people were busy on other work.

Six weeks later, Lakeisha called Abdur to check on progress. "The last project took me longer than I expected," Abdur said. "I've gotten into the graphics work and looked at your hardware requirements, and I've been working like crazy, but now that I have taken a closer look at it, I think there's at least a good three months of work on this job, particularly on that online training stuff you sent me—I will have to sub that out."

Lakeisha almost choked on that one. Her stomach told her she was in trouble and she murmured to herself. That would make the total development time for the Schneider project six months instead of four months. "Three months," she said, "you have to be kidding. I need the software code in two weeks to begin integration. You were supposed to be done by now! I am not paying your last invoice."

Abdur responded, "OK, but you already paid our last invoice—your accounting office has been very efficient. We just got the check, along with a nice holiday greeting." Lakeisha knew this was not her day.

"I'm truly sorry," Abdur said, "but this isn't my fault. There's more work here than we could have ever done, and more than you estimated in your schedule. We found that the software code doesn't work in your hardware, and we haven't been able to figure out why. And your team people aren't available to talk to us. I will finish it as fast as I can."

It turns out that Abdur delivered the software in three months, but the project took another month after that because of integration problems with the in-house team's code. In the end, the Schneider program took over seven months rather than the four month estimate. Lakeisha concluded that Bill and the company had sandbagged her by palming off a bad project that he was not able to handle and that had inherent big risks.

This story is about the fact that the new product innovation process is rarely a question of major breakthroughs and starting new concepts out of the clear, blue sky. Most innovations and new product developments are subtle product increments, improvements in design or development or processes, which generate major benefits in time to market. These incremental product developments require creative concepts and ideas that are no less important than brand new products. They evolve in organizations that encourage flexibility in process and procedures, allow redefinition of tasks during development, and encourage testing new materials and technologies *in development*.

As an example, in the development of an electronic instrument involving electronic, mechanical, and software engineering, new materials are often discovered in developing products that reduce production costs and improve the financial performance of the product in revenue use. An avionics firm recently found that when developing a radio control unit, a new low-voltage resistor package would produce major safety and performance benefits but would require a new mechanical housing design. The idea came from an experienced electronic engineer who thought through the role of the resistor in the instrument and searched for new designs by surfing the Internet. He found a new resistor design and, after simulating its performance, ordered the new resistor on his own and integrated it into the instrument. Results were startling in reduced cost and efficiency in the performance of the product.

The way configuration management is used is key when balancing the creative process with the control of product design. If new designs cannot be integrated into design because configuration management *locks in* a design too early, new ideas can be stopped prematurely. On the other hand, if new designs are not locked in until production is scheduled, information on alternative designs is lost and production can be delayed because the actual configuration of the product is not clear and not documented to support production.

People who participate in this product development process have to feel that the project allows flexibility while at the same time schedule milestones are taken seriously.

Another Story of New Product Development

A small (150 employees) avionics research facility reports to corporate headquarters in another state. The group is in charge of developing and producing new avionics products, electronic instruments used in aircraft for aircraft control, and subject to strict Federal Aviation Administration (FAA) regulations.

This particular story is about the development of a radio control unit with a new design and system requirements. The story demonstrates that creativity, innovation, and new product development decisions come at many points in the process, and not through a rational, controlled environment. These are not one-shot decisions or activities controlled by "gateway" decisions, but rather incremental and often personal—the tyranny of small decisions. The small decisions create the larger whole. To demonstrate, consider a typical development and manufacturing organization, and trace how new product development actually happens in many cases—at various levels of the organization.

The product is a new control unit that will manage the radio system in a business aircraft suite of avionics equipment. In the following example, it will be clear that the opportunities for new product and service ideas come up in very subtle ways and survive through very personal decisions affected directly by the tone of the organization.

Don Akers, a mechanical engineer in charge of the instrument housing, sees the new product as a challenge in housing the electronics in a new mechanical concept that he has been advocating. But his ideas have not been welcomed in the unit, which is dominated by electronic engineers. Akers feels he can make a mark through this new housing concept, but fears that if he goes too far, his ideas will be ridiculed and, worse, ignored. This new product offers a great opportunity to create a new mechanical concept, but there are organizational barriers to his idea and he has to be careful in promoting it.

Bill Verdon, an electrical engineer and functional head of the electronic engineering department, is the senior electrical engineer in charge. His focuses are his own expertise in this kind of avionics instrument and maintaining his role as the primary guru in the company. Verdon is skeptical of mechanical and software engineers, and does not welcome new ideas unless they come from him.

Buster Right, a software engineer in charge of embedded software design, is working on a new process of developing software code. This approach is a new and innovative software development concept that produces code in full increment stages, rather than in pieces and phases. He has advocated this approach with his software guru, Ben Branish, but his ideas are new to the company and not the way things are normally done. Branish is skeptical.

Ben Cartwright, the facility general manager in charge of the whole effort, is caught up in corporate and local management issues that have little to do with this new product. Cartwright, an electrical engineer, deals with a strong regional manager who started the whole business unit and who is known in the company as the corporate genius. However, management is not his strong point, and he often interferes with product development by intervening directly with engineers working on a given product with new assignments that divert their attention.

Bruce Hall, the test technician, is working with a new test system and new equipment, and is intent on showing that the new test process will speed up testing and reduce the cost of testing systems. However, the electrical engineers are skeptical of the process and generally look down on test technicians who do not have their level of education or experience. Technicians are technicians and should not be experimenting with test protocols, they would say.

Bart Strong, the project manager, is a physicist who backed into the new product process. Brilliant and gifted in many fields and a generalist, Strong is not accepted by the electronic engineer community because he does not have the training and education they have, and because he is loose in his management style. He actually enables staff to produce and expects them to take responsibility, while the electrical engineers often end up doing the work themselves. He is a manager in an organization that does not encourage management.

John Cherell, the head of the project management office, is trying to create order and consistency in the new product development process. But Cherell does not have an engineering degree and does not have the support of the general manager. Bruce Cartwell, the purchasing head, is not part of the new product team, reporting instead to the finance and procurement office. His interest is bulk buying to reduce costs in product components. He does not always purchase materials when the team needs them because he is driven by an incentive to wait until components and inventory can be procured in volume.

Sid Still is an electrical engineer in charge of putting the instrument together and testing it before it goes into test protocol. He has ideas about a new resistor for the product that will increase efficiency and reduce costs, but will take some time to procure. Still is faced with the issue of whether to push for the new resistor in the light of pressure to get the instrument into product, cheaper, better, and faster. Mildred Best, the configuration management director, is a strong personality who uses her specialized configuration management software to control product designs.

No product can go through new product development without dealing with configuration management. Best tries to capture designs early in the process to control them; engineers try to avoid locking in designs with configuration management (CM) because the CM change process is time consuming and difficult.

Joyce Cobb, the documentation manager, is assigned to the team and has taken the responsibility to document each step in the process. She is trying to organize the process so that documents are produced by engineers at each appropriate step, but is challenged by the propensity of engineers, especially software engineers, to avoid documentation until late in the process.

Bob Smith, the manufacturing manager, is interested only in scheduled production and making sure he makes the monthly quota of product units. Smith works closely with configuration management to ensure that products can be assembled and produced, pressing for early information on scheduling and inventory needs.

Each of these team members participates in new product and service development daily, as opportunities surface in their work. Some of these ideas will flourish as they are accepted, others will not.

What are the motivations and thought patterns of those who participate in new product development activity, and how does program management handle the unique dynamics of the new product environment?

Akers, the mechanical engineer, is motivated by the prospect of creating a new housing for the instrument that is innovative as seen by his peers, e.g., the new housing creates minimal tolerances with internal electronics and is designed to cost. "Design to cost" means that the design takes into consideration the costs of production. To Akers, this means that he will be recognized as really good at what he does if he minimizes the tolerances in his housing through robust design and at the same time reduces the production costs associated with the mechanical housing.

Verdon, the electrical engineer and functional head of electronics, has a different focus. He is motivated by the need to maintain his head guru status in the company by serving as a general consultant on the new design. He stays in his office and waits for engineers to come to him with questions. He has quiet empathy for his electronic engineers and less-quiet disdain for the mechanical and software engineers outside his functional department. He would like his people to receive the recognition they deserve for designing a truly new, state-of-the-art, low-voltage system for the new unit that will change the standard for the industry.

Buster Right, the software engineer, is a hardened software engineer who has bumped around the industry and knows coding and software development—but is relatively new to the embedded software challenges of this avionics instrument. Right is also an advocate of a new software development process that is at odds with the linear, sequential approach used by the company. He believes he is creating new product concepts through his innovative *process engineering*. His approach is called staged iteration, a process that designs and tests various iterations of the eventual software on the test platform rather than designing software and coding for one, final test on a platform. Right's approach requires the platform to be ready earlier than usual in the process. Right's functional software chief, Jim Barrangan, is skeptical of the new approach, but goes along with it. In holding back his personal and professional views until well into the project, Barrangan creates a major bottleneck by later holding up the new product development process because he decides the new approach is dysfunctional and puts Right back on the drawing board.

Cartwright, the general manager of the research and manufacturing staff and a well-respected senior electrical engineer, is caught up in satisfying the constantly changing whims of the local vice president, the corporate technical and sales guru in charge of several local operations. Between these distractions and more remote, corporate micromanagement, Cartwright fends off the outside, corporate world to protect his project staff. His interest in new product development is to try to leverage the corporate aircraft fleet office so that an appropriate test aircraft is available when his projects needs it, a typical bottleneck in the new product development process.

Bruce Hall is a local community college graduate with an electrical test technician degree who runs the test lab, which is always overscheduled and understaffed. Hall juggles many projects wanting testing services all at the same time. Engineers tend to treat Hall as a lower-level staff aid who does not understand the sophisticated world of electrical engineering.

Bart Strong is a physicist and project manager. He is in charge of the project team but has no authority to manage them. Strong is not an electrical engineer and has an open-ended and collaborative personality at odds with the head-down, structured approach of the other engineers and project managers. He is not accepted into the *fraternal* grouping of inside professionals.

John Cherell heads up the project management office that Cartwright created to accelerate the productivity of the group. He is a senior author in the field and teaches at the local business college. He finds that his attempts to standardize and control project management practice, his interpretation of his role, comes up against the need of the key staff *to do their own thing*. Cartwright eventually becomes disinterested in the function amidst his problems with corporate.

Cartwell, the purchasing head, supports the project team but is marching to different orders from the local vice president to buy in volume. Buying in volume means promising high volumes of orders of components to reduce prices. Also, Cartwell is not buying until demand builds up from several projects. The company continually fails to meet its volume predictions, which leads to unfulfilled orders and major vendor problems, all of which delay project activity.

Still is an electrical engineer who truly sought out innovative product enhancements, including a new resistor concept that he had generated himself. Generally focused on his profession, he actually creates brand new approaches that are later accepted as major state-of-the-art contributions. For his work, he is unceremoniously laid off along with several of his colleagues when company management recognized what everyone in the staff had known for months, that it had overstaffed the group given the project load.

Best is the configuration management specialist and managed her function with discipline and narrow perspective, often using the role to control designs and components before they were fully developed. Typical in the new product development environment, there is constant tension between the engineers who want to keep options and designs open, and configuration management staff who want to lock in designs to prepare for production.

Joyce Cobb, the documentation specialist, is organized and competent. Her approach to new product concepts is to generate a documentation process that quietly organizes the process by framing documentation needs early and getting documentation done during design and development.

Bob Smith is pressed to improve production in a facility characterized by changing assembly workforce and undocumented manufacturing processes. His interest in new product development is to meet expectations for defect-free production of prototypes and volume units.

The point of this story is to suggest that new product development teams are characterized by a wide variety of motivations as well as internal and external forces that shape the effort and its output. An effective new product development manager understands these dynamics and works with them to ensure that the team performs. There is a time and place for new product innovation, during design and early product planning. But when a product is in configuration management the issue is getting it produced and marketed quickly, not to second guess the product.

Chapter

2

Strategic Alignment and the New Product Portfolio

New Product Portfolio

Especially in middle to large companies (100 employees and up), new products typically evolve out of a company-wide, strategic process of generating ideas, conceptualizing them into opportunities, new products and new services, and conducting an evaluation process that qualifies them for the company's *new product portfolio*. A new product portfolio is a collection of new product projects that have been generated, evaluated, and selected on the basis of their estimated business value, financial performance, alignment with business strategy, and risk management potential. They may or may not be funded. This discussion of a new product portfolio will focus on the Eastern case study, and how the process works from initial business planning through to selection and funding of new product candidates. This case study will provide a business application of *new product portfolio development.*

A new product concept becomes a project when it is selected to the portfolio and then funded with resources and company support. As the new product or service concept is being analyzed for selection, it is still a concept, not a project. Project identification begins when the company gives it resources, establishes a project team, and authorizes it as part of the company-approved portfolio.

New project process

A business will typically articulate a strategic plan, and state strategic objectives. It then identifies and categorizes programs of initiatives and specific projects. Programs are collections of projects and products in a particular area of the company's internal operations, product delivery, or customer service. Each of these candidate projects will support at least one strategic objective. When a portfolio of candidate projects is generated, each is analyzed in terms of its estimated cost and schedule, business value, financial performance, alignment with strategy, and risk management potential. A composite index is then constructed to allow for prioritizing the projects in numerical order. Finally, the company provides funds for projects from available funding categories based on each project's alignment with the company's business strategy. The concepts of cost estimate and project budget are given special meaning here. A cost estimate is the bottom-up "resourcing" and overall price of a project from its work breakdown structure and outline, using Microsoft Project or some equivalent software. Budgeting is the process of allocating available business funds to a project, thus giving it company legitimacy.

What differentiates this new product process from more typical portfolio development is the level of detail in the analysis and the timing for a *concept* to become a *project*. New product and service concepts will not be detailed as early as more conventional ideas because they present a high degree of uncertainty. In fleshing out the concept in the first phase of new product development, more detail is provided that sheds more light on the product before it becomes a project. Thus portfolio analysis continues on into the development process with new products, and the project reviews at the end of each phase allow analysis of go and no-go options.

The generation and evaluation of new product concepts and ideas is the most critical process for ensuring new product success. It is the preliminary articulation of the product and early judgments that serve to *filter and bundle* ideas and concepts for the purpose of narrowing down the alternatives. The risk is that good ideas do not survive because they are not well understood and not well represented. This process is one of advocacy as product idea sponsors are afforded the opportunity to make their case to management for a given new product concept in regular project reviews after each phase is completed.

The Eastern Case

The Eastern case demonstrates a process of developing a portfolio, beginning with business planning and strategic objectives in the manufacturing industry, and then moving to programs, new products, and projects. This process helps ensure that the new products actually funded and implemented are aligned with the business plan, promise good financial performance, and are accompanied by good contingency planning should risk events occur.

The risk of inadequate business strategy integration and new product development is inherent in the nature of a business. Business planning aimed at developing a business strategy considers various new product risks and threats to its success. But often the work of developing and implementing a portfolio of projects to improve the business does not align with the business priorities and plans.

This discussion uses the case study approach by addressing how a typical company—the Eastern Company—handles integration of strategy and new products in its business planning process.

The Eastern Company is a global manufacturer and distributor of aluminum products. Typically, Eastern faces major competition and challenge in a global aluminum market, and from foreign manufacturers who regularly "dump" aluminum into western markets at artificially low prices. The company faces major increases in the price of electricity, one of its core manufacturing inputs. Thus there is continuous risk in the business from forces somewhat "out of the control" of internal company leadership and project management. To address the risks of integration failure inherent in its business, Eastern prepared a risk-based strategic plan.

Eastern faces eight integration risks, and has developed eight strategic goals to address them, as seen in Table 2-1.

Eastern recognized the need to directly take action to sustain its capability to successfully compete on a continual basis in the world aluminum marketplace. The assumption was that despite the fact that Eastern employees—in general—were dedicated to providing the highest quality products and services to the customer at a competitive price, as well as providing a positive return for the owner's investment, Eastern was heavily unionized. The company was committed to the principle that, "*We will not be able to step up to those challenges unless our employees—and the union—can see where we are going and why, and have the opportunity to 'buy in.'*" Through this strategic and communication

TABLE 2-1 Eastern's Risks and Strategies

Risk number	Risk	Strategy
1	Required electric power will not be available at an affordable price	Secure economically priced power to reduce the risk of power shortage
2	Cost increase in aluminum manufacturing will increase faster than margin	Secure other resources at reasonable costs to offset the risk of cost escalation
3	Customers will not be satisfied with Eastern's products	Cultivate customer awareness and promote customer satisfaction to avoid customer satisfaction risk
4	Eastern's working environment will prove to be unsafe and the company will experience substantial loss of workforce and finance as a result	Create a safe working environment to control the risk of worker injury and associated costs
5	The Eastern workforce will not grow with the technology available for continuous improvement	Build a responsible and knowledgeable workforce to avoid the risk of workforce instability
6	Eastern will not act to improve the technology of manufacturing in time to keep ahead of competitors	Improve technology and plant equipment to produce products more efficiently to control productivity risk
7	Pollution from Eastern facilities will lead to noncompliance with government environmental requirements	Improve Eastern's impact on the environment to avoid the cost of pollution and noncompliance
8	Increasing waste in the manufacturing process and workforce will lead to uncontrolled costs	Reduce waste and non-value-added costs to control the risk of wasted effort

plan, Eastern saw that it could accomplish alignment and reduction of considerable risk exposure.

Special meetings and focus groups were assembled to discuss and explain the strategic plan and appropriate new product projects and ensure that all employees understood it and could relate their work to achieve success. Employees were encouraged to document actions they or their teams took to accomplish or support particular projects. This process continued as the plan was updated annually, and policies, procedures, and organizational structures were realigned to accomplish the plan.

This strategic plan was developed by the directors of Eastern with support from area managers.

Commitment and partnership

Eastern management and United Steelworkers of America stated directly that they were committed to the strategic planning process for the organization. It was clearly recognized that by working together to accomplish this mission, the interests of all participants would be served. All management and employees benefited from long-term job security, job enrichment, and the monetary rewards that resulted from a successful business that was able to manage its risks. Eastern's stakeholders and owners benefited from the product recognition and profitability gained by producing superior goods and services. Eastern's customers benefited from the high quality and service levels delivered to them. Finally, the community enjoyed a stable revenue base from Eastern's success, and from the skills and services individual employees offered.

Stakeholder relations

The company stated that Eastern's "stakeholders" were people, organizations, and groups of people who had a vested interest in the success of the company. Major stakeholders included:

- Employees—who seek continued employment and income, quality of work life, and opportunities to learn and develop. Their perceived risk was related to job security as well as lack of growth, development, and marketability.
- Customers—who seek quality products at low cost and reliable delivery. Their perceived risk was related to product quality and timing, but mostly price. Cost was a major issue as competitors dumped quality aluminum at lower prices.
- Owners—who seek return on investment and continued viability. Their risk was grounded in stock value.
- Regulators—who seek compliance with laws and regulations. Their risk was in noncompliance with regulations and the cost of enforcement and litigation.
- Community—that seeks contributions through taxes and service, and minimal environmental impact. Its risk exposure was in losing the industry tax base, but having to pay pollution and environmental control costs.

- Suppliers—who seek to meet Eastern's requirements and continue business with Eastern. Their perceived risk lays in their inability to meet Eastern's contract requirements and sharing more of the risk in contracted work than they can handle.

To illustrate the documentation of a risk-based strategy, the following discussion contains an executive summary, situation analysis, and detailed description of eight key strategies and how they were translated to new projects and products. The situation analysis provides a framework for the strategies including mission and goals, management direction, SWOT analysis, and linkage to the parent company strategic plan. The eight strategies are supported by specific initiatives and a system to measure achievement of those initiatives.

This strategic plan for Eastern Aluminum Company covered a five-year period from 1996—2000 and thus helped guide the company and its employees into the 21st century. As the general long-term pathway to growth and profitability, the plan presented the company's approach to achieving Eastern's central strategic goal: to compete successfully on a continuing basis in the world aluminum marketplace. The plan served a wide variety of purposes, including support to ownership decisions; support to budgeting and resource allocation; guidance for management and employee planning, training and education; and support to long-term capital investment planning. A major element of the strategic planning process that produced this document was the communication of the plan and its underlying vision, assumptions, and values to the employees because the hope was that they would generate new concepts and new products themselves.

The plan explored Eastern's current strengths, weaknesses, opportunities, and threats, and presented and discussed eight basic key strategies, initiatives, and measures to accomplish the central strategic goal.

Eight strategies

Within the overall framework of the basic strategic goal to compete successfully in the world aluminum marketplace, and to remain consistent with the parent company's strategic objectives, eight key strategies were at the heart of this strategic plan:

1. **Secure economically priced power.** Eastern would find ways to lower its power costs through a variety of strategies, including building stronger partnerships with power companies and state and local governments, and through exploration of independent options for generating less expensive power. It was quite possible that if power were not procured cheaply, the company would have to generate a new product concept to produce it onsite.

2. **Secure other resources at reasonable costs.** As the cost of materials rises, Eastern planned to find low-cost sources for raw materials as well as explore approaches for using lower-graded materials. Eastern would take the initiative to ensure effective partnerships were built with quality suppliers.

3. **Cultivate customer awareness and promote customer satisfaction.** Eastern would work to educate employees about customers and their requirements,

and promote closer customer ties. Greater appreciation of customers would give employees more incentive for addressing future requirements, and connecting their daily work more clearly with the "value chain" to the customer.

4. **Create a safe working environment.** Eastern was working to improve its safety record through enforcement of safety and health rules and regulations. Employees would be better educated and trained to understand safety implications of their work. Safety compliance would be considered a major performance standard for all employees, and employees would be expected to generate new safety products and processes.

5. **Build a responsible and knowledgeable workforce.** Facing a major workforce turnover in the next five years, Eastern placed special emphasis on strategies to build a more responsible and skilled workforce, improve its partnership with the United Steelworkers, improve performance and productivity, lower labor costs, and find better ways to work together through teamwork. They recognized that if this strategy—grounded in the commitment to building a team-based organization—is not accomplished, Eastern could not thrive and grow, even if the other strategies were accomplished.

6. **Improve technology and plant equipment to produce products more efficiently.** Eastern prided itself on its leadership in technology and technical innovation, and planned to continue this industry leadership. Eastern was managing several capital improvement projects to make major breakthroughs in productivity and quality. Eastern felt it was demonstrating that major investments were being made in the plant to meet the challenges of the future global marketplace. These projects were in fact new products and would be managed as such. This meant they would need to be designed, produced, tested with customers, and marketed to their own employees as "new ways of doing business."

7. **Improve Eastern's impact on the environment.** Through strict compliance with federal, state, and local environmental standards, Eastern would continue to respond to and anticipate environmental impacts and address them. Special emphasis was being made to meet new clean air requirements.

8. **Reduce waste and non-value-added costs.** Eastern continued to pursue quality and process improvement initiatives to eliminate unnecessary costs due to accidents, rework and scrap, outdated positions and job requirements, and equipment damage. Employees continued to be trained and educated in process improvement, and reengineering streamlined the way work was accomplished.

Overview on integration issues

Eastern had already turned around from a high-cost swing plant with a confrontational labor atmosphere to a much more competitive operation, practicing effective and efficient management and supervision, worker empowerment, and self-directed team concepts. However, there was new urgency to integrate employee views with the business strategy, to ensure that all employees understood that the plant would grow only "by permission" from future customers, and only if it continuously improved its productivity, quality, internal cohesion, and teamwork across departments. The following discusses how they were positioned to compete in the future.

Strengths, weaknesses, opportunities, and threats

The following discussion covers Eastern's strengths, weaknesses, opportunities, and threats.

Strengths. Eastern had made a concentrated effort to retain its competitive position in the marketplace through technology. Its major strength was its capability to continuously produce quality products, focus on technology and capital improvement, and keep wages and salaries relatively high for its employees while controlling costs. Capital improvements and improved management and team practices made it possible to achieve record premium production in the recent past. Eastern continued to demonstrate its leadership in technological improvements and plant capital investment.

Eastern had experienced the longest run at full capacity in its history, remaining one of the few North American smelters not curtailed due to the recent metal surplus caused by the flow of aluminum into the world market from the Commonwealth of Independent States. Eastern continued to show resilience and responsiveness in the face of changing market conditions.

Eastern was working hard to empower its workforce by improving knowledge, skills, and responsiveness. It sought to align its incentive and reward programs, partnership practices with hourly employees, performance appraisal systems, and quality and process improvement initiatives with key long-term strategies. It faced the future turnover of the workforce with a strong commitment to use change to build a leaner, more integrated, and productive plant team.

At the heart of its strength was Eastern's traditional core competency to choose, operate, and improve process technology effectively; to produce a variety of difficult-to-produce new premium products; and to understand and meet new customer needs. Whatever initiatives Eastern undertook, it knew it had to continuously improve these success drivers.

Weaknesses. Eastern's products (called primary, slab, billet, tee, and foundry pig) were priced by the worldwide commodities market. High quality and excellent service of these products ensured a positive customer relationship, but Eastern could not control the selling price of the finished product.

Eastern knew that it was a high-cost plant compared to other producers, primarily because of the age of the facility and technology, and because of high wages, salaries, and fringe benefit levels. Because Eastern had little or no control over the market price, the cost of producing aluminum became a key determining factor in remaining globally competitive. In fact, 75 percent of all aluminum in the world was being produced at a cost lower than Eastern could produce it.

Eastern faced major challenges in turning over its workforce and creating a more energetic and knowledgeable workforce team. Past practices had not always inspired employees to align themselves with the best interests of the plant, and commit themselves to new product and service concepts and continuous improvement through teams.

Eastern needed to improve its capability to learn and document its successes, in short, to become a "learning" organization. Past practice had not always taken advantage of what the organization had already learned through the years.

Opportunities. Demand for aluminum was continuing to rise; supplies of aluminum had increased each year with primary aluminum products now sold on a worldwide basis. Eastern had the opportunity to position itself midpoint on the world cost scale, the point at which 50 percent of world production costs would be higher than Eastern. In achieving this position, Eastern could take more advantage of its high-quality products and services, and improve productivity.

A reduction of four cents per pound by 1999 would have placed Eastern in that competitive position, keeping mind that other aluminum plants were also attempting to reduce their costs.

Eastern's major opportunity was to improve its process efficiency and productivity through a combination of technology and capital improvement, building a more efficient workforce, and reducing labor costs. The four cents per pound cost reduction could be achieved by:

1. Conversion of potlines (production lines) to a new "point feed" technology, already underway, a major "new process" concept
2. Reduction of manhours per ton by 15 percent from 1996—2000
3. Reduction of non-value-added costs wherever possible through process improvements, total quality management, ISO 9000 certification, and other quality initiatives

Eastern had a major opportunity to improve its human resource practices and programs as the plant transitioned its workforce in the coming five years, both through better training and development of supervisory and hourly employees, and better, more effective assessment and hiring practices.

Threats and risks. If Eastern did not continually reduce costs, their position would worsen because:

1. New plants with lower costs would open
2. Existing competitive plants would reduce cost and improve their cost position
3. Other plants with higher costs would close, worsening Eastern's position

The most critical of these risks was the possibility that power costs would continue to rise beyond Eastern's capacity to absorb them. This scenario represented the most significant threat to Eastern's continued growth and had to be avoided. In addition to power costs, the long-term cost of coal could be another important threat to Eastern's growth, as well as unanticipated environmental regulations, particularly from the federal Clean Air Act.

In addition, although Eastern had made major progress in building a more team-based culture, the process could not be slowed by resistance to change and failure to be clear about new roles and functions. Therefore, one source of threat and risk came from within, that being the threat of slow deterioration of the momentum of teamwork and process improvement already underway. Such a step backwards could always happen as a result of neglect, as well as lack of trust and respect in the organization.

Eastern's Strategic Plan

Eastern's strategic plan was an integrated set of strategies, initiatives, and measures supporting an overall goal of competitiveness. Figure 2-1 presents a graphic depiction of the company's eight key strategies. Each strategy was seen as serving the central goal of world competitiveness, but each strategy was also inextricably tied to the others, indicating a strong interdependency of all plan elements. If any one strategy and risk reduction plan was not accomplished, overall achievement of the goal suffered.

The plan described plant strategies, initiatives, and measures of success. Initiatives were programs and projects underway or planned to help accomplish a particular strategy. Measures were indicators of progress and were used to monitor achievement of the eight key strategies.

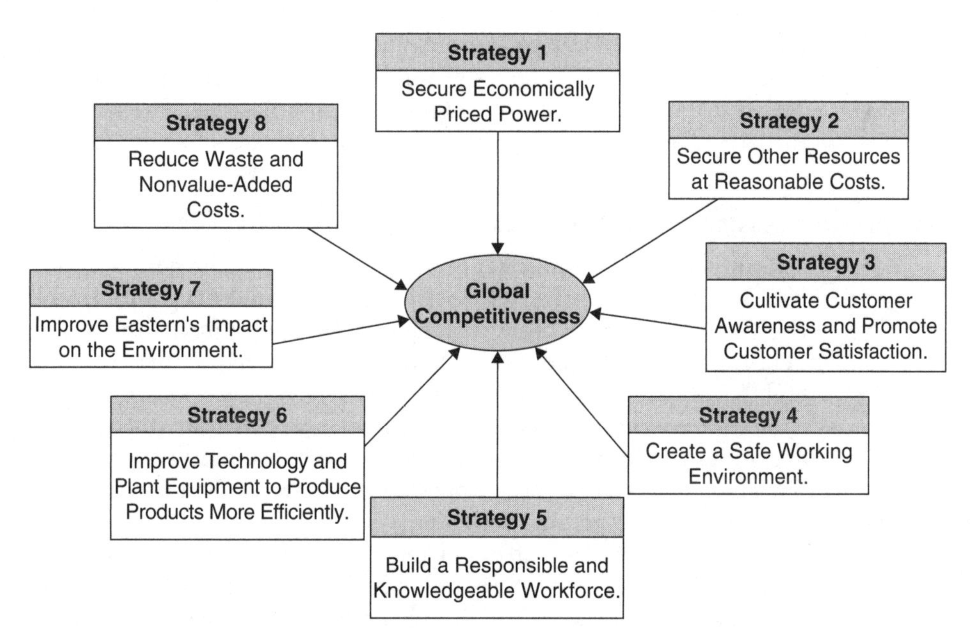

Figure 2-1 Eastern's business strategies.

Underlying Elements of the Risk-Based Strategic Plan

Five major elements formed the basis for this risk-based strategic plan: mission, commitment and partnership, driving forces, core competencies, and stakeholder relations.

Mission

Eastern's mission was to be the most cost-effective producer of the highest quality primary aluminum products, shipped on time to its customers, with optimum utilization of resources. They placed special emphasis on employees and their role in defining the mission, and on good community relations. Eastern recognized that accomplishing its mission involved a never-ending journey of continuous improvement.

Commitment and partnership

Eastern management and the United Steelworkers of America (Local 7886) indicated that they were committed to this mission for the organization. It was clearly recognized that cooperation with the union was a necessary ingredient of success.

Driving force: Production capability

An underlying element in this strategic plan was the single most important driver of company success: its capability to convert resources effectively into products, some new designs requiring new manufacturing processes, through highly organized and managed production processes. Their value added for the future will continue to be their capacity to produce products continuously.

Core competencies and risk contingencies

Three core competencies separated Eastern from its competitors:

1. **Its capacity to effectively choose, operate, and improve process technology.** Its capability to keep up with changing technology with new product and service concepts was rooted in its means to anticipate technology risk stemming from out-of-date technology.
2. **Its capacity to produce a variety of difficult premium products.** Its capability to change its production systems quickly was rooted in its wherewithal to anticipate the risks of change in new product requirements—including new products—and plan for them.
3. **Its capacity to understand and service customer needs.** Its capacity to understand its customers and especially to anticipate and manage customer services reduced the risk of failed customer service expectations.

Eastern would strive to maintain and build on these core competencies.

Eight Key Strategies

Eastern identified eight key strategies to carry out its central strategic goal of global competitiveness. Each strategy was being carried out through several initiatives, and was being monitored by the measures shown.

Strategy 1—Secure economically priced power

The cost of power was a major factor in Eastern's strategic plan. In its partnership with the community, power companies, and state and local government, Eastern would develop support for its efforts to continue to compete in the world aluminum marketplace. Eastern planned to negotiate lower power costs and explore independent options for generating less expensive power. These initiatives are shown in Table 2-2.

Power costs became the key factor in maintaining competitiveness because of impending major increases in costs from new sources. Eastern management was working closely with utilities, government officials, the community, and other power sources to ensure that it could achieve independence in power generation, should that be necessary. Power-wheeling sources and benefits were being pursued as well as self-generation options.

Although this issue was beyond the scope of any one employee, special attention was given to communicating the power cost issue to all employees so they could understand the urgency of the situation and help Eastern achieve power independence.

TABLE 2-2 Securing Economically Priced Power

Initiatives	Measures	Risks
Address power pricing issues by: maintaining relationships with Potomac Edison, Public Service Commission, People's Council, local and state government; investigate power-wheeling sources and benefits; develop alternative power sources, including self-generation; increase community support for reducing Eastern's power costs; eliminate power modulation	Reduced power costs by 2–4 mills per kilowatt hour (approximately $6–12 million/year); favorable public response and concern for Eastern's power pricing issues	Relationships with stakeholder agencies would deteriorate; power-wheeling sources (independent sources of power created by deregulation) would not provide lower prices; self-generation of power would fail either from technology problems or cost; the community would not support Eastern; power modulation (the practice of energy providers to reduce power) could not be anticipated

Strategy 2— Secure other resources at reasonable costs

The cost of materials continued to rise. This situation held the potential to erase savings created by increased productivity and reduced power costs. Eastern planned to find low-cost sources for raw materials as well as explore approaches for using lower-graded materials, as shown in Table 2-3. Eastern continued to manage human resources costs through employee attrition and retirements.

Key raw materials (alumina, aluminum fluoride, petroleum coke, and liquid pitch) were purchased for all parent company smelters by the same parent office. These costs were rising to a point such that Eastern's overall cost effectiveness was threatened. This issue challenged the company to find and use lower-graded materials, such as lower grades of petroleum coke. The company would continue to acquire both raw materials and supplies from the most efficient sources, all the while ensuring quality. This involved forming partnerships with suppliers to limit the number of such sources. This would accomplish two objectives: holding down costs, and minimizing purchasing and warehousing requirements.

As a major cost element, labor costs had to be controlled while productivity was enhanced through capital improvements, better management, and team and individual performance. Reduction of manhours per ton by 15 percent by the year 2000 was a major measure of success in reducing risk exposure.

Strategy 3—Cultivate customer awareness and promote customer satisfaction

Eastern continued to provide consistent and high-quality products and services to end-users and customers. As shown in Table 2-4, the company worked to ensure that all employees were aware of customers and their needs. Emphasis on the

TABLE 2-3 Securing Other Resources at Reasonable Costs

Initiatives	Measures	Risks
Obtain raw materials such as petroleum coke, pitches, alumina, and hardeners; secure high-quality supplies from the most economical sources	Maintain or decrease current raw material costs	Raw materials would not be available on a just-in-time basis
Manage human resource (labor) costs through attrition and retirements	Reduce manhours per ton by 15 percent by the year 2000; contribute to overall efficiency and productivity	Human resource costs would inflate and attrition goals would not be achieved
Explore innovative approaches to using lower-graded materials and lower grades of petroleum coke to produce new products	Maintain or decrease current raw material costs	Lower-graded materials would be acquired

customer would encourage the development of new products and services, and help Eastern establish a larger market niche. Eastern looked to external stakeholders to verify gains made in employee-customer awareness and customer satisfaction.

Eastern had to establish a market niche in high-quality, premium products to remain a viable company and successfully compete. To meet this demand, Eastern had to work closely with its ownership to identify future customer needs. Eastern continued to work with parent company marketing teams in the areas of initial order processing, customer team visits, and customer surveys.

Eastern would also make it easier for customers to deal with the plant. Increased use of bar code systems and electronic data interchange established a "seamless" electronic relationship with prospective customers. More attention would be paid to promoting the laboratory and metallography capabilities.

The continuing move to quality worldwide was having its impact on the company. More customer inquiries, such as from the automotive industry, were expected regarding quality standards. This development prompted efforts to maintain registration and refine documentation to both ISO 9000 and American

TABLE 2-4 Cultivating Customer Awareness and Promoting Customer Satisfaction

Initiatives	Measures	Risks
Enhance employee awareness about customer and end-product satisfaction	Third-party assessment of employees' customer awareness; recognition through accreditation and quality audits (American Association for Laboratory Accreditation, for example)	Employees were not able to connect their success with company success in end-product quality
Selectively diversify products and services to support market expansion—support new product development	Capacity to change products; number of customer assists through the Metal Quality Group	Its product mix could not be diversified
Support parent company marketing strategy; market services to make customers aware of Eastern's capabilities	Inventory Management System (IMS) data; customer team visit comments; customer satisfaction data	The parent company strategy was not consistent with Eastern's strategic plan and core competence
Focus on individual customer demands in metallurgy, product chemistry, packaging, and delivery requirements through process improvement; develop a long-term cast house plan and monitoring systems	International Standards Organization (ISO) 9000 registration; QS 9002 accreditation; monitor customer claims and contacts about technology services and products; review customer satisfaction survey results; improved product turnaround indicators	Eastern's process improvement efforts were not successful because of personnel and union disincentives
Set up cross-functional teams to increase awareness of internal customers	Internal customer satisfaction surveys; extent to which internal customer requirements are met	Cross-functional teams would not work because of internal conflicts and role definitions

Association of Laboratory Accreditation, and to attain QS 9000 and 14000 certification as well. Increased cycle time was becoming a major customer expectation, generating internal plans to develop systems to measure order entry, production scheduling, and shipping performance.

Because many employees did not have a direct relationship with customers and customer needs, the company was undertaking a program to enhance employee appreciation of customer needs and new product opportunities. This program included use of a third-party organization to monitor employees' understanding of these issues.

Strategy 4—Create a safe working environment

Eastern had significantly reduced accidents in recent years, and needed the support of employees and management to continue these safety efforts (see Table 2-5). In addition to developing and implementing state-of-the-art safety procedures and guidelines, the company needed to consistently enforce safety and health rules and regulations.

Eastern recognized its responsibility and accountability for the safety and health of each employee, and for the preservation of property and equipment. The company would continue to incorporate safeguards and procedures into the design and operation of all facilities that minimized risks of personal injury and loss of property and equipment. Management was responsible and accountable for the safety and safe work conduct of all employees who report to them. Employees were equally responsible and accountable for safe practices as well as assisting in the on-going safety program by reporting unsafe practices, procedures, or conditions when they were observed.

As indicated in the initiatives of this strategy, Eastern was giving special priority to upgrading engineering standards to reflect safety requirements and criteria. In some cases, this could have meant added cost and time constraints on planned capital projects, an expense well worth the investment in a safer working environment.

TABLE 2-5 Creating a Safe Working Environment

Project Initiatives	Measures	Risks
Eliminate safety and health hazards by:	Decrease accident incident rate	That safety initiatives would not be accepted and implemented by employees and managers
Upgrading engineering standards, safety features, and ergonomics consistently	Stay within accident and safety scorecard budget	That safety guidelines and regulations would shift substantially
Promoting employee awareness	Improve safety severity ratio index	
Update Joint Safety and Health Committee guidelines	Rate of completion of items on safety list and audits	
Enforcing rules and regulations	Improvements in efficiency and job performance	
Increasing team and employee accountability for safety and health	Improved plant safety performance record	
	Increased safety gainsharing payout	

Strategy 5—Build a responsible and knowledgeable workforce

By increasing the skills and abilities of individuals, teams, and supervisors, Eastern would be able to increase productivity, reduce operating costs, solve personnel problems, and increase teamwork across the entire plant (see Table 2-6). Initiatives in support of this priority included training and developmental opportunities in support of self-directed work teams.

Strategy 5 held the key to successful achievement of the other strategies—the building of a workforce and organization that: (1) was aligned with the strategic direction of Eastern; (2) was structured, capable, and motivated to improve performance; and (3) worked together across departments to provide a "seamless" process of production and quality and to generate new product concepts.

In building a flatter, more streamlined workforce, the company's strategy in the past had been to press for reduced manning and more teams and teamwork. As a result, many teams had been generated and trained to take responsibility for the problem solving and decision making necessary to keep their process operating at peak efficiency. Supervisory and hourly positions were reduced and roles and functions were changed.

In the spirit of building the total Eastern organization, the company's strategic emphasis went beyond reduced manning and generation of teams. The strategy was focused on organizational effectiveness: building the whole organization through a stronger linkage and alignment between management, supervisors, and bargaining-unit employees. The opportunity before company management was to build new

TABLE 2-6 Building a Responsible and Knowledgeable Workforce

Project Initiatives	Measures	Risks
Continuation of empowered, self-directed work team development (decision making and responsibility); new performance appraisal system for salaried employees; development planning; knowledge and skills training for bargaining unit employees; supervisory development program; strategic plan communication process conversion to parent salary structure; new bargaining unit job classification (stemming from the labor contract); HR strategic plan	Better communication and coordination within team members and between supervisors and teams; innovative, timely, and sound employee and team decision-making; better use of tools, equipment, and raw materials; employees will be prepared to assume new responsibilities as a result of developmental exposure; enhanced partnership agreement; increase in ideas and solutions from employees; reduce manhours per ton	Self directed teams would not work in the unionized work setting; employees would not act on incentives to train and develop new skills; strategic communication plan is not effective in improving employee support of company goals.
Offer developmental opportunities to sustain employee education and growth through mentoring, inside training, outside technical managerial training, and opportunities to manage	Successful development planning; track progress through training records; enhanced employee performance	The plant could not implement mentoring and training initiatives because of the company culture

supervisory roles and functions into the new team-based organization, requiring development of leadership skills, better business and productivity management and monitoring skills, and more support for technical supervision and cross-department process improvement. Support services such as human resources management helped to lead the effort. Organizational barriers to effective supervision were identified and eliminated. Organizational and training initiatives were underway to help supervisors function as the guiding force for day-to-day operations.

Development tools included business and productivity management, process improvement, facilitating and mentoring opportunities, inside training, outside development (technical and managerial), and management opportunities within the organization. To focus on incentives, Eastern reviewed its performance appraisal and gain-sharing structure to ensure that they were aligned with this strategic plan, and made improvements when necessary.

To ensure effective communication, quarterly plant communication meetings continued and more information was provided to employees "online," especially in the area of human resources.

Strategy 6—Improve technology and plant equipment to produce new products more efficiently

The company was managing several capital improvement projects to upgrade the condition of equipment and work processes at the plant. The company needed to continue these improvements while also employing sound capital project management skills. Eastern would work to speed up completion of these capital projects and to keep them within budget and quality requirements (see Table 2-7).

As evidenced in the partial list of capital improvements listed in Table 2-7, Eastern was heavily engaged in upgrading its technological and equipment base to maintain its leadership and core competency. The company was a front-runner in keeping pace with required capital improvements to an aging plant infrastructure. Improvements, essentially new products, were underway in the product production lines, carbon plant, cast house, substation, and laboratory, and in general plant functions such as emission and noise control, and information system management.

The focus for this strategic plan was the completion of new capital projects within budget, schedule, and technical requirements. This meant developing a stronger capital project management system and employing more effective project management practices.

Strategy 7—Improve Eastern's impact on the environment

The company would continue to monitor its impact on the local environment, as shown in Table 2-8. These efforts would be directed toward reducing environmental degradation and pollution.

TABLE 2-7 Improving Technology and Plant Equipment to Produce Products More Efficiently

Project initiatives	Measures	Risks
Complete capital program and budgets each year; conduct major maintenance projects and overhauls	Completion of capital improvements, including conversion of potlines to point feed technology, substation life extension, cast house continuous homogenizing furnace rod shop anode cleaner, ladle shop ladle cleaner, bake oven rebuild potline capacity expansion, rebuilt remelt furnace, developed stack filter systems for metal treatment, facilities expansion, completed stamper upgrades for billet and slab	Capital budgets would not be completed; maintenance projects are not completed for a variety of reasons
Conduct research to ensure that Eastern adapts or incorporates improved or emerging technologies	Completion of research and development (R&D) projects within budget	Necessary research on emerging technologies is not conducted
Develop a stronger capital project management system (CPARs) through training and other developmental assignments	Improved capacity to complete projects on time within budget and schedule	The SPARS system is not made operational

TABLE 2-8 Improving Eastern's Impact on the Environment

Project initiatives	Measures	Risks
Comply with federal, state, and local environmental regulations by providing proactive assistance to regulators, educating employees about regulatory requirements, promptly reporting non compliance and correcting any violations, and filing Title V air permit application	Eliminate incidents of non-compliance; monitor response time for identifying and fixing violations	New regulations would be enacted that Eastern could not respond to
Participate in voluntary activities on environment, safety and health issues, such as EPA Greenlights, reducing greenhouse gases and PFCs, and noise nuisance reduction	Eliminate environmental, safety or health complaints about the plant or its operations	Voluntary efforts do not improve community relations
Encourage environmentally sound industrial and agricultural growth	Partnerships with state and local agencies	Local growth objectives and dynamics would change substantially
Continue farm production	Farm production and maintenance of safe environmental practices	Company efforts at farm production (the company formed the area around the plant to improve its appearance) around the fringe of the plant were unsuccessful

This strategy addressed the company's environmental and community relations practices. Eastern continued to stay ahead of environmental requirements through two basic approaches: being proactive in assisting regulators at all levels in developing sound and cost effective regulations that both implement environmental legislation and meet the needs of community and the business, and planning and implementing capital improvements and operating measures to comply with environmental requirements, attempting at the same time to ensure that such improvements also contribute to overall plant productivity.

Costs of compliance increased as well in the administrative areas of record keeping, reporting, training, planning, and monitoring, and in acquisition of necessary monitoring equipment, creating the need to streamline these systems. Eastern continued to develop the capacity to prevent pollution through technology improvements and through a multi-media approach that addressed losses of material to air, storm water runoff, and solid or liquid waste streams.

Strategy 8—Reduce waste and non-value-added costs

Eastern continued to experience waste and non-value-added costs, such as safety and property costs related to accidents, rework and scrap, and equipment damage. Process improvement and problem-solving teams continued to focus on reducing these costs (see Table 2-9).

TABLE 2-9 Reducing Waste and Non-Value-Added Costs

Project initiatives	Measures	Risks
Involve Quality teams in identifying and resolving quality problems in key production processes	Record amount of rework and scrap by department on a monthly basis Stay within approved budget guidelines for rework and scrap costs	The company Quality teams were not successful in resolving quality issues
Minimize equipment damage by educating employees, monitoring equipment use, and enforcing rules for properly using equipment	Review monthly maintenance to ensure departmental accountability for responsible equipment use; stay within approved budget guidelines for equipment expenses	Equipment damage rates continued
Eliminate duplication of effort in administrative processes; process improvement/reengineering; encourage employee use of best practice techniques	Benchmark other processes; monitor process costs	Administrative redundancy and increase costs of operation continued
Improve inventory management of supplies and equipment (includes maintenance, production and raw material in-process)	Reduce inventory by at least 5%	Inventory management initiatives were not successful because of internal plant or supplier performance limitations
Minimize waste generation and increase recycling	Waste product reductions	Increasing rates of waste production continued

This strategy was in concert with Strategy 5—to build a knowledgeable and productive workforce. Both were required to improve overall productivity. This strategy was key to improving the overall productivity of Eastern by eliminating waste and unnecessary work, for example, reducing the cost of poor quality through process improvement and ISO and QS 9000 and 14000 documentation.

The company's quality and process improvement efforts started on the production floor, where quality was built-in through consistent practices and extensive use of statistical process control methods. Eastern was committed to being quality driven, not cost driven, thus the quickest route to elimination of waste and non-value added costs was "doing it right the first time." They looked to this strategy as a major factor in lowering the operating expenses by four cents per pound.

The Quality teams would continue to identify and resolve quality problems in key production processes; a new focus would be placed on administrative and support processes to ensure that they were under review in the context of process improvement as well.

Communicating Strategy and Risk

The company prepared a communication program to promote the company strategy and explain the risks inherent in the business and the local plant setting. The structure of that plan is presented in Figure 2-1. The strategic goal was to improve Eastern's capability to compete on a continuing basis in the world aluminium market place.

Programs and New Product Ideas: Generation of a New Product Portfolio to Implement Eastern Strategies No. 3 and 6

Designing programs of new product ideas

Once this process produced some sense of direction, ideas to implement these strategies were collected from leadership, management, and employees and filtered through a Product Portfolio Council using the *ideation* concept. Ideation is a process of generating and capturing product ideas and concepts and translating them to forms that will allow definition and evaluation. Ideation works only in companies that are characterized by a culture of innovation and creativity. Eastern faced considerable obstacles to innovation in the traditional nature of its unionized employees and the large capital investments in doing work the way they had always done it.

Now let's look at Strategies 3 and 6 and how individual projects were generated and framed in the new product portfolio. The program of projects for Strategies 3 and 6 involved technical and production process improvements, and would logically come from the technology and manufacturing departments, but not exclusively. Therefore the company asked the engineering department to head up the development of new product ideas in this program category. As is often the case, new internal products such as manufacturing process improvements must be linked with new products envisioned for marketing and sales. Working with marketing and operational people in production, they came up with the following programs and new product ideas and concepts.

After these strategies were articulated and communicated to the workforce, the company opened up the ideation process for ideas and new product concepts. As an example of what can come out of this process, two core strategies were selected to start the program and new product development:

Strategy 3: Cultivate customer awareness and promote customer satisfaction

Program Area 1: Customer Awareness

Program Goal: To generated new product ideas for making customers aware of plant operations and costs

New Product Idea 1-Develop new video on plant operations for customers

New Product Idea 2-Develop new onsite workshops for customers in aluminum production and product development

New Product Idea 3-Develop new function and role in plant management for customer awareness and education

New Product Idea 4-Develop new website for customers on plant operations

New Product Idea 5-Develop new "tour" of plant operations for customers

Strategy 6: Improve technology and plant equipment to produce products more efficiently

Program Area 1: Plant Equipment Products

Program Goal: To generate new product ideas for internal improvements to plant efficiency

New Product Idea 1-Design and build new plant mobile vehicle

New Product Idea 2-Design and build new plant communication system

New Product Idea 3-Design and build new plant production information system to allow customers to participate in designing new products and new production systems to deliver them

New Product Idea 4-Design and build new cost model for reviewing production costs

New Product Idea 5-Design and build new robotic hot metal conveyer system

Each idea was then developed into a *concept paper* that was circulated within the plant staff, marketing, and sales for initial responses. Each concept was analyzed for strategic alignment, financial performance, and risk by small work groups, and developed by a new product team through the process defined later in this book. No program management office was there to guide the process as it would likely be today. Later on we will describe the analytic process through which new products were reviewed for alignment, financial performance, and risk, and how they were translated into projects with schedules, resources, and teams.

A Special Case: New Product Idea 3:

New product idea #3 involved the design and building of a new plant production information and simulation system to allow customers to participate in designing new products and new production systems to deliver them. This new product idea contained elements of internal production and new consumer product at the same time. The idea was to provide options to customers to participate in the design, demonstration, and building of various product prototypes using Eastern equipment and manufacturing process in a simulated environment. The new product concept came from manufacturing engineers looking for ways to market new consumer aluminum products by allowing prospective customers to *virtually build* them through this product.

Since the product had internal production, marketing, and computer technology implications, a cross-functional product team was put together to explore the product concept and guide it through portfolio analysis and project selection and planning. The team was to act as a product advocate, of sorts, to make the business case for the product.

Postscript to the Strategic Plan

The Eastern strategic plan had been designed as a guidepost for the future, a way of realizing the vision of becoming more responsive to changes going on globally, more supportive to customers and employees, more cost effective in manufacturing processes, and more useful in guiding new product development and project management. However, it was not a "cookbook" for success. Eastern recognized that management and employees would continue to have to make informed judgments together each day to make the plan work. And they would have to learn better from their successes and mistakes.

Acquisition and merger

Although the strategic planning and risk reduction process was a focused and comprehensive process, and had measurable impacts on plant productivity and success in dealing with its costs and product problems, a major development was not anticipated in the process—acquisition.

During the process of developing and implementing the plan, the company was acquired by a competing parent company, creating a high degree of uncertainty and disruption in the process. Work in implementing the plan was delayed until the acquisition and merger process was completed, but the new company leadership endorsed the process and bought into its outcomes, thus guaranteeing that its product portfolio would survive.

Integration in Global and International Projects

Global, international projects inherently face difficult integration issues simply because the normal and typical integration barriers are compounded by political, national, and language issues. Projects to assist developing countries, for instance, are typically managed by the collective of non-governmental (NGO), national, donor country managers, and local team members, who often do not agree on the purposes and goals of the project. Opportunities for fraud, waste, and abuse are again compounded because of the many *players*.

Postscript on Integration and the Eastern Case

The Eastern case is an interesting application of new product development issues at the strategic level, e.g., how does a company integrate its strategic objectives with its program of projects, in this case the manufacturing, new product, and tooling improvements? And how does it integrate its strategies with its employees' views, in this case aligning key union leaders and its own plant management on the key issues? And then how do individual projects improve the business, e.g., producing a simulated manufacturing software program to allow customers to design their own products on company equipment, get designed and implemented?

Analyzing a New Product Portfolio—General Lessons fron Other Cases

Once a new product is legitimized and made part of the portfolio, it goes through at least three *filters*: (1) it is assessed for alignment using a weighted scoring model; (2) it its financial performance is forecast using net present value and cash flow estimates; and (3) its risks are analyzed using a risk matrix.

Weighted scoring model and net present value

How are candidate new product and service projects reviewed and prioritized in a typical company? Let's take another case, the Seitz Company, a manufacturing and distribution company facing the issue of whether to fund a new plant equipment project or build a completely new plant in Huntsville, Alabama.

First we will use the weighted scoring model to see how the two projects measure up to the strategic objectives of the company, and then we will use the net present value analysis to estimate the financial performance of the projects. These two tools are typically used to help make the business case for a new product; a high score in the weighted scoring model means that the new product or project will support the direction the company is going; a high score in the net present value analysis means that at the discount rate chosen, the so-called hurdle rate for investments in that firm, the project will produce a return above what could be returned from an investment that breaks even at that discount rate.

Let's say that the Seitz Corporation is in the process of selecting major investment programs for the 1996–1997 period. A total of $3,000,000 remains available for projects during this period. One selection remains and will be made from two candidate projects, which have been culled over from over a dozen new product and new project proposals. Two new product ideas are generated:

Project 1
Name: New Equipment
Submitted by:

- Steve Pokorski, Vice President of Operations
- Joe Downs, Director of Plant Engineering

Project description:
The project would focus on the replacement of the extrusion equipment in the West Milwaukee plant. The proposed new equipment is to be manufactured based on a prototype designed and built in Down's developmental facility. All existing equipment had been acquired from three major suppliers. Pokorski authorized the prototype because he felt that the suppliers had become unresponsive to Seitz's bid requests. The technical superiority of the new type of equipment will allow for significant improvements in productivity.

Project 2
Name: New Huntsville Plant
Submitted by:

- Janis Clark, Vice President of Marketing

Project description:
This project proposes to establish a new plant in Huntsville, Alabama. During the last three years, sales have been steadily increasing in the area to the point where Seitz has achieved third place in the market share. Clark claims that the market is extremely price sensitive and Seitz's more efficient production capacity would ensure quick attainment of the second position if transportation costs to markets in the southeast were lowered.

Part of the evaluation is based on the strategic objectives the leadership provided. Table 2-10 shows the weighted scoring of each objective—the weighted value of each objective and how each project performs against that objective. Each objective is first assigned a percent by making judgments on what relative weight to give each objective—assuming they are not all of equal value.

The second part of the evaluation is based on the net present Value Model that lists the return over the next five years utilizing a NPV factor at 12 percent. The first year cost is the cost of producing the project, e.g., the cost of plant equipment improvement in the case of project #1, and the cost of building a new plant for project #2. Then a forecast is made of net income that will be produced each year over the next five years; net income is the sum of all costs and revenues by year. Then each net income value is discounted by the chosen discount rate using a new present value table.

TABLE 2-10 Weighted Scoring Model

Priority	Company strategic objectives	Weight assigned to strategic objective	Score project 1 new equipment against objective	Score project 2 new plant against objective	Composite index project 1 (percent × score)	Composite index project 2 (percent × score)
1	Double total sales within the next decade	40%	8	10	320	400
2	Develop and market new products based on the company's plastics experience	15%	5	10	75	150
3	Reduce dependency on equipment suppliers	15%	10	0	150	0
4	Be first or second, based on market share in any regional	5%	5	10	25	50
5	Increase productivity	25%	5	8	125	200
	TOTALS				**695**	**800**

The result shows that the new plant, project #2, is more aligned with company strategic objects.

Table 2-11 shows the new present value analysis.

Risk matrix sample

Next, a risk matrix is developed for the two candidate projects. The risk matrix is a table capturing the following for each project:

1. The major risk inherent in that project, judged by a group of stakeholders and company management.

TABLE 2-11 Net Present Value Model

	Project 1			Project 2		
Year	Net income	NPV factor	NPV	Net income	NPV factor	NPV
0	($2,475,000)	0	($2,475,000)	($2,550,000)	0	($2,550,000)
1	700,000	0.89	624,999	550,000	0.89	491,071
2	800,000	0.79	637,755	600,000	0.79	478,316
3	800,000	0.71	569,424	900,000	0.71	640,602
4	800,000	0.63	508,414	1,150,000	0.63	730,845
5	800,000	0.56	453,941	1,300,000	0.56	737,655
TOTAL			**$2,794,535**			**$3,078,490**
Return			**$319,535**			**$528,490**

TABLE 2-12 Risk Matrix

Task	Risk definition	Probability	Impact	Severity	Contingency plan
Build new plant	New plant cannot compete because of high costs from expensive labor base in local area	25 percent	Major loss of revenue	Showstopper	Conduct simulation to see if new plant productivity can meet demand efficiently with local labor base
Upgrade current plant production system	New manufacturing equipment will not enhance manufacturing productivity	25 percent	Quality and schedule impacts	Showstopper	Conduct simulation to see if new equipment can enhance time to market and cut production costs

2. A definition of that risk in risk management terms
3. Probability, the probability that the risk will occur, from a best judgment of stakeholders or sponsors
4. Impact, the nature of the impact of the risk should a risk event occur, e.g., where and when will the risk "hit us"
5. Severity, how severe is the impact thought to be, e.g., will it be a *showstopper*
6. Contingency Plan, what can we do to prevent the risk and/or offset it with mitigation

The risk matrix (Table 2-12) does not score the projects; its function is to qualitatively assess each project so that risk and contingency can be included in the decision making or business case made for the new product or new project.

Funding New Product Projects

When a portfolio of new products or new projects is generated using the analyses above, the candidate projects enter the resource and budgeting process. They are typically prioritized and available funds are allocated to each project based on a weighting scheme of some kind using the results of the weighted scoring model used to select them.

For instance, see the example below (see Figure 2-2) showing how three new product projects are funded. First, the company vision and mission are stated and broken down into goals, or program areas, such as Consumer Products, Business-to-Business Products, and Government Sales. Then five strategic objectives are stated (as in the weighted scoring model), numbered in the figure 01 through 05. Each objective is given a priority percentage (using the ones developed in the weighted scoring model or updated ones), shown in the fraction below the objective number. For instance, objectives 02 and 04 rank high at .3, while 01 and 05 are low at .1.

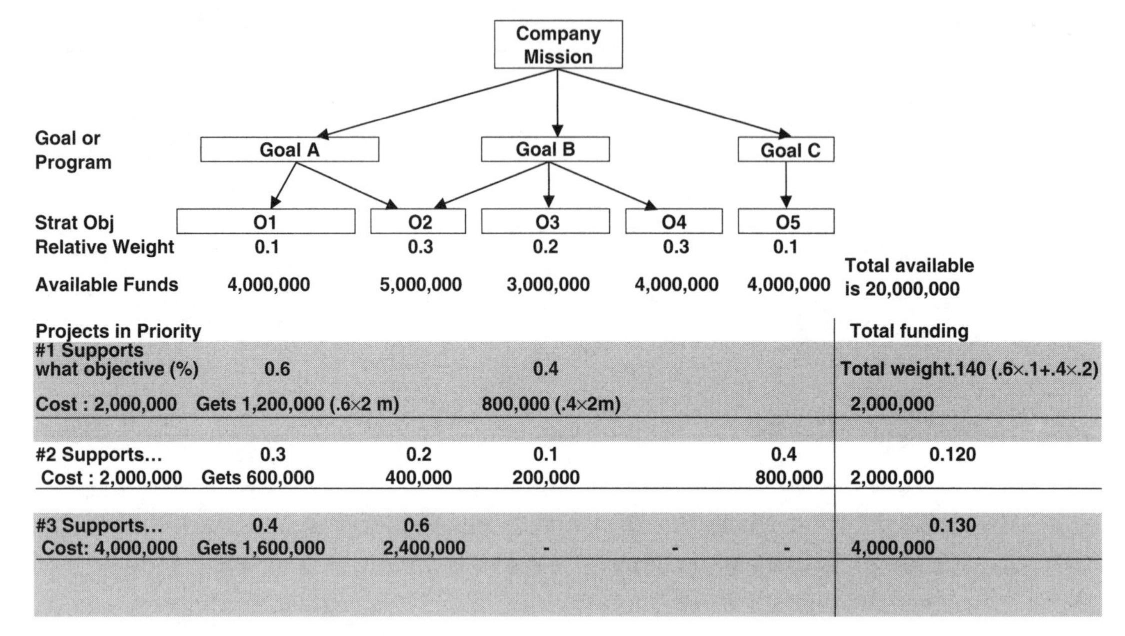

Figure 2-2 Project funding model.

The company's available funds budgeted for new product development, in this case $20,000,000, are then allocated to each strategic objective based on its priority and the estimated costs of projects in that area. Below candidate new product projects are listed with their estimated costs, 1-3.

Each project is now compared to each objective to see which projects support which objectives, using percent as an indicator. Based on a scale of 10, each project is aligned with objectives it supports in proportion to their relative importance. In project 1, it supports objectives 01 and 03. But it supports objective 01 more than it does objectives 03, so 01 gets a .6 and 02 gets a .4. To develop the project score, this figure is multiplied by the priority ranking of the objective it supports, producing its composite score. Project 1 was multiplied .6 by .1 for .06 and .4 by .2 for .08, for a total score of .140, which shows on the right column as the score for project 1.

If the objective is to fully fund each project, you can see that these three projects are fully funded, drawing from each strategic objective fund as appropriate.

The New Product Development Pipeline

In sum, new products and new projects go through a screening process that can be characterized as a *pipeline*, with funded, high-priority projects coming out of the pipeline for prioritization, funding, detailed project planning, and kickoff.

Figure 2-3 shows a graphic representing the pipeline new products go through as they are generated, analyzed, and selected. Concepts and ideas are generated

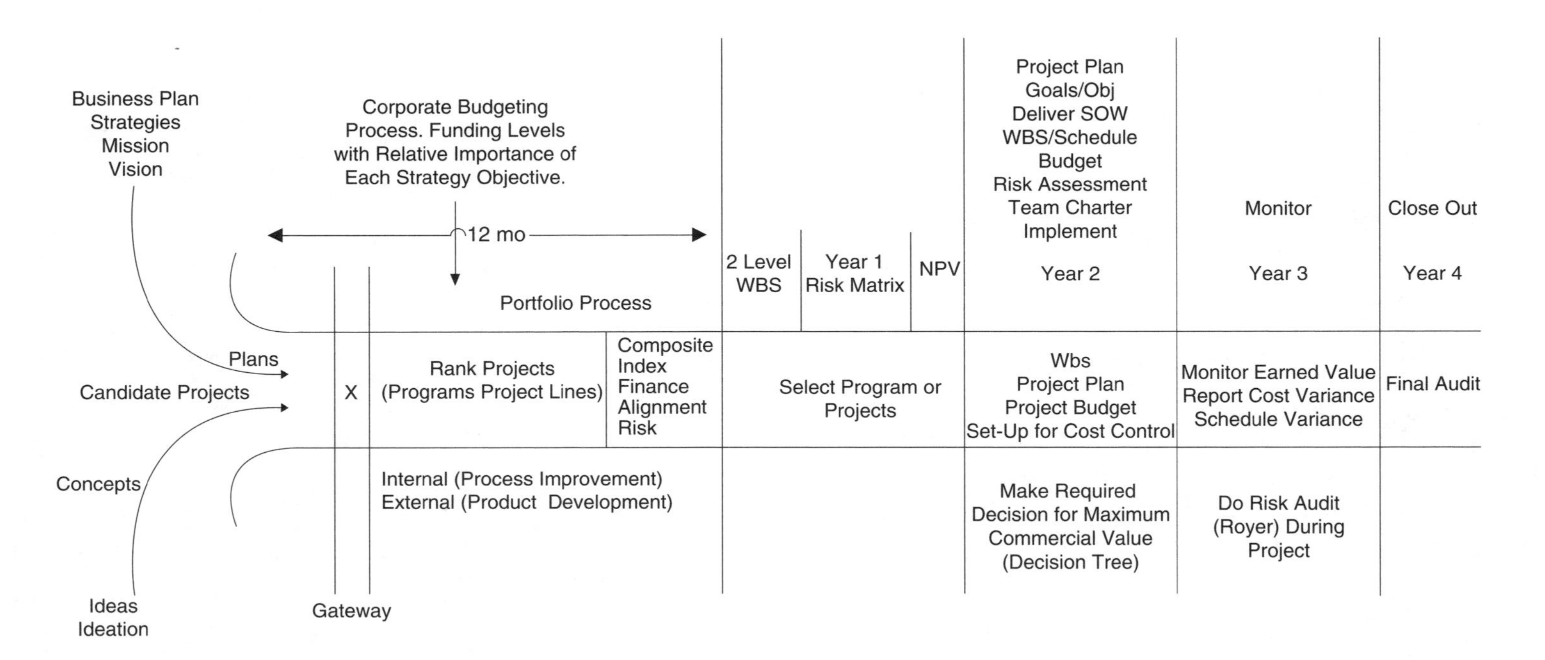

Select Priorities
Alignment (Weighed Score Model)
Financial Performance Tools: Rate of Return, Ash Flow (Npv), Risk (Project Level)
Budget - Funding (Capital Rationing) - Take Funding Levels and Prioritize Projects. Fund From Top Down Until Money Runs Out.

Figure 2-3 New product pipeline.

and aligned with strategy and plans, then proceed through an initial gateway, a go or no-go point. Projects are ranked using a composite index reflecting financial performance projections, alignment with strategy, and risk. The company funds the projects based on the process explained above. Selected projects are then defined and implemented.

Chapter

3

Project Integration and Setup

Project Management System

While new product development is often an open-ended process, the company must have a system in place for making new product decisions. A company can lose control of new product development projects unless there is a working and integrated project management system in place. New products raise critical risks and decision options during development and are frequently open-ended until development is nearly completed; therefore, control and analysis must be integrated into the process at every key juncture.

Project integration management, or integrated *program and project management* as it is sometimes called, involves selecting, coordinating, and synchronizing projects in a company or agency so that all the key factors for success are *optimized.* Program managers see both the big picture and the details of program and project work, all at the same time. The project review process is set up so that management can make *go or no-go decisions* at each key transition point from one phase to another.

Integration involves analyzing project business value at the high level, mobilizing team performance and dynamics, monitoring projects to ensure midstream adjustment and project recovery, resolving technical, resource, and interpersonal conflicts at every level, managing program interfaces and multitasking, identifying organizational constraints and *exploiting* them, keeping tabs on accountability, and reporting to avoid ethical and waste problems.

In an integrated project organization, program and project managers:

- See projects as capital investments made by the organization to achieve corporate business strategy and competitive advantage
- Analyze a portfolio of candidate projects using net present value techniques in combination with risk assessment and other tools
- "Read" team dynamics as they relate to team leadership, motivation, effectiveness, the decision-making process, and conflict resolution

- Formulate and evaluate alternative completion plans to optimize program quality, cost, and schedule. Select and apply appropriate project planning and tracking tools as well as project management leadership skills to recover the program
- Develop the appropriate program interface event management plan to integrate and manage the projects effectively, and resolve any complexities that result from multi-tasking within the networks or on the part of the project manager
- Plan the development and implementation of suitable project management systems and methodologies to support multiple projects. Identify best practices that lead to effective program management
- Develop and manage an integrated program schedule. This includes the elements of the schedule, types of schedules, schedule development, and the processes of schedule management
- Work with the structures of schedules based on the Integrated Master Plan (IMP), Integrated Master Schedule (IMS), supplier data integration, and the overall influences of the WBS
- Establish requirements and make the "make/buy" decision. The make/buy decision determines whether the product will be made by the company or bought from a supplier
- Work with networking basics and critical path tools (critical path identifies the linkage of interdependent tasks in a project that will take the longest time to complete, e.g., any delay in the critical path delays the project as a whole)
- Work with integrated product and process teams (IPT), their outputs, and their relationships to each other. They also recognize key product and process team tasks and their relationships in a program schedule, and are able to analyze the performance aspects of an IPT
- Work to transition from Integrated Master Plan to the Integrated Master Schedule
- Manage the integration of production recurring schedules with the nonrecurring program level schedules
- Manage reporting of integrated schedule data to customers
- Understand when and where to use Earned Value Management (EVM), and how this system is used to benefit the project management effort
- Design and manage an Assignment Matrix (RAM)
- Serve as an effective cost account manager, and understand how cost accounting supports the cost account manager
- Provide a working level definition of a project/program baseline schedule to include the basic objectives of that schedule
- Implement the planning and budgeting process within the resource loading activities of the schedule integration process

- Produce a variance analysis report, and understand the process of arriving at the conclusions cited in the report
- Define and integrate risk management, and elaborate on the process of risk analysis and risk management
- Provide the definition for low risk, medium risk, and high risk while being able to integrate the likelihood of the occurrence with the consequences of the occurrences onto the tailored risk grid, and incorporate necessary mitigating elements

Reviewing the current state of project management literature, it could be argued that the word "integration" is used in so many different contexts and applications that its usefulness is in question. Engineers integrate products and systems; managers integrate and coordinate their organizations; and planners integrate their plans. Yet, despite its prolific use, the term still defines a critical aspect of successful organizations, projects, and teams—working together toward program and project objectives. The integration of product and system means that the components come together to produce product performance and customer satisfaction. They come together because people make them come together; integration just doesn't happen, it must be proactively encouraged by the participants in the program management process.

Integration as a Leadership Function

Organizations effectively integrate their program and new product development project work when their leaders encourage it. Systems do not integrate unless key people at the working and project levels actually *think* integration. Thinking "integration" is a way of looking at your work as interdependent, as a part of the whole. Information is shared in an integrated organization simply because the key people know that shared purpose and shared information serve the customer better, faster, and cheaper. New product developers, however, often want to do their work in isolation, off to the side of the main stream of the company. Management needs to work against this tendency to stay "offline" by making new product development a key business system.

Leaders prepare their organizations for integration by loosening bureaucratic barriers and encouraging cross-functional training and work settings. Information systems encourage integration. For instance, an electronic time sheet system is tied into networked Microsoft Project software so that project managers can see actual costs in real time. Leaders insist on these supporting systems because they know the value of information and sharing in building products that work.

Integration as a Wide Ranging Quality and Process Improvement Standard

Integration is addressed in a wide variety of quality standards for corporate management, and for program and project management, including the Project

Management Institute PMBOK, the National Baldrige Quality Award, the PMI OPM 3 maturity model, and critical chain concepts. Along with increasing complexity in systems and projects, and the challenge of putting together the efforts of global outsourcing teams, the concept of integration becomes more and more important to achieve "cheaper, better, faster" project cycles.

For instance, the national Baldrige Quality Award criteria are used by many companies as benchmarks for best practice in integrating planning, operations, and project/process management. The Baldrige criteria address integration in terms of alignment and consistency of purpose and in measurement of outcomes. For instance, the 2005 criteria for health services organizations read:

> This item examines your organization's selection, management, and use of data and information for performance measurement and analysis in support of organizational planning and performance improvement . . . This performance improvement includes efforts to improve health care results and outcomes (e.g., through the selection of statistically meaningful indicators, risk adjustment of data, and linking outcomes to processes and provider decisions). The item serves as a central collection and analysis point in an integrated performance measurement and management system that relies on clinical, financial, and non-financial data and information. The aim of measurement and analysis is to guide your organization's process management toward the achievement of key organizational performance results and strategic objectives.
>
> Alignment and integration are key concepts for successful implementation of your performance measurement system. They are viewed in terms of extent and effectiveness of use to meet your performance assessment needs. Alignment and integration include how measures are aligned throughout your organization, how they are integrated to yield organizationwide data/information, and how performance measurement requirements are deployed by your senior leaders to track departmental, work group, and process-level performance on key measures targeted for organizationwide significance and/or improvement. (Baldrige Award Health Criteria 2004, p. 40).

Translated to the project management environment, the Baldrige criteria stress the importance of selecting projects that implement business goals and plans, and ensures outcomes of multiproject portfolios and business processes (such as project planning and control) are tied together through alignment with the business direction.

Tools in Building an Integrated Project Management System

The theme of project management systems is the tradeoff between cost, time, and value or performance. The objective of the system is to optimize all three since one can always be enhanced at the expense of the other two. Figure 3-1 shows time, cost, and value.

A new product project can be brought in early, but the process of crashing the project can increase the cost and create quality and value problems. On the other hand, the project cost can be reduced, but this action may produce schedule and timeline problems, and again quality impacts. If value is increased beyond the scope of work and customer requirements, the price will be paid in cost and time.

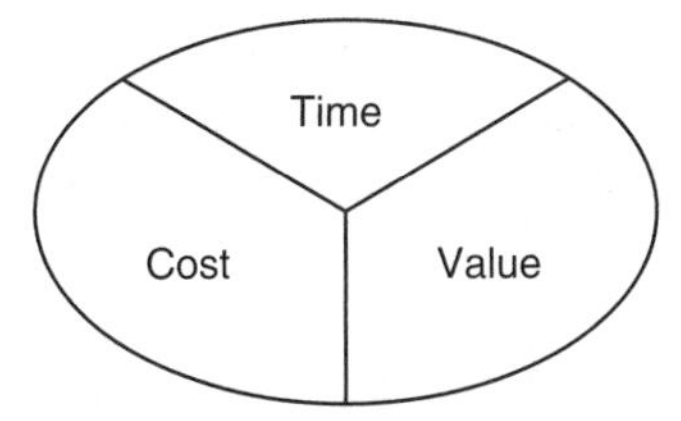

Figure 3-1 Time, cost, and value.

In building a company or agency system to support new product development project integration, there are ten areas for process improvement. These are: organization- or enterprisewide project management systems, program portfolio system development, integrated resource management systems, information technology, technical product development (including a stage-gateway review system), interface management system, project portfolio management, project monitoring and corrective action, change control, and program evaluation.

Organizationwide project management system

The following are the attributes of a fully mature project management system.

- **Integrated project management culture.** Leaders develop their organizations to accomplish integration through systems and communication. This system involves the development of a culture of defining and capturing work in terms of projects, e.g., all work of the organization outside of recurring production work is considered project work with a customer and deliverables. All training and development, and incentive systems, are built to encourage work to be accomplished through formal projects, plans and schedules that integrate cost, time, and quality.
- **New product project council.** A new product project council is made up of top management, marketing, functional departments, and project management. This council manages project reviews at the end of concept definition and full development, and make the go or no-go decisions.
- **Generic work breakdown structure.** A generic work breakdown structure (WBS) is a task outline in sequence, but not linked. The purpose of the generic WBS is to integrate the work, which is project coded to capture costs and task performance history, with the scheduling of any task the company takes on in any project. The generic WBS defines each task in a *data dictionary*, or task definition that covers what the task expectation is and what its deliverable is, for a safety task in a product development project.
- **Scheduling system.** A scheduling system places all work in a project schedule software, e.g., MS Project or Primavera, assigns resources, and estimates costs to control the work. Integration of all the work of the company is accomplished through scheduling, which is seen as a process of *committing resources to work.*

- **Resource assignment.** Resources are assigned to projects and tasks so that the workforce is integrated into the work that is authorized and sponsored by the company. Projects are seen as investments in the business plan; therefore, there is a major impetus to capture the work being performed in a resource assignment system.
- **Task linkages and interdependency.** Projects are consolidated and tasks are linked to stress the interdependency of project work. No piece of work in the company is left *unconnected* to ensure integration.
- **Matrix team structure.** The matrix structure ensures integration because functional departments and project teams are intermingled in every aspect of the company's work, from projects to process development and improvement. Project teams are staffed by functional departments that are in charge of the quality of the work and the development of technical processes and systems. Project managers manage assigned team members toward project deliverables and earned value.
- **Work authorization system.** Again, to ensure that work that goes on is authorized work, all work is authorized and directed by the project manager. The way work is approved is through the baseline schedule that defines the authorized work.
- **Guidelines for the project management plan.** The project management plan is defined in a company policy statement to guide for the definition and control of the work. Therefore, the plan must include control points, e.g., project reviews, when project management authorizes advancement of a project from one phase or stage to another. Reporting and monitoring strategies, including the use of earned value to integrate cost, schedule, and quality performance, should be made explicit.

The plan should also address accountability, particularly in view of the recent legislative and regulatory requirements of the Sarbanes-Oxley Act. This requirement of compliance with internal control and accounting standards is no longer optional for project managers. In fact, the price of disconnected and inconsistently applied efforts throughout a project and its interfaces, and lack of financial tracking systems that provide for audits, could be businesswide. Compliance with Sarbanes-Oxley, therefore, is not a choice but a requirement, and the plan should state standards for estimating costs, tracking the costs and relating costs to work performed, and monitoring the integrity of the closeout procedure and invoices to customers for work performed.

Program/portfolio planning and development system

The following are the attributes of a program/portfolio planning and development system.

- **Business planning system and strategic objectives.** The integrated company has a business and strategic planning process that produces a statement of strategic objectives to serve as a guide for all planning and budgeting. Such a system helps to shape the project portfolio and ensures that the company invests in projects that are integrated with the direction of the business and its ownership.

- **Decision process.** Some kind of decision process supports integration because open decisions, if prolonged, can lead to waste and ineffective work. Decision trees are used to assess the commercial value of various decision paths involved in defining the task structure and sequence of approved projects.
- **Budgeting system.** A capital rationing system, or some way to allocate company resources in line with the priority of relative strategic objectives, is part of integration. After budgets are identified to carry out business plans, projects are planned and prioritized in the portfolio system; then costs are estimated. Finally, projects are funded according to their relative merit against business plans and available budget.
- **Risk management system.** Some kind of risk management planning system that identifies and assesses risks, and generates risk contingency plans, is necessary in an integrated project management system. The risk matrix is the format for developing risk information that is used in scheduling and controlling the work.
- **Program definition.** Programs are sets of projects with similarities in process, product, and customer base. Definition of longer term *product lines* help clarify the boundaries of a given program over time.
- **Portfolio pipeline system.** A pipeline of approved projects is maintained so that as funds and resources become available, projects are quickly initiated. Project plans and schedules are produced for projects in the pipeline so that when authorized they can proceed quickly.

Resource management system

There needs to be a way to manage resources in an integrated project management organization simply because there is value in targeting all resources and equipment on the right project work. A resource pool can be established using MS Project that records all assignments to keep a running view of how people and equipment are being utilized.

- **Workforce planning.** A workforce planning system integrates the hiring and training of personnel with the needs of the program portfolio of projects. In others words, people, equipment, and systems are brought into the company to fill needs that are made explicit in the project resource allocation pool that reflects both current and planned work. Measures such as *person month needs* by project are used to predict resource needs.
- **Staffing planning.** A staffing system allocates staff to the priority project needs to fully integrate the core competence of the workforce with the priority needs of key projects. Staff is focused on assignments that are visible and reviewed regularly.

- **Financial and accounting control.** Financial and accounting control is ensured in a project management system that captures all project costs, both direct and indirect, and ensures internal controls on project costs and equipment inventories.
  - *Earned value.* Reports on work progress and costs are used to calculate earned value so that the company knows how each project is doing in terms of schedule and cost.
  - *Industry standards.* Industrial cost and work standards are used to control the estimated duration of scheduled tasks, e.g., using a trade association to schedule an industrywide activity on which there are work and industrial standards.

Program information technology system

A program information system that documents all project work in consolidated schedules and resource pools ensures that work is staffed, planned, and monitored in a uniform way. This allows comparison of project progress and supports decisions on where to focus resources.

- **Network system.** All program and project information, e.g., schedules, resource pools, project review, gate review data, and configuration management documents are kept on a company intranet to allow wide ranging visibility.
- **Accessibility to key information.** Accessibility to information is controlled and focused on need-to-know criteria. However, customers are regularly informed on program and project progress through MS Project Central Web–based reporting systems that allow review of schedules without parent software.
- **Reliability planning.** Reliability planning targets products with failure mode effects assessments and functional hazard assessments, along with risk matrix documents, to consistently design and test reliability of product performance to customer requirements and specifications.
- **Workforce training.** Workforce training is designed to meet project needs as evidenced in work performance feedback reviews and lessons learned exercises with project teams in closeout.

Product/service development process

Integrated project management cannot be accomplished without integrated product development processes with strong Stage-Gate milestones.

- **Project review process.** Project management is a process of managing time, cost, and quality, but the underlying strength of any project integration process is a strong, phased development process with clear controls on entry to the next stage. Project reviews are used to decide to proceed or terminate a project.
- **Technology support and testing.** Technical support that meets industry standards ensures that product integration and testing are verifiable. Designs

are tested against specifications, specifications are tested against scope of work, and scope of work is traced to customer requirements and expectations.

Interface management

The following are the attributes of interface management.

- **Matrix organization.** Interfaces between functional departments, e.g., accounting, engineering, project management, and testing, are ensured through strong interface management. Separate departments and functions are brought together constantly through information and reporting systems and face-to-face review meetings at key gates.
- **Program review meeting formats.** Review meetings are controlled by generic meeting agendas, and data and information support from a professional project management office or staff. This way review information is objective and consistent.
- **Procurement interface.** Because of the importance of contract and outsourced work, contractor personnel and processes are integrated with sponsor company personnel and processes. Common scheduling and reporting systems are designed.
- **Financial, accounting, and internal control interfaces.** New impetus for strong accounting and accountability reporting now requires that project managers capture costs and be able to relate costs to work performed and equipment purchased and in inventory.
- **Marketing and sales interface.** Integration of marketing, sales, and project work is accomplished by assigning marketing and sales personnel to project teams. They attend and input to the teams on customer developments, and learn what they can and cannot commit to customers.
- **HR interface.** The interface with HR is important to integrate personnel and HR policies and procedures with project work and priorities. Performance reviews are left flexible, yet they are important in assigning resources to future projects.

Portfolio management

The following are the attributes of portfolio management.

- **Top management visibility of programs and projects.** The whole set of projects in a multiple project system is managed consistently in an integrated project management system. All projects are monitored using common earned value and other measurement systems, e.g., balanced scorecard.
- **Uniform project management system.** A uniform approach to projects in the portfolio is ensured through a professional project management staff and PMO support system.

- **Pipeline management.** Pipeline management consists of:
  - *Generation of projects.* A systematic way of generating projects through brainstorming, budgeting processes, and business planning
  - *Evaluation of projects.* A way of reviewing portfolio projects using net present value and cash flows, weighted scoring models to score projects against business objectives, and risk management
  - *Selection of projects.* Projects are selected using a uniform set of measures

Program monitoring and control system

The following are the attributes of a program monitoring and control system.

- **Project management office (PMO).** Monitoring is based on earned value reporting, and quality is ensured by a task planning system that relates percent complete to defined milestones in the baseline schedule.
- **Corrective action/risk management process.** Contingencies and corrective actions are based on remaining work, and are forward-oriented. Contingencies are embedded in schedules to ensure that should risks occur, contingencies have already been scheduled and budgeted for.
- **Escalation system for decisions.** Conflicts and differences within project processes are reviewed regularly by top management to ensure that decisions are not delayed.

Change management system

The following are the attributes of a change management system.

- **Change management system.** A change management system is especially important in new product development because design changes are typical and frequent, and each change in configuration or design must be documented.
- **Change order system.** All changes to a scope of work are submitted by project team members or the sponsor/customer to ensure that changes are reviewed and managed.
- **Change impact system.** Change impact statements are prepared for all substantial changes, with risk, schedule, cost, and quality impacts specified.

Program evaluation system

The following are the attributes of a program evaluation system.

- **Document lessons learned.** Closeout includes a lessons learned meeting and documentation of the outcomes. The PMO is responsible for ensuring that lessons learned are integrated into future projects.
- **Financial auditing system.** A financial and program audit system is managed to ensure accountability and internal control of all assets.

Limitations of Integration Systems

Systems don't integrate new product projects—people do. Even if the organization is able to design and install compatible systems to help integrate projects, these systems will not work if the people who manage the work don't use them. Configuration management as an integrating function in product development between design and production cannot be effective if the configuration manager does not see both ends of that spectrum. Project managers who are obsessed with schedule and on-time delivery at any price and do not take costs into consideration *will not help the business succeed.* The lesson is this: Individual project success should not be at the expense of the business itself. To ensure that this does not happen, company leadership must continuously work toward an integrative vision and process at all levels of the organization—leadership personnel must daily walk the integrative talk.

The Critical Chain Concept

Critical chain theory borrows heavily from integration concepts as it links scope and time management to risk management. The critical chain approach to project planning emphasizes developing a WBS and project network, and focuses on identification of dependencies and starting tasks on time. The focus is on doing the work as quickly and correctly as possible, and not locking in on task durations that are often impossible to predict in a new product project. Dependencies require coordination and integration. From systems theory we recall that systems will go naturally into disorder, e.g., that the forces of system dynamics tend to push outward, away from the center. Integration acts in contrast to the normal centrifugal forces in a project and its environment. Critical chain focuses on the use of buffers, or allotments of time that are "tapped" up front by project managers and doled out as necessary to offset risk events and unanticipated problems. Because most networks are highly complex, a statistical analysis of all the inherent risks in starting tasks on time is usually impossible. Chains of tasks typically include a myriad of risks, many of which are the result of *disintegration*, the opposite of integration. Thus cost and schedule control tend to manage disintegration. For instance, as two components—one software and other electrical—of a product are being designed, each effort tends to make design assumptions independent of the other, only to find in downstream integration and testing that different assumptions made integration impossible. Thus the process of integration forces the electrical and software designers to share assumptions upfront, through concurrent work, constant cross-functional communication, information sharing, design reviews, and eIntegration (electronic) tools.

PMI OPM (Organizational Project Management) 3

PMI's OPM 3 maturity models (in fact, all such maturity models) actually measure the extent to which an organization is integrating its program and project work. OPM 3 integrates:

- Design and implementation of organizational strategic planning
- Identification of projects
- Determination of team and project chartering conditions
- Changes in priority and allocation of resources
- The process of managing the environment
- Management of the program and project portfolio

The overriding theme of OPM 3 is continuous improvement, the systematic and sustained improvement of business processes and products. This typically means linking information systems and teams to achieve standardization whenever possible.

Balanced Scorecard

Measures are important because people tend to do what is measured. The balanced scorecard encourages the measurement of four main areas—financial, process, learning, and growth—framed overall in the customer's perspective. Project managers have traditionally focused on time more than cost simply because the customer typically focuses on time as the priority. Now financial measures are more important, due to Sarbanes-Oxley reporting and accounting requirements. Thus financial reporting at the project level becomes important as a part of the earned value process, thus truly integrating cost and time.

eProcurement

The increasing use of electronic procurement systems has tended to tie businesses to businesses in an unprecedented network of collaboration and cooperation in the acquisition of materials and goods, and services. Product development teams now integrate contractors and vendors directly into the project through eProcurement systems that allow instant exchange of project and product information and documents.

The reason integrated project management is important is that we are increasingly aware that projects fail or "underperform" because of the lack of organizational and management support and the dysfunctional separation of key financial, human resource, marketing, and IT systems from project management in the typical company. In fact, projects naturally *disintegrate.*

In the integrated model, the basic tools of project management, e.g., work breakdown, scheduling and schedule variance control, chartering, resource management and cost variance control, project team development, portfolio project selection, product development, quality control and assurance, project review and performance monitoring, and interface management, are placed in a simple conceptual framework.

Integration: Concepts and Models

What does new product development and integration look like? One way to answer that question quickly is to look at two ways to achieve cost control.

The simplest way to control cost is to match actual spending to the spending plan, pure and simple. An accounting office might do control that way, lacking any other information or perspective. This is a classic mistake in cost control that does not integrate with work performance or value.

The integrated cost control approach, on the other hand, involves looking at cost from the standpoint of work performed and quality/value achieved, not simply in terms of costs incurred. Integrated cost control is a *forward integration tool* that points toward completion of the work, keys on current progress, and matches costs to quality output. Forward cost control involves looking at the variance between the work performed in project execution against *what it should have cost to do that work.* It also looks at the value or quality of the deliverable at any given time to ensure that the customer is getting value for the dollar spent. In other words, a good indicator of whether you are forward integrating your cost control is whether invoices for work performed are paid.

What does integration feel like? In other words, this overused term has a significant meaning in many fields, perhaps beyond its literal translation. Asking what integration feels like is not as superficial as it may sound. To explore when we have it is to explore how to get it. These indicators come to mind:

- When there is complete integration, a project deliverable reflects in its performance and value to the customer and stakeholders all of the project requirements and components outlined for it in the project plan and work breakdown structure, along with horizontal and lateral coordination. Further, the design and performance of the product facilitate the customer's performance because they are integrated *into the customer's systems.*
- When there is complete integration, all parties to a program and project—managers, team members, support people, suppliers, and customers—all are delighted with the project outcome and deliverables, and their roles in its success.
- When there is complete integration, all costs, schedules, quality, risk factors, and changes along the way are adequately reflected in the final outcome and due dates. The learning that occurs in a program or project is integrated into the product or service through *integrated* change controls.
- When there is complete integration, the professional and technical project staffs, e.g., administrative staff, support people, software engineers, mechanical engineers, electrical engineers, construction workers, who participated in defining specific components for a technical product deliverable feel they have made a significant contribution and gained new working relationships with their colleagues.
- When there is complete integration, there are no surprises and there has been an effective blending of cost, schedule, and quality considerations along the way in the project cycle; earned value has been maximized given project developments.

Understanding integration

What significance does integration have for program and new product project managers in tomorrow's business settings? Because the term *integration* is a key theme of this book, let's explore further what integration means. The concept of integration has many dimensions, individual, technological, organizational, interpersonal, and informational, but the core concept of integration is grounded in *connection and alignment.* But why will integration be more important in tomorrow's business organization; what makes integration key to organization and product performance?

Integration means *completeness* and *closure*, bringing components of the *whole* together in an operating system. Components of a larger system—increasingly global in nature—are brought together to create performance; but what is the process of integration and how does it work generically? The answer lies in systems theory; a system is a series of parts working together with a common objective. After the system is defined, the analysis function breaks down the whole into its components for purposes of understanding, building, and managing the system. Integration then puts the "built components" of a system back together to create a performance model that is aligned, e.g., all components work together as they were designed to do.

Projects must be internally and externally integrated. Internal integration means that project work packages, deliverables, and systems are connected; external integration means that the project interfaces with customer systems and produces value for the customer and the market/industry as a whole. Repeated internal and external project integration produces economic development in the larger community and societal framework.

The characteristics of integration that help to frame our understanding of program and project management and that underlie this book are as follows:

- Systems don't integrate; people do. The individual and project team member, working with an external contingent of support people and stakeholders, is the beginning of integration. The way people who work in a project environment think about their roles, responsibilities, and tasks creates the conditions for integration. *Integration thinking* means that as people perform their functions, their behaviors reflect an awareness of impacts on other team members and on other product components—and, most importantly, on the customer's satisfaction with the outcome. *Integration support systems* connect key aspects of project performance so that data is produced automatically on cost, schedule, and quality to allow informed decisions.

- *Forward integration* means that communication and connection are focused *forward* on producing deliverables and creating customer satisfaction, not necessarily to bring a project back to its original plan. Plans are estimates; real work performance serves as the basis for corrective action. Forward integration is a downstream concept in which work is performed to provide value downstream toward the deliverable; sequence means that integration occurs at the right moment in the process. This is a horizontal function, cutting across traditional functions to create synergy and cooperation.

- Top management builds the culture and mechanisms for successful connection and integration, and involves extensive coordination by a centralized program and project management function that works to avoid disconnected efforts throughout the enterprise.
- Integration means integrity. There is a connection between integrity, e.g., producing what you promise and doing it in a professional and ethical way, and integration, which is making sure required connections occur at the right times. The outcome, product, or service has integrity because it is integrated.
- Accountability requires integration; new requirements, including the Sarbanes-Oxley legislation, demand top management fiscal accountability, making financial and work performance integration imperative. The new requirement for internal accountability stresses internal control and checks and balances. Once seen as a low level accounting and audit requirement, this new mandate now requires integration at every level of the organization, including programs and projects.
- Integration begins at the business level. New forces require a new way of thinking about business itself, business strategy and operations, projects, and markets. These forces come about from developing changes in the landscape of business management, most notably at the global level. Integration now occurs across geographical, economic, political, and system boundaries as never before.
- The "regime" of business—the whole business enterprise system—is changing, too, as more and more middle- and small-sized businesses surface and disappear with the tides of business fortune. How does a business organization, designed as it is to grow and profit through serving customers, ensure that it "plays" in the regime of business fairly and with integrity. Such a business plays by the rules not just to avoid regulatory and government interference, but because *the business equates truth with integrity and success.*

Integration model

The following integration model captures the essential factors requiring a new level of integration in program and project management. Figure 3-2 shows program and project management integration.

People. People integrate, not systems, so people are trained to coordinate and interact with program and project participants, forming a true interdisciplinary team.

Projects. Projects become more cross-functional as project work is defined in terms of coordination and integration of work.

Technology. Complex products are managed at the interface, placing more emphasis on product and service integration.

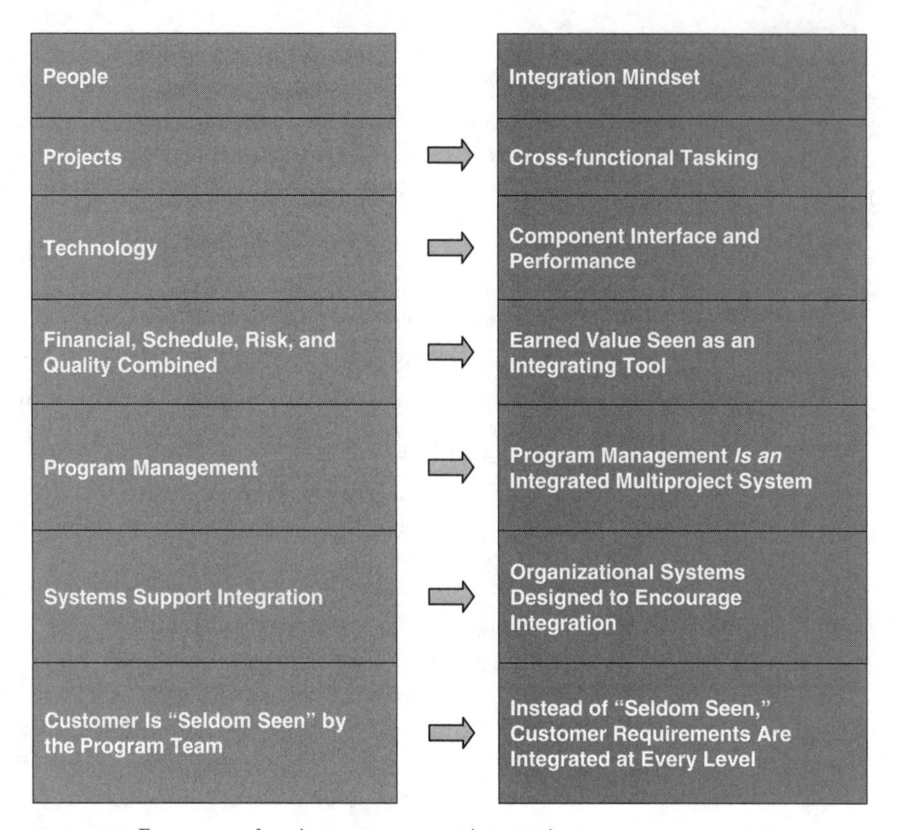

Figure 3-2 Program and project management integration.

Financial, schedule, risk, and quality combined. Through earned value and integrative tools, program and project progress is seen in terms of the combined impacts on financial, schedule, risk response, and quality issues.

Program management applications. Integration defines the program manager's role, working between top management and project managers. Program managers integrate projects with company plans and strategies, and work with enterprisewide resource management systems.

Systems support integration. Organizational and information technology systems are designed to interface with each other and to encourage integration.

Customer is "seen" by the program team. All program and project activity is performed with the customer in full view, integrating the work with the customer's expectations.

Project Integration Management: Organizational Issues

The organization model for successful project integration management starts with the steps shown in Figure 3-3.

Figure 3-3 Keys to successful project integration management.

Prepare the organization

As shown in the analysis of the Project Management Institute Project Management Body of Knowledge (PMI PMBOK) requirements for integration later in this chapter, the concept of organizational and technical coordination starts with a key business process focused on integration. The process of "putting things together" and recognizing the interdependencies of the project process is the first priority. It starts with preparing a culture and supporting mechanism for working together for a shared outcome and deliverable, and designing systems to integrate rather than diffuse organizational performance. Preparing the organization starts with a corporate and enterprise policy on integration, backed up by top managers who *walk the talk* that brings people and systems together, and treats the outcome as a shared vision.

Develop systems of integration

Systems of integration include business processes such as design reviews, project earned value analysis, configuration management, software compatibility, and interdisciplinary assignments. The more the organization provides mechanisms that in their very existence further integration, the more successful the integration process will be.

Develop integration skills

Integration skills start with a mindset that one's project tasks fit into a larger whole, and that the success of the project and the enterprise itself is dependent on the collective success and efficiency involved in producing the product or service.

Recognize integration success

What you recognize and measure is usually what you get; therefore, the success of integration in planning and project management begins in what the company measures and rewards. Measurement of integration success can be accomplished using various indicators of integration:

- Duration of integration tasks, looking for faster integration turnarounds
- Conflict intensity during integration
- Lessons learned

Integrate with the customer

The final integration is the alignment with customer expectations and systems, leading to a "perfect storm" of timing and performance with the customer's processes. This integration can be accomplished by keeping the customer involved and engaged throughout the project life cycle.

More Detail on the PMI PMBOK Standard for Project Integration

The PMI standard for project integration has fundamentally changed from its early form—a narrow focus on project-only issues—to a broader treatment of project integration from an organizationwide and global view. Project integration is now a Project Management Knowledge Area, published in 2005, that includes the processes and activities needed to identify, define, combine, unify, and coordinate the various processes and project management activities within the project management groups, e.g., initiating, planning, executing, monitoring, controlling, and closing. In the project management context, integration includes the characteristics of unification, consolidation, articulation, and integrative actions that are crucial to project completion, successfully meeting customer and other stakeholder requirements, and managing expectations. Integration, in the context of managing a project, is making choices about where to concentrate resources and effort on any given day, anticipating potential issues, dealing with these issues before they become critical, and coordinating work for the overall project good. The integration effort also involves making tradeoffs among competing objectives and alternatives.

What this means in simple terms is that integration has become the essential pulling together of project and organizational systems and processes for a multiproject, portfolio approach to project management. Integration is essentially the function of program management, running several projects at once using all the organization's support systems.

The need for integration in project management becomes evident in situations where individual processes interact. For example, a cost estimate needed for a contingency plan involves integration of the planning processes described in greater detail in the project cost management processes, project time management processes, and project risk management processes. When additional risks associated with various staffing alternatives are identified, one or more of those processes must be revisited. The project deliverables also need to be integrated with ongoing operations of either the performing organization or the customer's organization, or with the long-term strategic planning that takes future problems and opportunities into consideration.

Most experienced project practitioners know that there is no single way to manage a project. They apply project management knowledge, skills, and processes in different orders and degrees of rigor to achieve the desired project performance. However, the perception that a particular process is not required does not mean that it should not be addressed. The project manager and project team must address every process, and the level of implementation for each process must be determined for each specific project.

The integrative nature of projects and project management can be better understood if we think of the other activities performed while completing a project. For example, some activities performed by the project management team could be to:

- Analyze and understand the scope. This includes the project and product requirements, criteria, assumptions, constraints, and other influences related to a project, and how each will be managed or addressed within the project.
- Document specific criteria of the product requirements.
- Understand how to take the identified information and transform it into a project management plan using the planning process group described in the PMBOK guide.
- Prepare the work breakdown structure.
- Take appropriate action to have the project performed in accordance with the project management plan, the planned set of integrated processes, and the planned scope.
- Measure and monitor project status, processes, and products.
- Analyze project risks.

Among the processes in the project management process groups, the links are often iterated. The planning process group provides the executive process group with a documented project management plan early in the project and then facilitates updates to the project management plan if changes occur as the project progresses.

Integration is primarily concerned with effectively integrating the processes among the project management process groups that are required to accomplish project objectives within an organization's defined procedures. Figure 3-4 provides an overview of the major project management integrative processes. As seen in Figure 3-4, the integrative project management processes include:

a. Develop project charter—developing the project charter that formally authorizes a project or a project phase

b. Develop preliminary project scope statement—developing the preliminary project scope statement that provides high-level scope narrative

c. Develop project management plan—documenting the actions necessary to define, prepare, integrate, and coordinate all subsidiary plans into a project management plan

d. Direct and manage project execution—executing the work defined in the project management plan to achieve the project's requirements defined in the project scope statement

Project Integration Management

I Develop Project Charter
1. Inputs
   a. Contract
   b. Project Statement of Work
   c. Enterprise Environmental Factors
   d. Organization Process Assets
2. Tools
   a. Project Selection Methods
   b. Project Management Methodology
   c. Project Management Information System
   d. Expert Judgment
3. Outputs
   a. Project Charter

II Develop Preliminary Project Scope Statement
1. Inputs
   a. Project charter
   b. Project Statement of Work
   c. Enterprise Environmental Factors
   d. Organizational Process Assets
2. Tools and Techniques
   a. Project Management Methodology
   b. Project Management Information System
   c. Expert Judgment
3. Outputs
   a. Preliminary Project Scope Statement

III Develop Project Management Plan
1. Inputs
   a. Preliminary Project Scope Statement
   b. Project Management Processes
   c. Enterprise Environmental Factors
   d. Organizational Process Assets
2. Tools and Techniques
   a. Project Management Methodology
   b. Project Management Information System
   c. Expert Judgment
3. Outputs
   a. Project Management Plan

IV Direct and Manage Project Execution
1. Inputs
   a. Project Management Plan
   b. Approved Corrective Actions
   c. Approved Preventive Actions
   d. Approved Change Requests
   e. Approved Defect Repair
   f. Validated Defect Repair
   g. Administrative Closure Procedure
2. Tools and Techniques
   a. Project Management Methodology
   b. Project Management Information System
3. Outputs
   a. Deliverables
   b. Requested Changes
   c. Implemented Change Requests
   d. Implemented Corrective Actions
   e. Implemented Preventive Actions
   f. Implemented Defect Repair
   g. Work Performance Information

V Monitor and Control Project Work
1. Inputs
   a. Project Management Plan
   b. Work Performance Information
   c. Rejected Change Requests
2. Tools and Techniques
   a. Project Management Methodology
   b. Project Management Information System
   c. Earned Value Technique
   d. Expert Judgment
3. Outputs
   a. Recommended Corrective Actions
   b. Recommended Preventive Actions
   c. Forecasts
   d. Recommended Defect Repair
   e. Requested Changes

VI Integrated Change Control
1. Inputs
   a. Project Management Plan
   b. Requested Changes
   c. Work Performance Information
   d. Recommended Preventive Actions
   e. Recommended Corrective Actions
   f. Recommended Defect Repair
   g. Deliverables
2. Tools and Techniques
   a. Project Management Methodology
   b. Project Management Information System
   c. Expert Judgment
3. Outputs
   a. Approved Change Requests
   b. Rejected Change Requests
   c. Project Management Plan Updates
   d. Project Scope Statement Updates
   e. Approved Corrective Actions
   f. Approved Corrective Actions
   g. Approved Defect Repairs
   h. Validated Defect Repair
   i. Deliverables

VII Close Project
1. Inputs
   a. Project Management Plan
   b. Contract Documentation
   c. Enterprise Environmental Factors
   d. Organizational Process Assets
   e. Work Performance Information
   f. Deliverables
2. Tools and Techniques
   a. Project Management Methodology
   b. Project Management Information System
   c. Expert Judgment
3. Outputs
   a. Administrative Closure Procedure
   b. Contract Closure Procedures
   c. Final Product, Service, or Result
   d. Organizational Process Assets Updated

Figure 3-4 Project integration management overview.

e. Monitor and control project work—monitoring and controlling the processes used to initiate, plan, execute, and close a project to meet performance objectives defined in the project management plan

f. Integrated change control—reviewing all change requests, approving changes, and controlling changes to the deliverables and organizational process assets

g. Close project—finalizing all activities across all of the project management process groups to formally close the project or a project phase

Develop Project Charter

The project charter is the document that formally authorizes a project. The project charter provides the project manager with the authority to apply organizational resources to project activities. A project manager is identified and assigned as early in the project as is feasible. The project manager should always be assigned prior to the start of planning, and preferably while the project is being developed.

A project initiator or sponsor external to the project organization, at a level that is appropriate to funding the project, issues a project charter. Projects are usually chartered and authorized external to the organization by an enterprise, a government agency, a company, a program organization, or a portfolio organization, as a result of one or more of the following:

- A market demand (e.g., a car company authorizing a project to build more fuel-efficient cars in response to gasoline shortages)
- A business need (e.g., a training company authorizing a project to build a new substation to serve a new industrial park)
- A customer request (e.g., an electric utility authorizing a project to build a new substation to serve a new industrial park)
- A technological advance (e.g., an electronics firm authorizing a new project to develop a faster, cheaper, and smaller laptop after advances in computer memory and electronics technology)
- A legal requirement (e.g., a paint manufacturer authorizing a project to establish guidelines for handling toxic materials)
- A social need (e.g., a nongovernmental organization in a developing country authorizing a project to provide potable water systems, latrines, and sanitation education to communities suffering from high rates of cholera)

These stimuli can also be called problems, opportunities, or business requirements. The central theme of all these stimuli is that management must make a decision about how to respond and what projects to authorize and charter. Project selection methods involve measuring value or attractiveness to the project owner or sponsor and may include other organization decision criteria. Project selection also applies to choosing alternative ways of executing the project.

Charting a project links the project to the ongoing work of the organization. In some organizations, a project is not formally chartered and initiated until completion of a needs assessment, feasibility study, preliminary plan, or some other form of analysis that was separately initiated. Developing the project charter is primarily concerned with documenting the business needs, project justification, current understanding of the customer's requirements, and the new product, service, or result that is intended to satisfy those requirements. The project charter, either directly or by reference to other documents, should address the following information.

- Requirements that satisfy customer, sponsor, and other stakeholder needs, wants, and expectations
- Business needs, high-level project description, or product requirements that the project is undertaken to address
- Project purpose or justification
- Assigned project manager and authority level
- Summary milestone schedule
- Stakeholder influences
- Functional organizations and their participation
- Organizational, environmental, and external assumptions and constraints
- Business case justifying the project, including return on investment
- Summary budget

During subsequent phases of the multiphase projects, the "develop project charter process" validates the decisions made during the original chartering of the project. If required, it also authorizes the next project phase, and updates the charter.

The project charter commissions and challenges the project team as well as the business to work *with their eyes fully open to both the risks and opportunities of full integration.* This means that the team is charged to plan, design, and monitor the program and/or project with an eye *outward* to customer expectations and requirements; cost, schedule, and quality tradeoffs (using earned value tools); reaching out to supporting functions and systems (IT, functional departments such as accounting and purchasing/acquisition, technical test equipment and tool managers); the costs of production and manufacturing of the project products and outputs; the need for strong configuration management to document and preserve the product; and other projects underway as part of the company's portfolio of projects. The team will look at the project as an investment in the company.

The charter will identify periodic stage-gate review points, the need for interfaces in those reviews, and will challenge the team to make the business case for the project at every stage-gate review.

Develop project charter: Inputs

Contract—A contract from the customer's acquiring organization is an input if the project is being done for an external customer.

Project statement of work (SOW)—The statement of work is a narrative description of products or services to be supplied by the project. For internal purposes, the project initiator or sponsor provides the statement of work, based on business needs, product, or service requirements. For external purposes, the statement of work can be received from the customer as part of a bid document, for example, request for proposal, request for information, request for bid, or as part of a contract. The SOW indicates a:

- Business need—an organization's business need can be based on needed training, market demand, technological advance, legal requirement, or governmental standard.
- Product scope description—documents the product requirements and characteristics of the product or service that the project will be undertaken to create. The product requirements will generally have less detail during the initiation phase and more detail during later processes, as the product characteristics are progressively elaborated. These requirements should also document the relationship among the products or services being created and the business need or other stimulus that caused the need. While the form and substance of the product requirements document will vary, it should always be detailed enough to support later project planning.
- A strategic plan—all projects should support the organization's strategic goals. The strategic plan of the performing organization should be considered as a factor when making project selection decisions.

Enterprise environmental factors—When developing the project charter, any and all of the organization's enterprise environmental factors and systems that surround and influence the project's success must be considered. This includes but is not limited to items such as:

- Organizational or company culture and structure
- Governmental or industry standards (e.g., regulatory agency regulations, product standards, quality standards, and workmanship standards)
- Infrastructure (e.g., existing facilities and capital equipment)
- Existing human resources (e.g., skills, disciplines, and knowledge, such as design, development, legal, contracting, and purchasing)
- Personnel administration (e.g., hiring and firing guidelines, employee performance reviews, and training records)
- Company work authorization system
- Marketplace conditions
- Stakeholder risk tolerances
- Commercial databases (e.g., standardized cost estimating data, industry risk study information, and risk databases)

- Project management information systems (e.g., an automated tool suite, such as a scheduling software tool, a configuration management system, an information collection and distribution system, or web interfaces to other online automated systems)

Organizational Process Assets

When developing the project charter and subsequent project documentation, any and all of the assets that are used to influence the project's success can be drawn from organizational process assets. Any and all of the organizations involved in the project can have formal and informal policies, procedures, plans, and guidelines, and the effects must be considered. Organizational process assets also represent the organization's learning and knowledge from previous projects, for example, completed schedules, risk data, and earned value data. Organizational process assets can be organized differently, depending on the type of industry, organization, and application area. For example, the organizational process assets could be grouped into two categories: organization's processes and procedures for conducting work, and organizational corporate knowledge base for storing and retrieving information.

Organization processes and procedures for conducting work

- Organizational standard processes, such as standards, policies (e.g., safety and health policy, and project management policy), standard product and project life cycles, and quality policies and procedures (e.g., process audits, improvement targets, checklists, and standardized process definitions for use in the organization)
- Standardized guidelines, work instructions, proposal evaluation criteria, and performance measurement criteria
- Templates (e.g., risk templates, work breakdown structure templates, and project schedule network diagram templates)
- Guidelines and criteria for tailoring the organization's set of standard processes to satisfy the specific needs of the project
- Organizational communication requirements (e.g., specific communication technology available, allowed communication media, record retention, and security requirements)
- Project closure guidelines or requirements (e.g., final project audits, project evaluations, product validations, and acceptance criteria)
- Financial controls procedures (e.g., time reporting, required expenditure and disbursement reviews, accounting codes, and standard contract provisions)
- Issue and defect management procedures defining issue and defect controls, issue and defect identification and resolution, and action item tracking
- Change control procedures, including the steps by which official company standards, policies, plans, and procedures, or any project documents will be modified, and how any change will be approved and validated

- Risk control procedures, including risk categories, probability definition and impact, and probability and impact matrix
- Procedures for approving and issuing work authorizations

Organizational corporate knowledge base for storing and retrieving information

- Process measurement database used to collect and make available measurement data on processes and products
- Project files (e.g., scope, cost, schedule, and quality baselines, performance measurement baselines, project calendars, project schedule network diagrams, risk registers, planned response actions, and defined risk impact)
- Historical information and lessons learned knowledge base (e.g., project records and documents, all project closure information and documentation, information about both the results or previous project selection decisions and previous project performance information, and information from the risk management effort)
- Issue and defect management database containing issue and defect status, control information, issue and defect resolution, and action item results
- Configuration management knowledge base containing the versions and baselines of all official company standards, policies, procedures, and any project documents
- Financial database containing information such as labor hours, incurred costs, budgets, and any project cost overruns

Develop project charter: Tools and techniques

Project selection methods—Project selection methods are used to determine which project the organization will select. These methods generally fall into one of two broad categories:

- Benefit measurement methods that are comparative approaches, scoring models, benefit contribution, or economic models
- Mathematical models that use linear, nonlinear, dynamic, integer, or multi-objective programming algorithms

Project management methodology—A project management methodology defines a set of project management process groups, their related processes, and the related control functions that are consolidated and combined into a functioning and unified whole. A project management methodology may or may not be an elaboration of a project management standard. A project management methodology can be either a formal mature process or an informal technique that aids a project management team in effectively developing a project charter.

Project management information system—The project management information system (PMIS) is a standardized set of automated tools available within the organization and integrated into a system. The PMIS is used by the project management team to support generation of a project charter, facilitate feedback as the document is refined, control changes to the project charter, and release the approved document.

Expert judgment—Expert judgment is often used to assess the inputs needed to develop the project charter. Such judgment and expertise are applied to any technical and management details during the process. Such expertise is provided by any group or individual with specialized knowledge or training, and is available from many sources, including:

- Other units within the organization
- Consultants
- Stakeholders, including customers and sponsors
- Professional and technical associations
- Industry groups

Develop project charter: Outputs

One final point about the project team charter in terms of integration: Teams that work together and communicate outward about the project with internal stakeholders, such as finance, purchasing, and engineering, tend to integrate their projects more effectively than teams that work in isolation. Therefore, the major determinant of successful integration is not technical but rather social and organizational. The way the team sees the priority of integration at all levels is to *see it explicitly in the charter itself*. Therefore every charter should have a statement such as the following:

> "The project team will integrate project activities at all levels, including with business planning and marketing, finance and budget, functional departments, and customers, to ensure that the project outcomes reflect all the stakeholder interests to the extent possible. Project planning shall include a comprehensive project schedule that integrates cost, time, and quality factors to create an optimum outcome, cheaper, better, and faster."

The charter might address the following topics:

- Project manager
- Priority of project
- Date
- Owner/sponsor
- Mission
- Scope
- Objectives
- Assumptions
- Constraints
- Schedule and major milestones

- Cost/budget/financial assumptions
- Quality specifications
- Major risks and contingencies
- Project core team
- Subject matter experts
- Contractors

Develop Preliminary Project Scope Statement

The project scope statement is the definition of the project. This statement defines—what work needs to be accomplished. The developed preliminary project scope statement process addresses and documents the characteristics and boundaries of the project and its associated products and services, as well as the methods of acceptance and scope control. A project scope statement includes:

- Project and product objectives
- Product acceptance criteria
- Product or service requirements and characteristics
- Project boundaries
- Project requirements and deliverables
- Project constraints
- Project assumptions
- Initial project organization
- Initial defined risks
- Scheduled milestones
- Initial WBS
- Order of magnitude cost estimate
- Project configuration management requirements
- Approval requirements

The preliminary project scope statement is developed from information provided by the initiator or sponsor. The project management team in the scope definition process further refines the preliminary project scope statement into the project scope statement. The project scope statement content will vary depending upon the application area and complexity of the project and can include some or all of the components identified above. During subsequent phases of multiphase projects, the developed preliminary project scope statement process validates and refines, if required, the project scope defined for that phase.

Develop preliminary scope statement: Inputs. The inputs to the scope statement are, as previously described:

1. Project charter
2. Project statement of work
3. Enterprise environmental factors
4. Organizational process assets

Develop preliminary project scope statement: Tools and techniques

Project management methodology—The project management methodology defines a process that aids a project management team in developing and controlling changes to the preliminary project scope statement.

Project management information system—The project management information system is used by the project management team to support generation of a preliminary project scope statement, facilitate feedback as the document is refined, control changes to the project scope statement, and release the approved document.

Expert judgment—Expert judgment is applied to any technical and management details to be included in the preliminary scope statement.

Develop preliminary project scope statement: Output. The major output of the "preliminary project scope statement" process is

- Preliminary project scope statement

This statement provides a generalized discussion of the work to be done in the project. It provides the basis for the scope of work for any outsourced work as well.

Develop project management plan

The develop project management plan process includes the actions necessary to define, integrate, and coordinate all subsidiary plans into a project management plan. The project management plan content will vary depending upon the application area and complexity of the project. This process results in a project management plan that is updated and revised through the integrated change control process. The project management plan defines how the project is executed, monitored, controlled, and closed. The project management plan documents the collection of outputs of the planning processes of the planning process group and includes:

- The project management processes selected by the project management team

- The level of implementation of each selected process
- The descriptions of the tools and techniques to be used for accomplishing those processes
- How the selected processes will be used to manage the specific project, including the dependencies and interactions among those processes, and the essential inputs and outputs
- How work will be executed to accomplish the project objectives
- How changes will be monitored and controlled
- How configuration management will be performed
- How integrity of the performance measurement baselines will be maintained and used
- The need and techniques for communication among stakeholders
- The selected project life cycle and, for multiphase projects, the associated project phases
- Key management reviews for content, extent, and timing to facilitate addressing open issues and pending decisions

The project management plan is a broad document with several elements. It is typically composed of one or more subsidiary plans and other components. Each of the subsidiary plans and components is detailed to the extent required by the specific project. These subsidiary plans include, but are not limited to:

- Project scope management plan
- Schedule management plan
- Cost management plan
- Quality management plan
- Process improvement plan
- Staffing management plan
- Communication management plan
- Risk management plan
- Procurement management plan
- Milestone list
- Resource calendar
- Schedule baseline
- Cost baseline
- Quality baseline
- Risk register

Develop project management plan: Inputs

Inputs to the project management plan include:

1. Preliminary project scope statement
2. Project management processes
3. Enterprise environmental processes
4. Organizational process assets

Develop project management plan: Tools and techniques

Tools and techniques for developing the project management plan include:

Project management methodology—This methodology is a disciplined process of planning a project from initial requirements, task work breakdown structure, schedule, budget, and risk management plan.

Project management information system/configuration management system—The project management information system is the totality of information on the project, usually provided by a project management software program, e.g., Microsoft Project, a technical database on the product, and a network for exchanging of project data.

The configuration management system is a subsystem of the overall project management information system. The system includes the process for submitting proposed changes, tracking systems for reviewing and approving proposed changes, defining approval levels for authorizing changes, and providing a method to validate approved changes. In most application areas, the configuration management system includes the change control system. The configuration management system is also a collection of formal documented procedures used to apply technical and administrative direction and surveillance to:

- Identify and document the functional and physical characteristics of a product or component
- Control any changes to such characteristics
- Record and report each change and its implementation status
- Support the audit of the products and components to verify conformance to requirements

Change control system—The change control system is a collection of formal documented procedures that define how project deliverables and documentation are controlled, changed, and approved. The change control system is a subsystem of the configuration management system. For example, for information technology systems, a change control system can include the specifications (scripts, source code, data definition language, etc.) for each software component.

Expert judgment—Expert judgment is provided through subject matter experts on the new product.

Develop project management plan: Outputs

The output is a project management plan.

Project management plan—In practice, the project management plan is a guide for the definition and control of the work. Therefore, the plan must include control points, e.g., project reviews. These review points ensure that management can authorize movement from one phase or stage to another. Reporting and monitoring strategies, including the use of earned value to integrate cost, schedule, and quality performance, should be made explicit.

The plan should also address accountability, particularly in view of the recent legislative and regulatory requirements of the Sarbanes-Oxley Act. This relatively new legislative requirement, now backed up by regulatory provisions and stemming from the corporate abuses of the late 1990s, makes internal control a mandatory part of any new product project. In fact, the price of disconnected and inconsistently applied efforts throughout a project and its interfaces, and lack of financial tracking systems that provide for audits, could be businesswide. Compliance with Sarbanes-Oxley therefore is not a choice, but rather a requirement, and the plan should state standards for estimating costs, tracking the costs, and relating costs to work performed, and the integrity of the closeout procedure and invoices to customers for work performed.

Direct and Manage Project Execution

The direct and manage project execution process requires the project manager and the project team to perform multiple actions to execute the project management plan to accomplish the work defined in the project scope statement. Some of these actions are

- Perform activities to accomplish project objectives
- Expend effort and spend funds to accomplish the project objectives
- Staff, train, and manage the project team members assigned to the project
- Obtain quotations, bids, offers, or proposals as appropriate
- Select sellers by choosing from among potential sellers
- Obtain, manage, and use resources including materials, tools, equipment, and facilities
- Implement the planning methods and standards
- Create, control, verify, and validate project deliverables
- Manage risks and implement risk response activities
- Manage sellers

- Adapt approved changes into the project's scope, plans, and environment
- Establish and manage project communication channels, both external and internal to the project team
- Collect project data and report costs, schedule, technical and quality progress, and status information to facilitate forecasting
- Collect and document lessons learned, and implement approved process improvement activities

The project manager, along with the project management team, directs the performance of the planning project activities, and manages various technical and organizational interfaces that exist within the project. The direct and manage project execution process is most directly affected by the project application area. Deliverables are produced as outputs from the processes performed to accomplish the project work planned and scheduled in the project management plan. Work performance information about the completion status of the deliverables, and what has been accomplished, is collected as part of project execution and fed into the performance reporting process. Although the products, services, or results of the project are frequently in the form of tangible deliverables such as buildings, roads, and so on, intangible deliverables such as training can also be provided.

Direct and manage project execution also requires implementation of:

- Approved corrective actions that will bring anticipated project performance into compliance with the project management plan
- Approved preventive actions to reduce the probability of potential negative consequences
- Approved defect repair requests to correct product defects found by the quality process

Direct and manage project execution: Inputs

The input to project execution is a project management plan.

- **Approved corrective actions**—Approved corrective actions are the documented, authorized directions required to bring expected future project performance into conformance with the project management plan.
- **Approved preventive actions**—Approved preventive actions are the documented, authorized directions that reduce the probability of negative consequences associated with project risks.
- **Approve change requests**—Approved change requests are the documented, authorized changes to expand or contract project scope. The approved change requests can also modify policies, project management plans, procedures, costs or budgets, or revise schedules. Approved change requests are scheduled for implementation by the project team.

- **Approved defect repair**—The approved defect repair is the documented, authorized request for product correction of a defect found during the quality inspection or the audit process.
- **Validated defect repair**—Notification that the re-inspected repaired items have either been accepted or rejected.
- **Administrative closure procedure**—The administrative closure procedure documents all activities, interactions, and related roles and responsibilities needed for executing the administrative closure procedure for the project.

Direct and manage project execution: Tools and techniques

- Project management methodology
- Project management information system

Direct and manage project execution: Outputs

- **Deliverables**—A deliverable is any unique and verifiable product, result, or capability to perform a service that is identified in the project management planning documentation, and must be produced and provided to complete the project
- **Requested changes**—Changes requested to expand or reduce project scope, to modify policies or procedures, to modify project cost or budget, or to revise the project schedule are often identified while project work is being performed. Requests for a change can be direct or indirect, externally or internally initiated, and can be optional or legally/contractually mandated.
- **Implemented change requests**—The approved change requests that have been implemented by the project management team during project execution.
- **Implement corrective actions**—The approved corrective actions that have been implemented by the project management team to bring expected future project performance into conformance with the project management plan.
- **Implemented preventive action**—The approved preventive actions that have been implemented by the project management team to reduce the consequences of project risks.
- **Implemented defect repair**—During project execution, the project management team has implemented approved product defect corrections.
- **Work performance information**—Information on the status of the project activities being performed to accomplish the project work is routinely collected as part of the project management plan execution. This information includes, but is not limited to:
  - Schedule progress showing status information
  - Deliverables that have been completed and those not completed

- Schedule activities that have started and those that have been finished
- Extent to which quality standards are being met
- Costs authorized and incurred
- Estimates to complete the schedule activities that have been started
- Percent physically complete of the in-progress schedule activities
- Documented lessons learned posted to the lessons learned knowledge base
- Resource utilization detail

Monitor and Control Project Work

The monitor and control project work process is performed to monitor project processes associated with initiating, planning, executing, and closing. Corrective or preventive actions are taken to control the project performance. Monitoring is an aspect of project management performed throughout the project. Monitoring includes collecting, measuring, and disseminating performance information, and assessing measurements and trends to affect process improvements. Continuous monitoring gives the project management team insight into the health of the project, and identifies any areas that can require special attention. The monitor and control project work process is concerned with:

- Comparing actual project performance against the project management plan
- Assessing performance to determine whether any corrective or preventive actions are indicated, and then recommending those actions as necessary
- Analyzing, tracking, and monitoring project risks to make sure the risks are identified, their status reported, and that appropriate risk response plans are being executed
- Maintaining an accurate, timely information base concerning the project's products and their associated documentation through project completion
- Providing information to support status reporting, progress measurement, and forecasting
- Monitoring implementation of approved changes when and as they occur

Monitor and control project work: Inputs.

The input to the monitor and control project phase is a project management plan.

- **Project management plan**—(See description of the project plan)
- **Work performance information**—Work performance information provides data on how the project is progressing based on the original plan.
- **Rejected change requests**—Rejected change requests include the change requests, their supporting documentation, and their change review status showing a disposition of rejected change requests.

Monitor and control project work: Tools and techniques. Tools and techniques include:

- **Project management methodology**—The whole process used to plan and implement a project using accepted project management tools.
- **Project management information system**—A system for providing data and information on the project.
- **Earned value technique**—The earned value technique measures performance of the project as it moves from project initiation through project closure. The earned value management methodology also provides a means to forecast future performance based upon past performance.
- **Expert judgment**— (As explained above)

Monitor and control project work: Outputs

- **Recommended corrective actions**—Corrective actions are the documented recommendations required to bring expected future project performance into conformance with the project management plan
- **Recommended preventive actions**—Preventive actions include all those contingencies to prevent risks from occurring.
- **Forecasts**—Forecasts include estimates or predictions of conditions and events in the project's future, based on information and knowledge available at the time of the forecast. Forecasts are updated and reissued based on work performance information provided as the project is executed. This information is about the project's past performance that could impact the project in the future; for example, estimate at completion and estimate to complete.
- **Recommended defect repair**—Recommended defect repair is a contingency plan addressing the process of repairing potential product defects.
- **Requested changes**—Requested changes are all those authorized change requests from internal and external (customer) sources, preserved in the configuration management system.

Integrated Change Control

The integrated change control process is performed from project inception through completion. Change control is necessary because projects seldom run exactly according to the project management plan. The project management plan, the project scope statement, and other deliverables must be maintained by carefully and continuously managing changes, either by rejecting changes or by approving changes so those approved changes are incorporated into a revised baseline. The integrated change control process includes the following change management activities in differing levels of detail, based upon the completion of project execution:

- Identifying that a change needs to occur or has occurred
- Influencing the factors that circumvent integrated change control so that only approved changes are implemented
- Reviewing and approving requested changes
- Managing the approved changes when and as they occur, by regulating the flow of requested changes
- Maintaining the integrity of baselines by releasing only approved changes for incorporation into project products and services, and maintaining their related configuration and planning documentation
- Reviewing and approving all recommended corrective and preventive actions
- Controlling and updating the scope, cost, budget, schedule, and quality requirements based upon approved changes, by coordinating changes across the entire project; for example, a proposed schedule change will often affect cost, risk, quality, and staffing
- Documenting the complete impact of a requested change
- Validating defect repair
- Controlling project quality to standards based on quality reports

Proposed changes can require new or revised cost estimates, schedule activity sequences, schedule dates, resource requirements, and an analysis of risk response alternatives. These changes can require adjustments to the project management plan, project scope statement, or other project deliverables. The configuration management system with change control provides a standardized, effective, and efficient process to centrally manage changes within a project. Configuration management with change control includes identifying, documenting, and controlling changes to the baseline. The applied level of change control is dependent upon the application area, complexity of the specified project, contract requirements, and the context and environment in which the project is performed.

Projectwide application of the configuration management system, including change control processes, accomplishes three main objectives:

- Establishes an evolutionary method to consistently identify and request changes to established baselines, and to assess the value and effectiveness of those changes
- Provides opportunities to continuously validate and improve the project by considering the impact of each change
- Provides the mechanism for the project management team to consistently communicate all changes to the stakeholders

Some of the configuration management activities included in the integrated change control process are:

- Providing the basis from which the configuration of products is defined and verified, products and documents are labeled, changes are managed, and accountability is maintained
- Configuration status accounting: capturing, storing, and accessing configuration information needed to manage products and product information effectively
- Configuration verification and auditing: establishing that the performance and functional requirements defined in the configuration documentation have been met

Every documented requested change must be either accepted or rejected by some authority within the project management team or an external organization representing the initiator, sponsor, or customer. Many times, the integrated change control process includes a change control board responsible for approving and rejecting the requested changes. The roles and responsibilities of these boards are clearly defined within the configuration control and change procedures, and are agreed to by the sponsor, customer, and other stakeholders. Many large organizations provide for a multitiered board structure, separating responsibilities among the boards. If the project is being provided under a contract, some proposed changes would need to be approved by the customer.

Integrated change control: Inputs

Inputs to the change control system include:

- **Project management plan**—The plan for managing the project.
- **Requested change**—Any changes requested by the customer, documented.
- **Work performance information**—Data on how the project is progressing in terms of time, cost, and quality.
- **Recommended preventive actions**—Contingency actions to prevent risks identified.
- **Recommended corrective actions**—Contingency actions to offset risk events that have occurred.
- **Recommended defect repair**—Recommendations to correct product defects identified in testing.
- **Deliverables**—All project management documents produced to date.

Integrated change control: Tools and techniques

The integrated change control system includes:

- **Project management methodology**—Project management tools and processes.
- **Project management information system**—The data and networked information on the progres of the project, including technical documents.
- **Expert judgment**—Use of subject matter experts on special problems uncovered in development.

Integrated change control: Outputs

Outputs include:

- **Approved change requests**—All change requests approved by the project manager.
- **Rejected change requests**—All change requests rejected by the project manager.
- **Project management plan**—The plan for managing the project.
- **Project scope statement**—The statement of the project work to be completed.
- **Approved corrective actions**—The statement of the project work to be completed.
- **Approved preventive actions**—A preventive actions approved by the project manager.
- **Approved defect repair**—All corrective actions aimed at fixing defects.
- **Validated defect repair**—Reported defects validated by technical subject matter expert in testing.
- **Deliverables**—All output documents.

Integrated change control requires, in practice, a clear understanding of the scope of work and what lies inside and outside the boundaries of the scope. Changes must be reviewed by top management and the customer to avoid new work generated by the team or by customer representatives. The project is seen as a contract, of sorts, and a change to the contract is considered "negotiable" but not given.

Cost control is driven by work done, not by budgets out of context from work performed. In other words, *cost control is seen as the process of aligning actual costs to the planned budget associated with the work performed.* This approach allows changes to be seen during a project *in terms of their impacts on remaining work,* not on the original budget and work schedule.

Close Project

The close project process involves performing the project closure portion of the project management plan. In multiphase projects, the close project process closes out the portion of the project scope and associated activities applicable to a given phase. This process includes finalizing all activities completed across all project management process groups to formally close the project or a project phase, and transfer the completed or cancelled project as appropriate. The close project process also establishes the procedures to coordinate activities needed to verify and document the project deliverables, to coordinate and interact to formalize acceptance of those deliverables by the customer or sponsor, and to investigate and document the reasons for actions taken if a project is terminated before completion. Two procedures are developed to establish the interactions necessary to perform the closure activities across the entire project or for a project phase:

- **Administrative closure procedure.** This procedure details all the activities, interactions, and related roles and responsibilities of the project team members and other stakeholders involved in executing the administrative closure procedure for the project. Performing the administrative closure process also includes integrated activities needed to collect project records, analyze project success or failure, gather lessons learned, and archive project information for future use by the organization.
- **Contract closure procedure.** Includes all activities and interactions needed to settle and close any contract agreement established for the project, as well as define those related activities supporting the formal administrative closure of the project. This procedure involves both product verification (all work completed satisfactorily and correctly) and administrative closure (updating of contract records to reflect final results and archiving that information for future use). The contract terms and conditions can also prescribe specifications for contract closure that must be part of this procedure. Early termination of a contract is a special case of contract closure that could involve, for example, the inability to deliver the product, a budget overrun, or lack of required resources. This procedure is an input to the close contract process.

Close project: Inputs

1. **Project management plan**—The plan for managing the project.
2. **Contract documentation**—Contract documentation is an input used to perform the contract closure process, and includes the contract itself, as well as changes to the contract and other documentation (such as the technical approach, product description, or deliverable acceptance criteria and procedures).
3. **Enterprise environmental factors**—All factors that might affect the project in the company organization.
4. **Organizational process assets**—All key business systems supporting the project.
5. **Work performance information**—All data and information on how the project is progressing.
6. **Deliverables**—All output documents.

Close project: Tools and techniques

1. **Project management methodology**—Project management tools and processes
2. **Project management information system**—All data and information on project progress
3. **Expert judgment**—Subject matter expert inputs.

Close project: Outputs

1. **Administrative closure procedures**—This procedure contains all the activities and the related roles and responsibilities of the project team members involved in executing the administrative closure procedure. The procedures to transfer the project products or services to production and/or operations are developed and established. This procedure provides a step-by-step methodology for administrative closure that addresses:
   - Actions and activities to define the stakeholder approval requirements for changes and all levels of deliverables
   - Actions and activities that are necessary to confirm that the project has met all sponsor, customer, and other stakeholder requirements, verify that all deliverables have been provided and accepted, and validate that completion and exit criteria have been met
   - Actions and activities necessary to satisfy completion or exit criteria for the project
2. **Contract closure procedure**—This procedure is developed to provide a step-by-step methodology that addresses the terms and conditions of the contracts and any required completion or exit criteria for contract closure. It contains all activities and related responsibilities of the project team members, customers, and other stakeholders involved in the contract closure process. The actions performed formally close all contracts associated with the completed project.
3. **Final product, service, or result**—Formal acceptance and handover of the final product, service, or result that the project was authorized to produce. The acceptance includes receipt of a formal statement that the terms of the contract have been met.
4. **Organizational process assets (updates)**—Closure will include the development of the index and location of project documentation using the configuration management system.
   - **Formal acceptance documentation.** Formal confirmation has been received from the customer or sponsor that customer requirements and specifications for the project's product, service, or result have been met. This document formally indicates that the customer or sponsor has officially accepted the deliverables.
   - **Project files.** Documentation resulting from the project's activities, for example, project management plan, scope, cost, schedule and quality baselines, project calendar, risk registers, planned risk response actions, and risk impact.
   - **Project closure documentation.** Project closure documents consist of formal documentation indicating completion of the project and the transfer of the completed project deliverables to others, such as an operations group. If the project was terminated prior to completion, the formal documentation indicates why the project was terminated and formalizes the procedures for the transfer of the finished and unfinished deliverables of the canceled project to others.

- **Historical information.** Historical information and lessons learned information are transferred to the lessons learned knowledge base for use by future projects.

In practice, closeout is often discounted by project managers and management in general as an *administrative* process, not worth much attention by *substantive* team members. Things have changed on that note. Now closeout is seen as the process of complying with Sarbanes-Oxley and the audit process to make sure that project work and expenditures are documented and traceable to customer requirements and project related expenditures.

Case Study of PMBOK Implementation: Integrated Transportation System

A Microsoft Project schedule for the case serves to illustrate how the work breakdown structure is translated to project tasks, schedule, and budget and resource assignments. This transportation program, entitled the Integrated Transportation System, involves the design and development of a product, a new transportation vehicle concept that will incorporate a controlled, line haul highway system for urban transportation. The case is scheduled into Microsoft Project and will be discussed, task level by task level, to illustrate integration issues and opportunities in a typical such project. This product development project is organized in the stage-gateway framework with entry from one stage to another controlled by an integration gateway review. This integration gateway review assesses progress, change, and impacts from each stage and provides the basis for a go or no-go decision to proceed to the next stage.

The program is being developed at ITS, Inc., a research and development firm concentrated on transportation systems and tools. Program stages are defined as separated projects to be *integrated forward* to the final program product. Forward integration is the process of looking forward to delivery rather than backwards to align with original baseline plans. Forward integration is based on earned value that is taking where we are in a project in both schedule and cost terms, and planning for remaining work.

The scheduled tasks serve as the basis for this analysis (see Table 3-1).

Integration gateway 1: Global interface

The first step in integration is a business planning function (often called environmental scanning) looking outward to integrate with global partners and customers. The *first line of integration is global.* According to Thomas Friedman in the *World is Flat*, the "playing field" for resources, partnerships, and systems is now global, beyond traditional political and economic jurisdictions. Thus ITS must first position itself to be successful in this program area by building interfaces with key customer groups, supplying partners, and, perhaps, with governmental agencies globally because of the "public" nature of transportation and the need to take advantage of worldwide supply and design sources.

TABLE 3-1 Integrated Transportation System Schedule

ID		Task Name	Duration	Start	2007 Q2	Q3
1		**INTEGRATED TRANSPORTATION SYSTEM**	**343 days**	**Mon 6/25/07**		
2		**Integration Gateway 1: Global Interface**	**7 days**	**Mon 6/25/07**		
3		International Partnering and Organization	7 days	Mon 6/25/07		
4		**Integration Gateway 2: Business Planning**	**12 days**	**Wed 7/4/07**		
5		Develop strategic Objectives	12 days	Wed 7/4/07		
6		**Integration Gateway 3: Organizational De**	**10 days**	**Fri 7/20/07**		
7		Prepare Organization for Integration	8 days	Fri 7/20/07		
8		Prepare for Internal Control	2 days	Wed 8/1/07		
9		**Integration Gateway 4: Global Team Com**	**11 days**	**Fri 8/3/07**		
10		Develop Team and Select Program and	11 days	Fri 8/3/07		
11		**Integration Gateway 5: Support Systems**	**4 days**	**Mon 8/20/07**		
12		Review all suppor Organizational Assets	4 days	Mon 8/20/07		
13		**Integration Gateway 6: Portfolio Develop**	**14 days**	**Fri 8/24/07**		
14		Business Strategy	6 days	Fri 8/24/07		
15		Portfolio Development	4 days	Mon 9/3/07		
16		Project Selection Criteria	4 days	Fri 9/7/07		
17		**Integration Gateway 7: Market and Custo**	**5 days**	**Thu 9/13/07**		

ID		Task Name	Duration	Start	2007 Q2	Q3
18		Demand studies	5 days	Thu 9/13/07		
19		**Integration Gateway 8: Program Integration**	**105 days**	**Thu 9/20/07**		
20		Six Sigma Goals	4 days	Thu 9/20/07		
21		Business Plan	2 days	Wed 9/26/07		
22		**Develop Project Charter**	**39 days**	**Fri 9/28/07**		
23		Program Plan	29 days	Fri 9/28/07		
24		Statement of Work	2 days	Thu 11/8/07		
25		Environmental Scan	2 days	Mon 11/12/07		
26		Organizational Process Assets	2 days	Wed 11/14/07		
27		Schedule Baseline	2 days	Fri 11/16/07		
28		Project Charter	2 days	Tue 11/20/07		
29		**Develop Preliminary Project Scope St**	**4 days**	**Thu 11/22/07**		
30		Draft Scope Statement	2 days	Thu 11/22/07		
31		Integration Review Cycle	2 days	Mon 11/26/07		
32		**Develop Project Management Plan**	**38 days**	**Wed 11/28/07**		
33		Project Scope Management Plan	2 days	Wed 11/28/07		
34		Generic Work Breakdown Structure	2 days	Fri 11/30/07		

ID	ℹ	Task Name	Duration	Start	2007	
					Q2	Q3
35		Schedule Management Plan	2 days	Tue 12/4/07		
36		Cost Management Plan	2 days	Thu 12/6/07		
37		Quality Management Plan	2 days	Mon 12/10/07		
38		Process Improvement Plan	2 days	Wed 12/12/07		
39		Staffing Management Plan	2 days	Fri 12/14/07		
40		Communication Management Plan	2 days	Tue 12/18/07		
41		Risk Management Plan	2 days	Thu 12/20/07		
42		Procurement Management Plan	2 days	Mon 12/24/07		
43		Milestone List	2 days	Wed 12/26/07		
44		Resource Calendar	2 days	Fri 12/28/07		
45		Schedule Baseline	2 days	Tue 1/1/08		
46		Cost Baseline	2 days	Thu 1/3/08		
47		Quality Baseline	2 days	Mon 1/7/08		
48		Risk Register	2 days	Wed 1/9/08		
49		Configuration Management Plan	2 days	Fri 1/11/08		
50		Change Control System Concept	2 days	Tue 1/15/08		
51		Contract Management Plan	2 days	Thu 1/17/08		

ID	ℹ	Task Name	Duration	Start	2007	
					Q2	Q3
52		**Direct and Manage Project Execution**	**4 days**	**Mon 1/21/08**		
53		Authorize and Supervise Work	2 days	Mon 1/21/08		
54		Assign Work	2 days	Wed 1/23/08		
55		**Monitor and Control Project Work**	**10 days**	**Fri 1/25/08**		
56		**Assess Earned Value**	**6 days**	**Fri 1/25/08**		
57		Cost Variance	2 days	Fri 1/25/08		
58		Schedule Variance	2 days	Tue 1/29/08		
59		Implement Corretive Actions	2 days	Thu 1/31/08		
60		**Integrated Change Control**	**4 days**	**Mon 2/4/08**		
61		Review and Approve Change Req	2 days	Mon 2/4/08		
62		Prepare Change Impact Statement	2 days	Wed 2/6/08		
63		**Plan for Project Closeout**	**6 days**	**Wed 2/6/08**		
64		Administrative	2 days	Wed 2/6/08		
65		Contract	2 days	Fri 2/8/08		
66		Financial	2 days	Tue 2/12/08		
67		**Integration Gateway 9: Systems Safety a**	**80 days**	**Thu 2/14/08**		
68		System Architecture	3 wks	Thu 2/14/08		

(Continued)

TABLE 3.1 Target Cost: Project Life Cycle (2 Years) (*Continued*)

ID		Task Name	Duration	Start	2007 Q2	Q3
69		Test Requirements	6 wks	Thu 3/6/08		
70		System Design Review (SDR)	0 days	Wed 4/16/08		
71		System Requirements Specification (SRS)	7 wks	Thu 3/6/08		
72		Test Architecture	4 wks	Thu 4/17/08		
73		Functional Hazard Assessment to Deter	15 days	Thu 5/15/08		
74		**Integration Gateway 10: Chassis, Mechan**	**96 days**	**Wed 6/4/08**		
75		**Power Assembly**	**1 day**	**Thu 6/5/08**		
76		I/O Drawing	1 day	Thu 6/5/08		
77		Performance spec	1 day	Thu 6/5/08		
78		Test requirements	1 day	Thu 6/5/08		
79		Full drawings	1 day	Thu 6/5/08		
80		**Chassis Concept**	**92 days**	**Wed 6/4/08**		
81		Layout	40 days	Thu 6/5/08		
82		Layout drawing	2 days	Thu 6/5/08		
83		**Chassis Design**	**91 days**	**Wed 6/4/08**		
84		Assy Design	29 days	Thu 6/5/08		
85		Approval	10 days	Thu 6/5/08		

ID		Task Name	Duration	Start	2007 Q2	Q3
86		Interpret Guidelines and Condition	0 days	Wed 6/4/08		
87		Release	1 day	Thu 6/5/08		
88		Procurement	15 days	Thu 6/5/08		
89		Procurement Check	18.2 wks	Thu 6/5/08		
90		**Computer Package**	**68 days**	**Wed 6/4/08**		
91		Systems design	0 days	Wed 6/4/08		
92		Hardware	0 days	Wed 6/4/08		
93		Software	16 days	Thu 6/5/08		
94		Embedded integration	5 days	Thu 6/5/08		
95		Special safety and security issues	1 day	Thu 6/5/08		
96		Dedection Systems	2 days	Thu 6/5/08		
97		Interface with diagnotic systems	3 days	Thu 6/5/08		
98		Software code	2 days	Thu 6/5/08		
99		Software integration	3 days	Thu 6/5/08		
100		Mechanical housings	13.4 wks	Thu 6/5/08		
101		In engine sensors	0 days	Wed 6/4/08		
102		Brake systems	20 days	Thu 6/5/08		

ID		Task Name	Duration	Start	2007	
					Q2	Q3
103		Automatic systems	6.8 wks	Thu 6/5/08		
104		**Exhaust and Emission Control**	**20 days**	**Thu 6/5/08**		
105		Standards	20 days	Thu 6/5/08		
106		International Issues	3 wks	Thu 6/5/08		
107		Harware	3 days	Thu 6/5/08		
108		Computer controls	17 days	Thu 6/5/08		
109		Self reguation system	3 days	Thu 6/5/08		
110		**Supply Chain Contracts**	**8 days**	**Wed 6/4/08**		
111		Systems	3 days	Thu 6/5/08		
112		Vertical Integration	2 days	Thu 6/5/08		
113		International consortia	0 days	Wed 6/4/08		
114		Supply interfaces	8 days	Thu 6/5/08		
115		**Configuration Management**	**5 days**	**Wed 6/4/08**		
116		Software tailoring	5 days	Wed 6/4/08		
117		Data Entry	1 day	Thu 6/5/08		
118		Change management	1 day	Thu 6/5/08		
119		**Tooling**	**5 days**	**Thu 6/5/08**		

ID		Task Name	Duration	Start	2007	
					Q2	Q3
120		Pre Manufacturing inspection	1 wk	Thu 6/5/08		
121		Safety system	1 wk	Thu 6/5/08		
122		Drawing	1 wk	Thu 6/5/08		
123		Alignment	1 wk	Thu 6/5/08		
124		**Electrical Components**	**4 days**	**Thu 6/5/08**		
125		Component designs	4 days	Thu 6/5/08		
126		**Chassis Assembly**	**65 days**	**Wed 6/4/08**		
127		Panels	*65 days*	Thu 6/5/08		
128		Insolation	5 days	Thu 6/5/08		
129		Welding	24 days	Thu 6/5/08		
130		Trunk	2 days	Thu 6/5/08		
131		Hood	11.4 wks	Thu 6/5/08		
132		Doors	6 wks	Thu 6/5/08		
133		Windows	5.6 wks	Thu 6/5/08		
134		**Computer Systems**	**24 days**	**Wed 6/4/08**		
135		Sensors	5 days	Thu 6/5/08		
136		Network	3 days	Thu 6/5/08		

(Continued)

TABLE 3.1 Target Cost: Project Life Cycle (2 Years) (*Continued*)

ID		Task Name	Duration	Start	2007	
					Q2	Q3
137		Wiring	0 days	Wed 6/4/08		
138		Embedded software	2 days	Thu 6/5/08		
139		Safety and Security	0 days	Wed 6/4/08		
140		Fail proof systems	3 wks	Thu 6/5/08		
141		Lighting systems	4 wks	Thu 6/5/08		
142		Engine injection	4.8 wks	Thu 6/5/08		
143		**Electric Assembly Builds**	**5 days**	**Thu 6/5/08**		
144		Prototype	5 days	Thu 6/5/08		
145		Systems design	1 day	Thu 6/5/08		
146		Wiring	1 day	Thu 6/5/08		
147		**Prototype Assembly**	**10 days**	**Thu 6/5/08**		
148		Molding	10 days	Thu 6/5/08		
149		Mold Development	1 day	Thu 6/5/08		
150		Shaping	2 days	Thu 6/5/08		
151		**Customer Requirements**	**54 days**	**Wed 6/4/08**		
152		Comfort	0 days	Wed 6/4/08		
153		Safety	10.8 wks	Thu 6/5/08		

ID		Task Name	Duration	Start	2007	
					Q2	Q3
154		Security	14 days	Thu 6/5/08		
155		Seating	2 wks	Thu 6/5/08		
156		Visibility	1 day	Thu 6/5/08		
157		Controls	2 days	Thu 6/5/08		
158		Dashboard	3 days	Thu 6/5/08		
159		Trunk	10 days	Thu 6/5/08		
160		Hood	5 days	Thu 6/5/08		
161		Tires	3 days	Thu 6/5/08		
162		Capacity	5 days	Thu 6/5/08		
163		**Six Sigma Supplier management**	**########**	**Thu 6/5/08**		
164		Key Processes	11.5 days	Thu 6/5/08		
165		Process Performance Indicators	8 wks	Thu 6/5/08		
166		Process improvement strategy	3 wks	Thu 6/5/08		
167		Measures	3 days	Thu 6/5/08		
168		Data collection	3 days	Thu 6/5/08		
169		Data analysis	2 days	Thu 6/5/08		
170		Process improvement teams	15 days	Thu 6/5/08		

ID	ℹ	Task Name	Duration	Start	2007	
					Q2	Q3
171		Review of Competition	3 days	Thu 6/5/08		
172		Benchmarking	10 days	Thu 6/5/08		
173		Internal reviews	15 days	Thu 6/5/08		
174		Supplier Quality	5 days	Thu 6/5/08		
175		Documentation	5 days	Thu 6/5/08		
176		**Performance Requirements**	**95 days**	**Wed 6/4/08**		
177	▦	Requirements analysis	15 days	Thu 6/5/08		
178	▦	Data analysis	19 wks	Wed 6/4/08		
179		Simulation Studies	14 days	Thu 6/5/08		
180		**Interface Integration**	**70 days**	**Thu 6/5/08**		
181		Electrical	16.5 days	Thu 6/5/08		
182	▦	Mechanical	5 wks	Thu 6/5/08		
183		Software	4 wks	Thu 6/5/08		
184	▦	Powertrain	*14 wks*	Thu 6/5/08		
185	▦	Wheels	0.6 wks	Thu 6/5/08		
186		**Outsource Controls**	**5 days**	**Thu 6/5/08**		
187	▦	Contracts	0.1 wks	Thu 6/5/08		

ID	ℹ	Task Name	Duration	Start	2007	
					Q2	Q3
188		Supplier negotiations	2.5 days	Thu 6/5/08		
189	▦	Collaboration and partnering	1 wk	Thu 6/5/08		
190		Outshoring	3 days	Thu 6/5/08		
191		**Integration Gateway 11: Software Design**	**10 days**	**Wed 4/23/08**		
192		**Iteration 1**	**10 days**	**Thu 4/24/08**		
193		Inception	1 wk	Thu 4/24/08		
194		Elaboration	2 wks	Thu 4/24/08		
195		**Radio and Communication**	**10 days**	**Thu 4/24/08**		
196		Learn environment	2 wks	Thu 4/24/08		
197		Write to requirements	1 wk	Thu 4/24/08		
198		Draw line	1 wk	Thu 4/24/08		
199		Draw character	4 days	Thu 4/24/08		
200		Research radios	1 wk	Thu 4/24/08		
201		**Transition**	**5 days**	**Thu 4/24/08**		
202		Architecture document	1 wk	Thu 4/24/08		
203		**Iteration 2 - S/W Application**	**10 days**	**Wed 4/23/08**		
204		Inception (Planning)	10 days	Thu 4/24/08		

(*Continued*)

TABLE 3.1 Target Cost: Project Life Cycle (2 Years) (*Continued*)

ID		Task Name	Duration	Start	2007 Q2	Q3
205		**Elaboration (Analysis & Design)**	**5 days**	**Wed 4/23/08**		
206		Write Use Cases	1 wk	Thu 4/24/08		
207		Class and Sequence Diagrams	1 wk	Thu 4/24/08		
208		Preliminay software requirements	0.8 wks	Thu 4/24/08		
209		Software Requirements Review	0.2 wks	Thu 4/24/08		
210		High-Level Requirements Comple	0 wks	Wed 4/23/08		
211		Refine Class Diagrams	1 wk	Thu 4/24/08		
212		Collaboration Diagrams	1 wk	Thu 4/24/08		
213		**Integration Testing**	**5 days**	**Thu 4/24/08**		
214		Test Protocol	1 wk	Thu 4/24/08		
215		Test the Test	1 wk	Thu 4/24/08		
216		Conduct Test	1 wk	Thu 4/24/08		
217		**Iteration 3 - Hardware Integration**	**10 days**	**Thu 4/24/08**		
218		Inception (Planning)	80 hrs	Thu 4/24/08		
219		**Application Development**	**10 days**	**Thu 4/24/08**		
220		Hardware setup	0.8 wks	Thu 4/24/08		
221		Learning curve	2 wks	Thu 4/24/08		
222		Design for software support layer	1 wk	Thu 4/24/08		
223		Message Manager	1 wk	Thu 4/24/08		
224		Radio Control	1 wk	Thu 4/24/08		
225		**Hardware Support**	**9 days**	**Thu 4/24/08**		
226		Design for hardware support	1.8 wks	Thu 4/24/08		
227		**Integration Gateway 12: Test Equipment**	**50 days**	**Thu 4/24/08**		
228		**Integration Test Equipment**	**50 days**	**Thu 4/24/08**		
229	▦	Requirements Definition	2 wks	Thu 4/24/08		
230		Detailed Design	10 wks	Thu 4/24/08		
231		Procurement	7.5 wks	Thu 4/24/08		
232		Assemble	4 wks	Thu 4/24/08		
233		Test	0.5 wks	Thu 4/24/08		
234		Documentation Release	1 wk	Thu 4/24/08		
235		**Integration Gateway 13: Integration of S**	**20 days**	**Wed 4/23/08**		
236		**Subassembly Availability - Prototype**	**0 days**	**Wed 4/23/08**		
237		Part 1 Available	0 days	Wed 4/23/08		
238		Part 2 Available	0 days	Wed 4/23/08		
239		Kit Available	0 days	Wed 4/23/08		
240	▦	Chassis Available	0 days	Wed 4/23/08		
241		**Prototype Full Integration**	**20 days**	**Thu 4/24/08**		
242		Build	20 days	Thu 4/24/08		

Customer groups would include a set of public transportation companies and agencies that would represent stakeholders in a new transportation system product. Integration in this case would involve understanding the strategic and business plans of these agencies and "building in" requirements and conditions as the new product is developed.

International partnering and organizational integration. This integration step involves setting up an internal business staff to provide liaison with global partners and customers. Completion of this step "rolls up" in the project schedule to "Global Interface." Organizational integration means actually serving to study and understand the system dynamics of candidate transportation agencies and systems in order to fully integrate product prototypes with customer expectations.

The concept of stage-gateway project management ensures that integration is achieved during each stage or phase of the project. Integration implies that there has been coordination and collaboration across functions and among participants *at the scale of the particular stage*. For instance, in the global interface stage, the gateway integration issue is, "Has the business positioned itself globally to exploit supplier, customer, and stakeholder values worldwide?" Thus integration is seen here on a global scale and targets the forces and factors that will create risk and opportunity for the business and its developing portfolio of projects.

Integration gateway 2: Business planning

Business planning is an integrative function and gateway because it involves the integration and combination of business objectives and strategies with candidate projects, resulting in a company strategy and aligned program and project portfolio. Here the business integrates its core competency and global partnerships with its internal business objectives and plans. Candidate projects are designed to carry out business investment goals. The integration of business interest and direction, with projects such as this ITS case study, is the second line of interface in fully integrating the project. Integration of business planning involves the use of the weighted scoring model and other integrative tools to score projects against business objectives.

In the case of ITS, one of its strategic objectives is to develop and test the feasibility of a variety of new transportation systems that would provide future customers and passengers with new levels of efficiency and effectiveness. The ITS project would score highly against this objective simply because it is directly connected to that business objective. ITS is "aligned" and thus integrated with the business itself, its capacity, its competencies, and its intended future.

Develop strategic objectives. The development of strategic objectives is a "roll up" task to business planning. Strategic objectives are developed after an audit of the business capacity and competencies, and the interests of stakeholders and management in directing business growth. Strategic objectives are developed as part of a brainstorming process, and then confirmed by top management and the board of directors. Objectives are then weighted based on their relative importance, providing a basis for later assessment of candidate programs of projects against weighted objectives.

The integration of business strategy and programs is the second line of integration following global interface. Integration ensures that all business owner and management expectations are embodied in documented strategic objectives, and that the global opportunities, both for markets and suppliers, are integrated into business decisions.

Integration gateway 3: Organizational development

If the organization cannot handle the challenges and risks associated with a given business purpose and direction, this stage ensures that work does not proceed until it can. The stage addresses the core competency of the company or agency to design, develop, and produce in a given program area, in this case transportation systems development. In effect this is an audit of company capacity, and this stage involves taking stock of company capacity in terms of people, support system infrastructure, information technology, facilities, business processes, and, perhaps most importantly, energy and commitment.

This stage is, in effect, an assessment of the company's *integrative capacity*, its proven competence in pulling together its people, systems, and processes to overcome risks and create opportunities in a given program area. If the gateway review uncovers weaknesses in this capacity, e.g., a proven tendency to deliver project deliverables late and over budget, the business undertakes improvement projects to increase its capacity to produce on time and within budget before entering the next stage.

Integration gateway 4: Global team composition and development

This stage involves setting up partnerships and team relationships in all areas necessary to achieve the business global objectives and to support candidate programs and projects. The composition of the team must be aligned with the business direction. Teams are composed of team members who possess the competencies and interests globally, across functions, suppliers, and customers that support the broad business plan. Teams are set up at each gateway level of the program, beginning with worldwide partnerships down to program and project teams. The stage is important because integration at every level requires commitments and trusting relationships between key company stakeholders.

The concept of integration is not often associated with people in a work setting as much as it is with systems and equipment. In a social or political setting, integration has meant many things to many people, e.g., racial and religious integration, economic integration, and so on. But the concept has rarely been applied to organizational dynamics. Other concepts, e.g., team harmony and team alignment, refer to teams where team members contribute according to their capacities and support each other.

However, team integration as used in this book is more than a simple team concept. It is not the question of whether the team members get along. It is, rather, grounded in the need for the team to be integrated across the organization and across the global marketplace in which it is working. Team members reach out to various support systems and stakeholders of the organization and to the customer base and *bring value* into the team. There is true integration when team members create team value because they are connected to key forces and factors outside the team. For instance, an engineer on the ITS team is in direct contact with governmental regulations that govern the dimensions and performance characteristics of the ITS. Thus his or her engineering contribution is informed by the external constraints and conditions that the project product must meet.

Virtual team tools are utilized to integrate people across geographical boundaries, thus integration is not spatial; it is informational. The integration of ideas is facilitated by online working relationships and teleconferencing, and customers and suppliers are integrated into team deliberations through online tools. This is true integration of the team into the global fabric of the business and its constituency.

Develop team and select key program and project managers. The selection of the program and project managers in an integrated project organization is keyed to training and cross-functional experience. Managers who have played many functional roles in a project management process will likely coordinate and collaborate more effectively than managers who have not. Thus, rotating talent across functions builds project management competency in integrating and delivering project outcomes.

Integration gateway 5: Support systems audit

This stage involves assurance that the organization or agency can support a complex program of projects; this is the "positioning" of the infrastructure of the organization, much like an engineering study of the readiness of a bridge for heavy traffic. Termed organizational process assets by the Project Management Institute in the PMBOK, the target here is key business processes, e.g., functional and technical support, procurement, contracting, human resources, accounting and financial cost capture and reporting, configuration management, information technology, and a project management office (PMO) that can *make or break* a program of projects. In these organizations, integration of the project, during both the planning and implementation stages, is facilitated by responsive and integrated support systems and services. These are the so-called tools of program management. No project can be successful in isolation from its parent business, and no project can be successful without a strong support system in the key *asset* areas. Integration is made possible by integrating systems.

This basic system audit task provides for reviews of the company's support systems and capacity to support program management in the following areas. Each system helps to integrate a particular business system support with project team needs.

- **Functional, technical support**—the presence of a matrix of functional capability, e.g., engineering, facilities, technical processes that support project management
- **Procurement and contracting**—the capacity to develop a supplier community and work with suppliers and contractors in a seamless, e-procurement environment
- **Human resources**—the support of an HR department that looks after the welfare of employees and the morale and well being of the workforce
- **Accounting and financial reporting**—a project oriented accounting and reporting system that captures costs at the project code level and is able to report in real time actual costs with 24 hours to project managers
- **Configuration management**—a staff and software that preserves and documents the project deliverables and their components, with disciplined numbering and filing systems, to serve transition to manufacturing and production
- **Information technology**—a company intranet system for facilitating the visibility of program and project information, e.g., schedules, plans, budgets, status reports, project review agendas, and a Web-based customer reporting system that allows project managers to share project information with stakeholders and customers
- **Project management office**—a staff function to program managers and project managers that provides analytic and administrative support to projects through generic work breakdown structures, tools and techniques for scheduling, and monitoring guidelines to ensure consistent management across all company projects

Integration gateway 6: Portfolio development and management

The process of producing and managing a portfolio of projects is an integrative function from beginning to end. Portfolio development, vertical integration process, involves aligning new projects to align *up the organization* with top-level strategic and business plans. The integration of projects with business direction is a key high-level activity that begins the portfolio process. In this case, integration means that projects are made part of the fabric of the business; projects are seen as instruments of the business plan, focused on business growth and expansion. The way integration is accomplished at this level involves the following steps:

1. Review of business strategic objectives
2. Weighting business objectives relative to each other
3. Scoring each candidate project against each strategic objective
4. Integrating weights of the objectives and project scores

This is the weighted scoring model, a tool to ensure that projects are seen in terms of how projects "rack up" against the business strategies, weighted to reflect their relative importance to the business. The tool is flexible, so weighted objectives can be changed as the business changes, and scoring can be enhanced by modification of project plans.

The development of a portfolio and scoring of candidate projects is largely a subjective system performed through a group meeting of key business management and experts. Integration in this case occurs in the thinking process of the participants as well, reflecting the views and insights of thoughtful participants.

The gateway decision here is whether projects fit the organization's direction and purpose, and serve as capital assets to grow the business. Projects enter the portfolio pipeline and are budgeted and funded by a *capital rationing* technique. Capital rationing integrates the company's allocation of available program resources with its strategic objectives through a weighted scoring model. Through the budgeting process, funds are allocated to strategic objectives according to their relative importance (triggered by their relative weights), then project costs and scores against objectives are used to fund projects until funds run out.

Integration gateway 7: Market and customer interface

New product development projects are most effectively managed when marketing is linked directly to project management from beginning to end. The integration of new product–oriented projects with the company's market and customer base is accomplished *through an organizational process*, not a technical one. The organization is vertically integrated to tie marketing and sales to project planning and management. This is accomplished by linking the marketing and sales departments in a matrix relationship with program and project management. Marketing and sales representatives are tied virtually to project team status and review meetings, and project design and development can be "interrupted" by marketing and sales inputs *at any time* through the change control process.

Integrated project management opens the change control process to its own marketing and sales people through this linkage. This is not easily done in many companies because of the traditional separation of project, marketing, and sales departments. This separation is functional because marketing and sales are outward bound activities, while projects are inward bound activities. Cultural differences stem from different workforce and incentive systems. Sales is *incentivized* by actual bottom line sales transactions; projects are typically unbound by customer acceptance transactions, with the exception of user acceptance tasks.

This institutional integration of marketing, sales, and project dynamics is a key ingredient to flexible project management, focused on better, cheaper, and faster. Although traditional project deliverables are constrained by generic processes and requirements, under this system customized changes in real time from marketing and sales are *injected* seamlessly into project processes at any stage gateway. Sales transactions on the customer's site can be made on the spot as salespeople and project people are in real-time communication on project outcomes and design specifications. New marketing research information can *be blended into project designs as they are surfaced and interpreted.*

Integration gateway 8: Project integration management

This stage gateway is the key management action, the *change intervention* that empowers the company to carry out project integration. The stage actions facilitate consistency with PMI's Project Body of Knowledge section on project integration management earlier in this chapter. Here the organization *shows* the face of coordination, integration, and teamwork through its processes. Good program planning is reflected in employees who are disciplined to do what is necessary to integrate, even when such activity may seem "bureaucratic" and overly administrative in nature. In other words, engineers may be uncomfortable with preparing a project charter when the ingredients of such a charter are already in their heads, until they see that the charter is necessary to *mobilize* the whole team.

An integrated environment also involves setting up the work of the organization at every level to allow monitoring for earned value. The concept of integration begins at the task level. Interface factors such as key earned value milestones and cost factors in a project should be built into individual task structures in the project planning process. If the task is structured correctly, project integration is practical and routine; if not, project integration becomes difficult because there is no objective set of indicators for project performance against time and cost.

The major stages of PMI's project integration management are

1. **Develop project charter**—a full and detailed discussion of the charter of the project team
2. **Develop preliminary scope statement**—a statement of the work of the project and how the staged, gateway system would work, including criteria for passage from one stage to another
3. **Develop project management plan**—an overall project plan including plans for scope, generic WBS, schedule, cost, quality, staffing, risk, procurement and contracting, resources, configuration management, and change control
4. **Direct and manage project execution**—the statement of the administrative and managerial actions that will be triggered to authorize work, to assign work, and to generate teamwork in the project team
5. **Monitor and control project work**—includes the approach to monitoring progress and remaining work, through the use of earned value (cost and schedule variance) tools
6. **Plan for project closeout**—includes procedures to terminate the project, including financial, contract, and resource disposition

Integration gateway 9: Systems safety and reliability

This is the technical design stage, the process of linking the requirements document with the design of the system and product prototype. System safety and reliability sets the design boundaries for making sure the ITS system is safe, e.g., that it performs to minimum "mean time between failure" standards, and that it is reliable, e.g., that it performs to minimum "mean time between maintenance standards."

Integration at the systems level changes the context of integration to the system and product being produced by the project. But again, technical integration does not occur without organizational integration. A quality electronic product requires that mechanical, software, and electronic engineers consult with each other in defining, designing, and building system components that must perform in the system being produced.

Outputs of this stage also include system architecture, test requirements, system design review criteria for gateway reviews, system requirements specifications, test architecture, and functional hazard assessment requirements. The gateway go or no-go decision is based on a direct tracing of design to functional requirements.

Integration gateway 10: Chassis, mechanical, and electronics design and development

This stage covers the design and development of the prototype product following designs produced in the previous stage.

Here the hardware prototype components are produced with the support of the manufacturing engineering staff. Integration of manufacturing factors and design factors is ensured here by involving manufacturing engineers in producing prototypes for testing and user approval. Components include power assembly, chassis, computer package, exhaust and emission control, supply chain contract partnerships, electrical components, and necessary drawings and graphics. Components are tested against six sigma quality standards set in design.

Integration gateway 11: Software design and development

Software design is seen as a high-risk activity because of the high probability of failure in integration, thus this activity has its own gateway. This stage involves inception and elaboration phases of software design and development, a separate stage to ensure control over the design and development of software design.

Integration gateway 12: Test equipment and testing

Test equipment is a separate stage because of the key role of testing equipment in integration. Test equipment must be aligned with testing requirements to allow for the next stage—integration of software and hardware.

Integration gateway 13: Integration of software and hardware

Full integration of software and hardware is accomplished in this stage. In a product development project, this is where software, electrical, and mechanical engineers work together to integrate their components into a fully functioning prototype for testing and acceptance.

This stage is not primarily a technical one, but rather a teaming activity where success is determined by past work between functional specialists. Integration is a coming together of trained specialists who have been working together while producing their components.

Another Case Application: Integration Issues in Portfolio and Project Planning Life Cycles

The purpose of this case is to illustrate, using a real project, how integration is achieved at each stage of a portfolio and project life cycle. The process described as follows includes many of the traditional project planning and control steps, simplified and "demystified" for practical use. This discussion is followed by a more in-depth treatment of integration, including technology and technical integration issues that typically characterize technical projects.

Integrating a project involves 13 basic steps:

1. Business and strategic planning
2. Portfolio development and project selection
3. Definition, work breakdown structure, and scope
4. Task list, with estimated durations, linkages and interfaces, and resources
5. Network diagram
6. Time-based network diagram
7. Baseline Gantt chart (schedule interfaces)
8. Resource integration
9. Project initiation
10. Project work assignment and performance
11. Team integration
12. Project monitoring
13. Project closeout

The Case: QUICK-TECH building systems

The case is a project involving QUICK-TECH, a development company in the commercial real estate and building industry. The case involves the selection, planning, and implementation of a complex building project involving a high-level technical component of building support systems. QUICK-TECH employs a new automated system of building architecture and planning that accelerates the completion of commercial buildings for specialized technical tenants who require tailored facility support and security.

Business and Strategic Planning

QUICK-TECH uses a classic portfolio development process, first identifying broad business plans and then developing a strategic plan using SWOT (strengths, weaknesses, opportunities, and threats) analysis. The company's

SWOT analysis has identified key SWOT issues in all four areas (strengths, weaknesses, opportunities, threats), concentrating on contingencies and integration challenges.

Business and strategic planning integration issues

We can define the process of integration as "the process of bringing the parts of a system together to produce a deliverable" through a proactive management approach that encourages people to worry about whether the left hand knows what the right hand is doing. Integration is made necessary because to design and build a project deliverable, that deliverable must be broken down into its "configuration" parts and produced from the bottom up. The "rolling up" of a deliverable is called integration because it requires both vertical and horizontal activity to bring pieces together from work package all the way up to the final project outcome.

We can further describe the integration process as coordinating:

1. The work of the project with the plans and ongoing operations of the performing organization
2. Product and project scope

Integration at the highest level, business and strategic planning, involves the coordination of company and business planning with project selection and financial planning. This means that the beginning of project integration is making sure the project itself is aligned with the company's strategy and plan for growth, whether written or simply a shared vision of company leadership.

Analysis of weaknesses in the strategic planning process, for instance, would be performed in the light of past company performance and market analysis. Integration of business plans with marketing plans means that a project is selected with a full view of how the project will serve the company in its growth objectives. From a personal standpoint, this means that the company leadership coordinates and communicates with program and project managers, and has access to information that facilitates project integration management.

The portfolio: Procedures in development

The business portfolio is coordinated with key company processes and workforce capacity. Resources are reviewed against the demands of the company project portfolio.

Procedures in development integration issues. The early integration of the portfolio with other company processes means that the financial and resource implications of the portfolio are reviewed by company leadership across the board. Each top company manager has veto authority over proposed projects, and a portfolio council is organized to resolve differences and move the portfolio on through development.

This building project would first appear in the business plan as a marketing concept or program objective to enter into a commercial building project consistent with the company's growth plans. Integration of portfolio issues with other company processes requires a planning information system with the following characteristics:

1. Provides a database on current workforce commitments
2. Provides for a business impact statement for each new project proposed for the portfolio

The database can be handled through a networked Microsoft Project "resource pool," a database of all current projects and committed resources against a calendar.

Integrated portfolio business impact statement. The business impact statement is a tool to ensure integration of the new portfolio with key business processes and plans. The impact statement is completed and shared widely at the top of the company to flesh out issues before further project commitments are made. This is the first and highest level integration tool tailored to weed out projects that would have adverse impacts on other company commitments or plans.

The impact statement format is shown in Table 3-2, complete with the building case now under consideration.

TABLE 3-2 Integrated Portfolio Business Impact Statement

Proposed portfolio project	Impact on current workforce	Impact on current financial condition	Impact on current risk posture of company	Impact of current market and customer base	Impact on current company technology and processes
Commercial "technical" building support project engineer staff	Current proposed project team will require hiring three draftsmen with technical building support background	Upfront financing of proposed project will require finding a venture capital source to earmark support of this highly risky project; however, company does not now have project level accounting and cost capture system to record costs	Current portfolio balance is toward high risk; this project will add to the current imbalance of highly risky projects with no committed customer contract	Company relies on technical clients that need this kind of building and support system, thus project is aligned with strategic objective of growing in this industry sector	Current company tooling is already there; capacity to perform with little new investment makes this an attractive project

The value of this statement is that each functional department completes their portion of the statement after review of the project proposal. This achieves the integration of the key perspectives in the company—many of which may be at odds—at its highest level. Thus integration activity may be completed by electronic forms, which are made part of the company business planning process.

Definition: Work breakdown structure

Before the process begins, the company defines a generic work breakdown structure (WBS). The generic WBS defines the company's prescribed technical and project process cycle for a given family or programs of projects. This is the standard for every project life cycle designed to deliver product and services and includes a set of generic tasks, linkages, and definitions, which represents the company's learning system based on past performance.

First level: The deliverable. The first step in defining the work necessary to produce the deliverable is to complete a work breakdown structure from the top (the deliverable) down to the third or fourth level of tasks. We will do this in outline form. The WBS is like an "organization chart of the work." At the top is the product or output of the project, and everything below it builds "up" to the product. The product at the top represents the final product or service outcome of the project, performing to specification and accepted by the sponsor, client, and/or user. In our case, it is the building itself. The "building" at the top of the WBS implies a finished product accepted by the user or customer.

Second level: Summary tasks. The second level across the organization chart of the deliverable includes the five or six basic "chunks" of high-level work that serve as the basic components of the project, the summary tasks that are integrated at the end of the project to complete the job. For our building project, these chunks of work might include the architectural drawing, building supplies, ventilation systems, water, and electrical systems. (For a software project, these chunks might include hardware platform, software, interfaces, training program, and financing. For a health management system, they might include the clinic population, health information system, medical personnel, space, and equipment.)

Third level: Subtasks. The third level includes a breakdown of the summary tasks already outlined into two or more sub-tasks that would be necessary to complete to produce the second-level summary task. For our building project, under the summary task "architectural drawing," this might include three tasks—get an architect, prepare preliminary blueprint, and check against standard blueprint template.

Fourth level: Work package. The fourth level is another level of detail at the real tasking level—the work package. These are the individual tasks assigned to team members. For instance, breaking down the summary task "get an architect" to the fourth level, we identify nine work packages: build list of candidate architects, develop criteria for selection, screen candidates, interview candidates, conduct reference checks, compile candidate information, distribute candidate information, convene meeting, and conduct process of selection.

It is this last level that is used to create the task list, the actual work assignments necessary to identify and schedule before the work can begin. These are the "schedulable" tasks.

The resultant WBS outline (can be shown as an organization chart) looks similar to the following.

1.0 The Building
- 1.1 Architectural drawings
  - 1.1.1 Get an architect
    - 1.1.1.1 Build list of candidate architects
    - 1.1.1.2 Develop criteria for selection
    - 1.1.1.3 Screen candidates
    - 1.1.1.4 Interview candidates, conduct reference checks
    - 1.1.1.5 Compile candidate information
    - 1.1.1.6 Distribute candidate information
    - 1.1.1.7 Convene meeting
    - 1.1.1.8 Conduct process of selection
  - 1.1.2 Prepare preliminary blueprint
  - 1.1.3 Check standard blueprint template
- 1.2 Building supplies
- 1.3 Ventilation system
- 1.4 Water system
- 1.5 Electrical system

Of course, in this case, we have filled out only one summary task to the fourth level; all levels are filled out in a real project. The concept is that all the project tasks at the lowest level of the WBS "roll up" to produce the deliverable.

Work breakdown structure integration issues. The work breakdown structure is the first integration challenge after the alignment of the project with the business strategy is ensured. The tool for integrating a project work breakdown structure is a generic WBS. The generic WBS is a standard business process that defines the company's core project and product development process and project codes. All WBSs are checked against the generic WBS to ensure alignment and workforce competency.

This is the way it works. First the generic WBS is developed by a cross-company team representing all key functions and support systems. A project management office is typically in charge of the effort, but the main thrust is to develop a "standardized" WBS and task list on each product deliverable the company produces. Key coordination steps across the company are built into the generic WBS, thus ensuring all perspectives are reflected.

The following outline is an example of a generic WBS and how it works.

2.0 Technical Building Design and Support Systems
- 2.1 Architectural drawings
  - 2.1.1 Get an architect
    - 2.1.1.1 Build list of candidate architects
    - *2.1.1.2 Check with building system department for past architect performance records*
    - 2.1.1.3 Develop criteria for selection
    - *2.1.1.4 Get customer input on criteria*
    - 2.1.1.5 Screen candidates
    - 2.1.1.6 Interview candidates, conduct reference checks
    - 2.1.1.7 Compile candidate information
    - 2.1.1.8 Distribute candidate information
    - 2.1.1.9 Convene meeting
    - 2.1.1.10 Conduct process of selection
  - 2.1.2 Prepare preliminary blueprint
  - 2.1.3 Check standard blueprint template

Plan tasks for earned value

To be able to interpret progress reports on tasks in terms of percent complete (as a precursor to earned value, schedule and cost variance analysis), tasks are structured so that key milestones align with percent complete determinations. For instance, if *conduct process of selection* is a two-week activity, one week deliverable could be *meet and discuss*, and the other could be *vote and decide*. Each costs the same in project labor costs. Thus 50 percent complete would coincide with meet and discuss, 100 percent complete would indicate that a decision has been made. This is an integrated approach to task structuring that enables meaningful earned value assessment later; without it, progress reports cannot be related to real deliverables.

Task list. The task list includes the fourth level of the building project previously referenced. This step defines the "work" for several purposes:

- To serve as the basic definition of the "work" of each task, consistent with the definition of work in MS Project (Work = duration × resource)
- To serve as the basis for the network diagram, each task will be an arrow in the network diagram
- To serve as the basis for identifying risks, the first opportunity to identify high risk tasks

In Table 3-3, tasks are listed and durations for each estimated by the task manager who will be accountable for that task, along with dependencies (predecessors) and assigned resources.

TABLE 3-3 Task List

ID	Task	Duration (total estimated elapsed time) (weeks)	Predecessor (linkage or dependencies)	Resources
A	Build candidate list	6	0	HR Specialist Plans Department
B	Define criteria for selection	3	0	Project Manager and Architectural Drawing Task Manager
C	Screen candidates	50	A	Architectural Drawing Task Manager
D	Interview candidates	30	A,B	Project Manager
E	Reference checks	25	B	HR Specialist
F	Compile information	35	C	HR Specialist
G	Distribute information	3	F	HR Specialist
H	Conduct selection process	3	E, G	Project Manager

Network diagram. Having identified the basic tasks of this summary task, you now build a network diagram of this summary task, which is later integrated with other summary tasks diagrams to create the whole project network, as follows.

Start with a network template. Always start your network diagramming with a template or model of the "typical" network, and adjust it to the project you are planning. A typical template looks like this, with three paths and parallel activities, ending in one task (See Figure 3-5).

Then tailor your model to your project as shown in Figure 3-6.

Time-based network diagram. Here you simply place the network diagram on a time-based graph. Draw the length of the arrows representing each task to equate with their actual durations as aligned with the bottom calendar of 97 days. Note that Figure 3-7 shows "float," or "slack," the dotted lines that represent the flexibility in what "time slot" you determine to do the non-critical

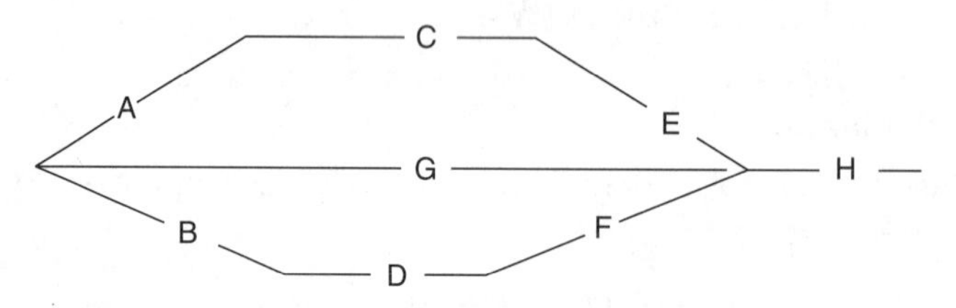

Figure 3-5 Network diagram template.

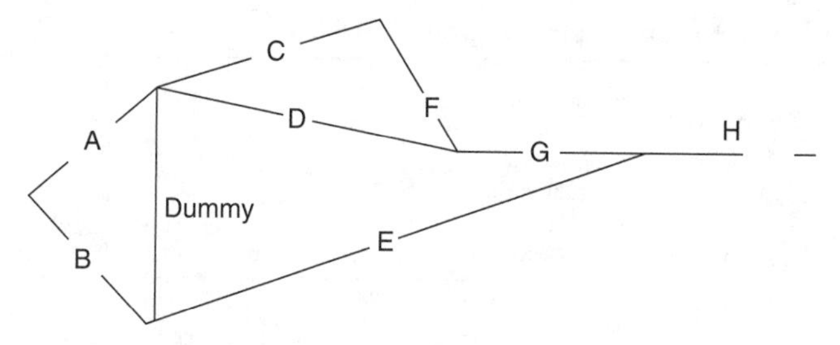

Figure 3.6 Network diagram for this project.
(The Dummy arrow connecting A and B is not a task; it's a link. This arrow shows that A and B are interdependent with D. D is dependent on both A and B, not just A).

path tasks. Note also that the critical path, A, C, F, G, and H, a continuous arrow with no breaks, represents the critical path.

Analysis of early and late starts. To determine what slack you have in the project plan to move tasks that are not on the critical path, do an analysis of early and late starts, early and late finishes, and slack, as shown in Table 3-4.

Project Paths: A, C, F, G, H = 6 + 50 + 35 + 3 + 3 = 97 weeks (critical path)

A, D, G, H = 6 + 30 + 3 + 3 = 42 weeks

B, E, H = 3 + 25 + 3 = 31 weeks

Gantt chart. The final Gantt chart (MS Project) takes the task list information entries and builds a bar chart representing the whole project graphically, based on linkages and durations. This is shown in Table 3-5.

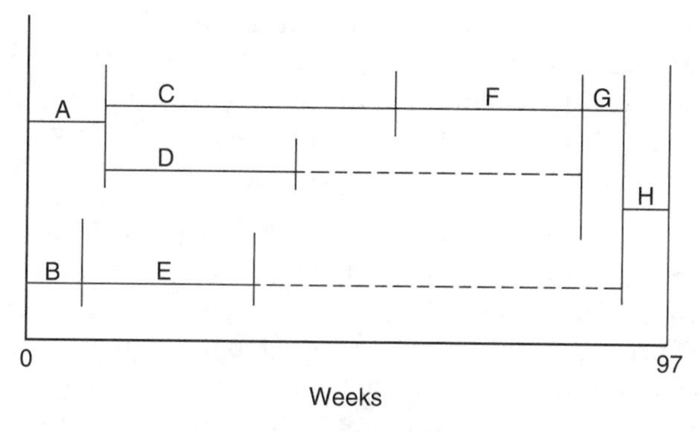

Figure 3-7 Time-based network diagram.

TABLE 3-4 Analysis of Early and Late Starts and Slack

ID	Task	Duration (total estimated elapsed time) (weeks)	Early start (week)	Late start	Early finish	Late finish	Slack
A	Build candidate list	6	0	0	6	6	0
B	Define criteria for selection	3	0	66	3	69	66
C	Screen candidates	50	6	6	56	56	0
D	Interview candidates	30	6	66	36	96	60
E	Reference checks	25	3	69	28	94	66
F	Compile information	35	56	56	91	91	0
G	Distribute information	3	91	91	94	94	0
H	Conduct selection process	3	94	94	97	97	0

Resource integration. Resource integration involves interfacing project planning with workforce and manpower planning. A successful resource integration system characterizes the workforce and makes it possible to know:

- What resources are required to accomplish selected projects?
- What resources are available in-house, when and where?
- What resource outsourcing or contracting is needed, when and where?

The integration of project planning with resource planning ensures that the company does not take on more than it can handle, and can avoid the kind of resource constraints or bottlenecks characterized in the critical chain concept.

Scale and integration. Integration occurs at different scales, or different perspectives. This makes decision making difficult in a complex company facing many tradeoffs in its management of multiple projects.

Integration occurs at five levels: global, industry, company, program, and project.

- **Global.** The global scale of integration works in the regime of global markets, supply chains, and new product development. Thomas Friedman's book, *The World is Flat: A Brief History of the Twenty First Century*, reminds us that we are passing the business-based, global economy and moving into a new era of individual-based information sharing and open source partnering that has no limits. Project management will be different in this new flat world, and project integration will be increasingly instantaneous and virtual.

TABLE 3-5 Gantt Chart

ID		Task Name	Duration	Start	Finish	1st Half		2nd Half		1st Half		2nd Half		1st Half		2nd Half		1st Half		2nd Half	
						Qtr 1	Qtr 2	Qtr 3	Qtr 4	Qtr 1	Qtr 2	Qtr 3	Qtr 4	Qtr 1	Qtr 2	Qtr 3	Qtr 4	Qtr 1	Qtr 2	Qtr 3	Qtr 4
1		A Build Candidate List	6 wks	Mon 7/28/03	Fri 9/5/03																
2		B Define Criteria for Selection	3 wks	Mon 7/28/03	Fri 8/15/03																
3		C Screen Candidates	50 wks	Mon 9/8/03	Fri 8/20/04																
4		D Interview Candidates	30 wks	Mon 9/8/03	Fri 4/2/04																
5		E Reference Checks	25 wks	Mon 8/18/03	Fri 2/6/04																
6		F Compile Information	35 wks	Mon 8/23/04	Fri 4/22/05																
7		G Distribtute Information	3 wks	Mon 4/25/05	Fri 5/13/05																
8		H Conduct Selection Process	3 wks	Mon 5/16/05	Fri 6/3/05																
9																					
10																					
11																					
12																					
13																					
14																					
15																					
16																					
17																					
18																					
19																					
20																					
21																					
22																					
23																					
24			1 day	Mon 7/28/03	Mon 7/28/03																

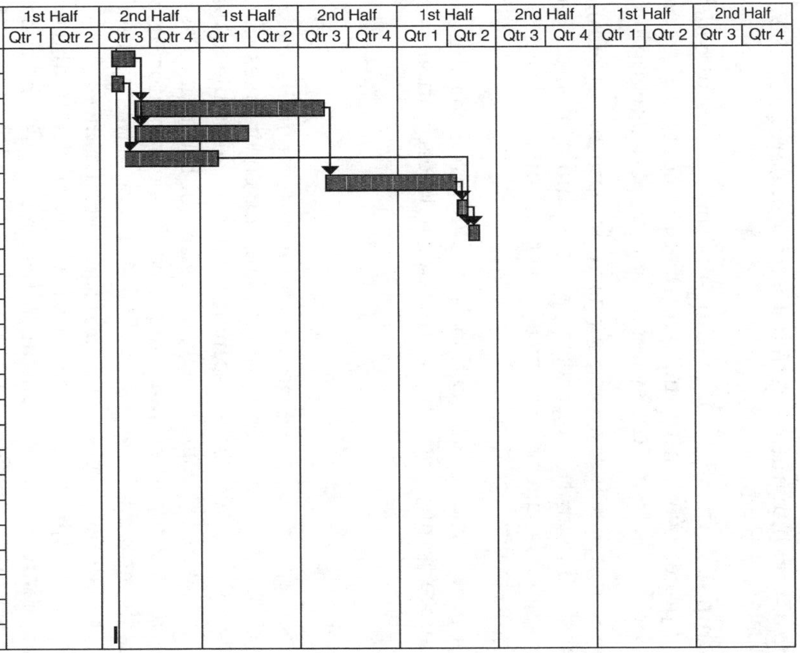

- **Industry.** Integrating with industry standards and the state of the art in the markets served.
- **Company.** Company integration is the successful combination of program and project work with the assortment of supporting organizational "assets," as PMI calls them, the systems that support integration.
- **Program.** Because a program is an integrated set of related projects and/or phases, integration of project work within a program is accomplished by a program manager to ensure the system outcome is consistent with customer requirements.
- **Project.** Project integration is the internal coordination and integration practice of the project team, working with each other in a seamless and open system.

Integrated monitoring

The concept of an integrated monitoring system best exemplifies integrated project management. Integrated planning encourages the joint assessment of impacts of various project scenarios on cost, schedule, quality, customer satisfaction, corporate competency, and growth. But regardless of how well project plans are put together and integrated, it is in monitoring that integration produces its most effective results. This is because the "system" as described in the project plan rarely plays out in actual performance.

Monitoring project performance requires tradeoffs, and tradeoffs are an application of integration. For instance, trading off cost and time requires the project manager to see *time and cost as a whole—and both in the context of risk.* Rather than simply looking at the "burn rate" of the project and making decisions on cost based on rate of expenditure, an integrated approach requires looking at *both cost and schedule.* As cost variance from an original plan is positive—that is, as the project is able to produce to its milestones at lower than expected costs—the schedule implications of positive cost variance must be viewed at the same time. Thus integration of cost, risk, and time is achieved. As cost variance becomes more positive, it is quite possible that reductions in cost are at the expense of time.

Integration gets more complicated when three or more variables are being viewed and traded off. For instance, as time and cost variances are more positive, it is possible that potential impacts on product quality are being sacrificed. Substantial variances in cost and time suggest either that the baseline estimates are incorrect, or that achievement of a quality deliverable is affected. Cost and time efficiencies could suggest skipping important quality processes, such as testing, and overages suggest that quality standards are coming at unexpectedly high costs and schedule.

Real integration during the project cycle is even more complicated when internal variables are added to the integration mix, such as company profit margins or resource efficiencies. Thus, at the same time the project manager is keeping tabs on cost, time, and quality, the dynamics of the project may be having unexpected impacts on the business itself. Here we see that picture of the project perhaps doing well, but using valuable resources that are needed elsewhere to achieve the company's overall strategic plan. This means that integration occurs now only at the deliverable, process, and project team levels, but also importantly at the business level. Tradeoffs at the company level may result in reversing decisions made for fully integrated rational reasons at the project level. For instance, a project doing well in terms of budget and time and likely to satisfy a customer may be negatively impacted or even terminated for "bigger picture" issues, such as marketing, company finances, or acquisition and merger. Thus one expects that top management will "integrate" at a different level than the program or project manager; a company that generates respect for upper management is more likely to be able to resolve this built-in tension than one that ignores the potential of it happening.

"Reading" the project as an integrated whole

Gareth Morgan made the concept of "reading" an organization popular in *Images of Organization.* His breakthrough analysis of organizations using metaphors provides a useful backdrop for viewing the challenges of program and project integration management. The higher the level of visibility in the organization, the more important it is to be able to *read* the project and the parent organization. "Reading," in this sense, means seeing the full dynamic of many projects and decisions in perspective, judging the collective results and impacts of the "tyranny of small decisions." Good project managers make decisions on the basis of integrating the factors and variables *they see in the larger context, requiring them to think terms of integration.*

Integration mixes art and science, intuition and data. Managers must make decisions with incomplete information, thus integration is both analytic and interpersonal. Analysis provides data on real impacts, but sometimes a full picture of the complete dynamics of a project cannot be provided with data and information. Thus integration usually requires communication with key people who can "read" the impacts at their accountability level, e.g., how will this project decisions affect project finances.

Integration of cost, schedule, risk, and quality

Project integration in practice in a real work setting often means "juggling" several factors and forces during the project planning and control process. Contrary to academic and "professional association" rubric that "segments" project management as if it were pieces of a pie, project planning and management does not occur in distinct phases nor in convenient "buckets" called planning, risk management, scheduling, scope, cost estimation, and so on. In fact, these functions and activities occur all at once as a project takes shape and is implemented.

Integration then becomes first a mindset, an approach that requires a project manager to keep her eyes on several factors at once. This discussion places this "multitasking" environment of integration into perspective. Let us try to integrate risk, schedule, cost, quality in a typical project setting.

First, planning *is* risk management. Every project plan approaches work structure and tasks in terms of overcoming uncertainty and barriers to project completion. So risk is inherent in a project simply because projects are usually new and different from past work and because there is a level of uncertainty and risk involved in every aspect of the project. The impact of failure because a risk is not "managed" well is felt in project outcomes such as cost, schedule, and quality. Bad risk management is bad cost and schedule control.

Integration involves making sure that all aspects of a project are attended to during the whole project life cycle.

- Work and task planning is structured so that earned value (schedule and cost variance) can be assessed at milestones built into the task definition and schedule. In other words, the way you build cost and schedule control into planning is to define the work in "segments" that correspond with percent complete determinations.
- Work and task planning is reviewed in terms of potential failures, e.g., risks, and high-risk tasks are presented in a risk matrix that includes a contingency plan *that is built right into the project schedule.*
- Quality is integrated into project planning by not only shaping scope and work around customer requirements, but by structuring work around quality. This means that earned value determinations, e.g., percent complete reports, assume that when a team member reports a piece of work completed that this means *completed according to user or customer requirements.*
- Projects are seen as decision systems, not just task planning and control. This means that the project manager anticipates decisions that have to be made and the expected value of these decisions downstream. This ensures that the schedule is flexible and includes alternative decision paths and scenarios when appropriate. Project plans become more than a single, linear look at the process; they start to reflect alternative directions as key decisions are made.

Project integration is first a personal and professional issue, not a process or technical issue. Project managers must be able to see the "big picture" as they proceed through the project process; they lead and direct the team to keep their eyes on all the major factors that need to be integrated to ensure success. They know the key players who can reflect different perspectives on the project, accountants, technical specialists and engineers, human resource people, contract and purchasing people, and top management. They are "plugged into" customers and users during the process to ensure integration of customer learning and change into the project.

The indicator of good project integration is that there are no surprises in the project that have not been addressed in planning and project design. Thus, if project integration is unsuccessful, we see that unanticipated costs appear, schedule delays occur that were not planned, quality problems are seen too late to resolve without costly change, and "outside" forces" intervene in project decisions.

Steps in the cost/schedule/risk/quality integration process

The following are four steps in integration. These steps are part of the normal project life cycle, not separate from them.

1. **Project alignment.** Projects are aligned and integrated with company strategy and financial goals. Projects that are not aligned with where the company or agency is going are destined to fail. Integration with the "home" competency and culture is key to success. Tools used to ensure such integration include a weighted scoring model, portfolio process, and top management interface.
2. **Interface management.** All stakeholders in the project are involved in project planning through effective "interface" management.

   Projects are continually "interfaced" with the following perspectives and functions:
   - Accounting
   - Purchasing
   - Engineering
   - Human resources
   - Marketing
   - IT
   - Manufacturing or production
   - Customer
3. **Project tradeoff management.** Cost, schedule, risk, and quality are built into project work and task definition so that tradeoffs can be made between them. Work is defined in terms that work can be monitored; if the monitoring system is attuned to the way work is planned, progress reporting can be successful.
4. **Earned value control.** Project progress is assessed through indicators of schedule, cost, and quality variance. In addition to the traditional focus on schedule and cost variance, a new concept—quality variance—is introduced. Quality variance is a distinct assessment of the gap between customer expectations and product or service delivery.

Integration Skills of the Program and Project Manager

We explore here the progression from project manager to program manager in an organization, and the new skills and competencies required to make the transition from leading a single project to leading a longer-range, multiproject program. While there are many gray areas in the transition from project to program, it is clear that new competencies and skills are "called up" as the focus becomes a long-term program and the broader scope needed to monitor many projects at once.

Single project management

The management of single projects requires several analytic and leadership skills in addition to proficiency with project management software. Project managers must be capable of working with a customer to develop the scope of work for the project and prepare a work breakdown structure that captures the deliverable in an organizational chart or outline. The project manager must be familiar with the technical field and changes in technology. The manager must be able to put a team together, assigning team members to activities and task areas in the work breakdown structure, and lead the team through the project life cycle. The manager must have the capacity to use project management software to prepare Gantt charts, assign resources, estimate costs, produce reports, and make presentations on the project.

There is an active debate in the field on the extent to which the project manager must be familiar with the technology of the project and technical aspects of design, development, testing, and product delivery. Some say the manager need only have a cursory sense of the deliverable and the technology involved, relying on team members and subject matter experts for technical assistance. Others say the manager must know enough "not to be snowed" by the customer, team members, or suppliers, and subcontractors. They indicate that the manager will have to interpret technical progress reports of team members and be able to communicate with technical counterparts in the customer organization. In any case, it is clear that the project manager must have a good grasp of the deliverable and be comfortable in the field, if not an expert. If the field is changing rapidly, it is especially important for the project manager to grasp the implications of change for the current project.

Single project managers must be wholly focused on the day-to-day dynamics of the project because the project environment is always changing and shifting unexpectedly. The horizon is short in project management. This requires that the project manager manage through the 80-hour rule; that is, the focus in each weekly review is the current status of projects, indicated in earned value analysis, and anticipating the next 80 hours of work. This way the team is reminded of the next two weeks' work, challenges that can be anticipated, and key milestones in that 80-hour period. The single project manager is typically wrapped up in the project at hand.

Project managers must have several key skills: the ability to lead a team and resolve team problems; the ability to communicate and report effectively to a wide variety of customers and stakeholders on technical and project issues; the ability to manage a number of technical assignments all at once; the capacity to deploy project management tools, such as Gantt charts and schedules; a full understanding of the project life cycle; and proficiency in project management software. In addition, the project manager must have judgment skills to make tradeoffs between cost, schedule, and quality during the progress of the project and to make difficult decisions quickly to keep a project moving.

Finally, project managers are expected to be advocates of their projects and to make effective arguments for resources and priorities based on their project needs and their projects' "critical path." They are not necessarily expected to see the big picture or make decisions on sharing resources with other project managers. They are expected to be narrowly focused on making their project goals, objectives, and deliverables regardless of what else is happening in the company. If they start compromising their focus in the context of the needs of other projects, they do a disservice to their customers and to their project team.

Program (or multiproject) management

Program management is a different kettle of fish altogether. Program management is the process of managing and integrating a "portfolio" of projects, some of which are going on at the same time and some of which are linked in a sequence of product enhancements over a longer time period. In either case the program manager's span of control is wider and broader than the project manager's responsibility. Program managers can be responsible for a long line of products and product enhancements in one program area over time, say over a ten-year period, transitioning from one project to another based on customer and market feedback. Program managers also can be responsible for many projects and project managers across programs or product areas, thus complicating the management process. Program managers often have responsibility for broad corporate product lines and markets across wide technical boundaries.

Program managers typically hire and supervise other project managers. They are responsible for developing the company's project manager workforce, building and advancing project managers into higher levels of responsibility. They coach project managers. Program managers serve as the interface between projects and broader company strategies and business plans, thus they are called on to communicate broad purpose to individual project managers and to report individual project results to corporate executives in high-level program reviews.

The complex, multiproject environment that program managers face requires different skills in managing information. The program manager cannot get lost in the details of one project or one project schedule or report, but must have an "enterprisewide" perspective. Information on many projects must be managed so that the program manager can see the big picture and make the necessary tradeoffs between projects, if necessary, to resolve resource and priority problems. Program managers must be able to step away from project details, see the broader implications, and make decisions on the basis of corporatewide considerations and impacts.

BuildIt: A Sample Integrated Program Structure

The selected course project business is a small, 100-employee, construction firm, BuildIt, Inc., that builds high-density housing around transportation hubs and transit stations. The business has built several complexes in the Atlanta area aimed at middle-income families, and specializes on small, compact cluster housing.

Organization

The firm does business through projects, with project managers leading teams of architects, engineers, electricians, carpenters, concrete specialists, and other home-building specialists. The organization is a matrix with functional managers for architecture, engineering, electrical, concrete, and drywall departments, who assign employees to various project teams.

Strategic statement

BuildIt's overall strategy is to capture the market in the Atlanta area for high-density home building and development around transportation hubs, particularly MARTA (Metropolitan Atlanta Rapid Transit Authority) stations. The strategy is to work with the area wide planners and MARTA to get early information on where stations will be built and improved, and use that information to acquire land and build developments adjacent to stations.

One- to five-year strategic objectives

The following are the three objectives related to the one- to five-year strategic objectives.

Objective 1: To build home developments around current MARTA stations

Objective 2: To research plans and programs for future MARTA sites

Objective 3: To build home developments around the intersections of major Interstate highways

Program of projects

The program of projects for BuiltIt is a broad description of a series of short-term projects and other support activities necessary to carry out the strategic intent of the company to increase its market share in transit-related community development. Let's take one program area, near term community development, and work through to specific projects:

Program Area 1 Near term community development

Program Goal: To build home developments around current MARTA stations.

This program of projects includes all of the program planning activity that will produce a set of short-term projects to design, plan, and build home developments around current MARTA stations.

1. **Marketing program:** The Marketing Program of Projects includes all program and project activity focused on identifying opportunities and markets for current MARTA-oriented housing development. The marketing program involves gaining a better understanding of not only what potential there is in building high-density homes around stations, but also what the capacity of BuildIt is to implement such a program. Marketing will dimension the needs

as well as the challenges of designing and building complexes in terms of the company's strengths, weaknesses, opportunities, and threats. Included in this program of projects are:

- Project 1. Perform market research on potential construction planning and engineering studies for Sandy Springs Station.
- Project 2: Perform market research on potential construction proposal for major downtown development around Underground, including major land redevelopment.
- Project 3: Perform feasibility study on options for high-density homes and development around the Atlanta airport station.
- Project 4: Contract out a survey and produce a comprehensive report to describe resident needs and services around MARTA stations.
- Project 5: Develop a model home design and build near the MARTA station to market housing development concepts.
- Project 6: Design an Internet marketing program and campaign to identify needs for high-density residences near MARTA stations.

2. **Project plans:** Projects from these programs would then be selected using three tools: (1) a financial model of the project that estimates revenues and net income from the project over five years and then discounts the estimates using new present value (NPV), (2) an analysis of alignment with strategy (weighted scoring model), and (3) a risk assessment using the PERT analysis in Microsoft Project. These tools will be illustrated later in this chapter.

Project cost accounting systems (PCAS)

PCAS integrate work packages, cost accounts, and project schedules into a unified project control package. They permit cost and scheduling overruns to be identified and causes to be quickly pinpointed among numerous work packages or cost accounts. Two elements common to most of these systems are the use of work packages and cost accounts as basic data collection unites, and the concept of earned value to measure project performance.

A program management manual for integrated project management

One of the most effective ways to integrate good project planning and control tools into new product development is to publish and implement a new product project management manual. The following discussion addresses what such a manual would look like.

Program management principles

It is company policy that all programs and projects will be planned and managed in a way consistent with project integration standards. The process will be managed by a program manager who is responsible for integrating and producing products that satisfy customer requirements. Customer requirements are documented in a System Requirements Specification (SRS). Products are produced through a product development process that meets customer requirements. It is company policy that in collaboration with department managers, program managers will follow this guide in planning, scheduling, and tracking programs through the product development cycle. The following principles underlie the program management process.

Meet customer requirements

We are a customer-driven company, striving to meet or exceed customer requirements. The customer's technical requirements are embodied in an SRS prepared by the project team. The customer's schedule and resource requirements are embodied in a baselined program schedule prepared by the program manager and approved by the Director of Product Development. In addition, the program manager maintains close contact with the customer to ensure integration at every level required.

Follow integrated, generic WBS—Product development process

The program management process will ensure that products are managed through the integrated product development process, tailored to particular product requirements. Program managers use the generic work breakdown structure (WBS) in that policy as the basis for scheduling a program, with exceptions for special programs.

Standard work breakdown structure

The standard WBS for schedules is specified as follows:

- **Program:** This is the display product line incorporating a basic set of features and functionality
- **Project phase:** This is the particular set of features and functionality for a program, based on particular customer need
- **Stages:** These are the steps in the product development process, e.g., requirements, detailed design, prototype development, design validation, verification, and manufacturing transition
- **Functions:** This is the task level within a stage, e.g., mechanical/optical, electrical, and software design within detailed design
- **Tasks:** This is the operating component level where work is achieved through individual or small team activity

Teamwork

Program managers and functional managers establish integrated teams to carry out the work with the objective of building an environment of high-performance teamwork and collaboration. To the extent possible, staff will be assigned tasks that are consistent with their backgrounds and expressed professional interests. Job descriptions will be written to encourage interface and integration of each individual's work. Program teams will be composed of professionals who are suited to the work they are expected to perform. Team staff members will be oriented and trained as necessary to enable them to perform assigned tasks.

Define and communicate the scope of work and assignments clearly

Product requirements and job assignments will be defined and communicated to the program team clearly through the program schedule and individual assignments. This is to empower staff to understand how their work contributes to the overall customer requirement, and prepare for work assignments with appropriate training and development.

Collaboration across the organization

Collaboration between the program managers, department managers, and the program team is the essential ingredient to the success of program management. The company encourages continuous, professional communication and information exchange among program team members and department and system managers in a concurrent engineering framework. The objective is to create both individual team member and joint team accountability for particular tasks and a broad support system to ensure individual success.

Work will be quality and schedule driven

Maximum emphasis will be placed on preparing tight program schedules that incorporate all the work necessary to meet requirements on time. Program schedules will be planned and "scrubbed" in a collaborative process that ensures all necessary work is included and all task durations, interdependencies, and resources are tightly planned and estimated. Schedule baselines will be established and work initiated only after schedules have been tightened through this process.

Ensure timely procurement of product components

Program managers pay special attention in early program scheduling to ensuring the availability of required product components and test equipment. Hardware specifications, parts, test equipment, and supply items will be included in initial program schedules and appropriate lead times established. Procurement actions will be generated in a timely way to avoid schedule delays attributable to lack of components.

Change will be managed

The company will administer the engineering change notice and configuration management processes to ensure that requirements and product component changes are managed and controlled. A systematic change management process ensures that specifications can be met within schedule and resource constraints.

Program progress will be tracked and periodically reviewed

Program managers will track program progress, prepare weekly reports, and prepare for weekly program reviews conducted by the Director of Product Development. Department managers and selected team members will participate in tracking and program reviews as appropriate.

Program management: Roles and responsibilities

The organizational framework for program management is a collaborative, team-based organization requiring cooperation between program managers and functional department managers. In that framework, the program manager has responsibility for delivery of the product within quality, schedule, and resource requirements. Department management is responsible for supervising staff and maintaining the technical capacity of their departments to support program management.

Program management office (PMO)

The program management department includes all program managers and the program administrator/planner. The PMO is responsible for ensuring consistency in the application of this program management guide throughout the product development process. The PMO tracks performance of the overall program management process.

Within the department, program managers will consult regularly with each other on scheduling and resource plans and potential conflicts, and ensure consistent approaches to scheduling details, work breakdown structure, budgeting, and sharing schedule information. On the initiation of new programs, program managers will consult with each other on potential resource impacts and issues.

Program manager role

The program manager is ultimately responsible for meeting customer requirements and delivering the program within schedule. The program manager provides leadership to the program team, ensures that the program meets product specifications, delivers the product on time and within resource constraints, and in general controls the "what and when" of the project. The program manager produces time-phased schedules for each program, tracks progress and anticipates future impact, and ensures linkages with related programs and projects.

The program manager requires that design reviews at every phase are conducted and documented to ensure integration, and all actions resolved.

The program manager has the primary responsibility for creating a program plan for each program and a program schedule composed of tasks and milestones. The program plan is created with support from the customer, program team members and department managers. After the program is underway, the program manager is required to keep the program schedule current, track progress, and incorporate changes as required. The program manager utilizes project management software to produce and update schedules and resource reports, and is expected to be proficient in the use of such software for control and presentation purposes.

The program manager has the primary responsibility to create and maintain a detailed program schedule that meets all program objectives. The schedule must be consistent with the generic WBS, and should include:

- Summary tasks and task structure and key milestones that correspond to all major program objectives contained in the program plan
- All product development activities and tasks required to execute a given program, including systems design, detailed design, certification, test equipment, reliability, safety, design reviews, manufacturing, procurement, test assets, and so on
- Tasks detailed to the lowest practical level; activities and tasks should generally be built four levels down
- Resources assigned to activities and tasks and leveled to reflect a realistic workload

Departmental manager roles in the matrix

Department managers for systems engineering, mechanical/optical design, electrical design, software design, and certification are responsible for building and maintaining the resource and technical capacities of their departments to support the product development process. The following are some key functions of department managers in the program management process:

- Assign staff to programs and support assignments
- Ensure technical processes and systems are in place to accomplish programs
- Prepare and maintain department schedules for each program, identifying department level assignments
- Attend program review meetings and provide advice and support to program managers

Each program team member is responsible for understanding his or her individual tasks and for general support to overall team performance. Team members are accountable for keeping technically proficient and performing their assigned tasks in a timely way, consistent with the schedule. Team members are responsible for communicating with their program managers and functional managers on issues or problems encountered in their team tasks. Team members collaborate with each other and the program manager, promptly attend program team meetings, and report to their program managers and department managers on schedule and technical issues, respectively.

Role of the program administrator/planner

The role of the program planner in the PMO is to promote consistent best practice in ID program management. The administrator/planner provides administrative support to program managers and departments with scheduling, resource planning, and reporting service, and prepares analyses of resource impacts to identify and resolve conflicts. In addition, the program administrator/planner prepares program management guidelines, provides training, develops program evaluation metrics, and maintains individual program schedules for the Director of Product Development and/or program managers.

Program planning, scheduling, and resource management

The product development process is primarily schedule driven. Effective and disciplined scheduling and tracking of work and resources is directly related to customer satisfaction because customer expectations always include timely delivery as a key priority.

The scheduling process begins with a program plan that describes the overall program in general terms. It is a reference source for all documents that impact on the program. The program plan includes:

- Program overview
- Program strategy
- Customer identification
- Program objectives
- Measures of program success
- Program scope and requirements (Summary)
- Program management, including team roles, schedule, resource plan, milestones, program review, and risk management
- Program development and review process
- Reference documents

After a program plan is approved, good scheduling is at the heart of the program management process. Program schedules, created in Microsoft Project, and linked to a central resource pool file and posted on the network, constitute the basis for program development, tracking, and review. A good scheduling process provides adequate time to ensure that the work breakdown is comprehensive and responds to the customer requirement and product functionality, that scheduled task durations and predecessors are as accurate as possible, that key linkages are made, and that people and resources, once assigned, understand interdependencies and are available and committed to the program when they are needed.

Before a schedule is drawn up, the work itself must be clearly defined in a work breakdown structure. Thus scheduling provides for an SRS (in the requirements stage) that defines the *what and why* of the program or product, e.g., a description of the product and its functionality. The scheduling process helps to flesh out requirements as individual features are programmed into various iterations of the schedule, leading to the baseline.

After the work scope is understood and signed off by the customer, scheduling defines *when and how* the work is going to be done, key interdependencies, *when* the deliverable will be produced, and *who* will do the work. The scheduling process assigns staff to scheduled work, and commits staff to do the work within the time constraints in the schedule. Scheduling is a resource planning tool, providing a high degree of discipline to the assignment of staff because each task is specifically described and time-constrained. Thus scheduling requires that those who are actually going to do the work—*those who are being scheduled*—also be part of the process. Because scheduling "signs up" and mobilizes staff to fit new work into their schedules, which typically include other program work, it requires that there be a clear picture of staff availability, e.g., the current resource picture. New program schedules are phased in based on the timelines and resource impacts of current work.

The integrity of a schedule is only as good as the description of the work, the processes in place to do the work, a good picture of interdependencies and resource availability, and the commitment of the people who are slated to do the work. This process becomes more complicated when there are several programs or projects operating at the same time and where staff time is always limited by previous commitments driven by earlier or concurrent projects, and by anticipated and unplanned work. In the end, it is the quality of the planning and communication process and the capacity, commitment, and motivation of the individuals actually doing the work that drives a successful schedule.

Project management software makes it easier to accomplish the scheduling process by capturing important planning and scheduling data and making it available to a wide cross section of people, and facilitating presentations and progress tracking. The following process assumes access and proficiency in Microsoft Project as the support software for a network-based planning and communication tool.

Scheduling tailors the product development process to real time and available resources, and flags conflicts and new resource needs. Scheduling provides time-phased and linked tasks and milestones, assigns resources to complete the work, and supports the monitoring of performance, resource allocations, budget, and earned value.

Table 3-6 shows an example of a program schedule (GANTT chart),but without the bar chart at the right.

Table 3-7 shows the resource usage view of that same schedule. The program manager and the project team can see from this view what resources are assigned to the project and the level of assignment in terms of hours, against a real calendar.

TABLE 3-6 Program Schedule

ID	Task Name	Duration	Start	Finish	Predecessors	Resource Names	Jun
1	**Huntsville Project**	**44.4 wks**	**Thu 4/17/97**	**Fri 2/20/98**			
2	Select Architect	2 wks	Thu 4/17/97	Wed 4/30/97		Project Manager[12%],Corp. Personnel[0%],Facility Specialist[73%]	
3	Recruit & Train Managers	6 wks	Thu 4/17/97	Wed 5/28/97		Project Manager[21%],Facility Specialist[0%],Corp. Personnel[80%]	
4	Select Real Estate Consultant	2 wks	Thu 5/1/97	Wed 5/14/97		Project Manager[10%],Facility Specialist[50%]	
5	Pre-Production Plan	2 wks	Thu 4/17/97	Wed 4/30/97		Production Specialist[80%],Manufacturing Engineer[85%], Marketing Specialist[0%],Project Manager[20%]	
6	Create Production Design	4 wks	Thu 5/1/97	Wed 5/28/97	5	Production Specialist[60%],Manufacturing Engineer[80%], Marketing Specialist[20%]	
7	Building Concept	2 wks	Thu 5/1/97	Wed 5/14/97	5,2	Facility Specialist[40%],Architect[70%]	
8	Building Design	6 wks	Thu 5/15/97	Wed 6/25/97	7	Building Design,Facility Specialist[20%],Architect[10%]	
9	Site Selection	3 wks	Thu 5/15/97	Wed 6/4/97	4	Real Estate Consultant[70%],Site Cost,Marketing Specialist[20%], Facility Specialist[30%]	
10	Select General Contractor	2 wks	Thu 5/15/97	Wed 5/28/97	7	Project Manager[10%],Facility Specialist[25%], Production Specialist[25%],Architect[20%]	
11	Permits and Approvals	3 wks	Thu 6/5/97	Wed 6/25/97	4,9,10	Project Manager[10%],Real Estate Consultant[30%], Architect[10%],General Contractor[15%],Permit Costs	
12	Building Construction	24 wks	Thu 6/26/97	Wed 12/10/97	11,8,9	Project Manager[10%],Facility Specialist[20%], General Contractor[35%],Building Costs	
13	Plant Personnel Recruiting	8 wks	Thu 5/29/97	Wed 7/23/97	3,6	Facility Specialist[10%],Production Specialist[10%],Project Manager[10%],Marketing Specialist[10%],Personnel Director[80%]	
14	Equipment Procurement	24 wks	Wed 5/28/97	Wed 11/12/97	6	Manufacturing Engineer[40%],Purchasing Agent[25%], Building Costs[0%],Equipment Costs	
15	Raw Material Procurement	8 wks	Thu 5/29/97	Wed 7/23/97	6	Production Specialist[10%],Manufacturing Engineer[10%], Accounting Director[10%],Purchasing Agent[20%]	
16	Equipment Installation	4 wks	Wed 12/10/97	Wed 1/7/98	14,12,13	Production Specialist[10%],Manufacturing Engineer[30%], Facility Specialist[40%],Equipment Costs[0%],Equipment Installation	
17	Product Distribution Plan	2 wks	Thu 5/29/97	Wed 6/11/97	6	Marketing Specialist[45%],Traffic Manager[40%]	
18	Landscaping	3 wks	Thu 12/11/97	Wed 12/31/97	12	Facility Specialist[15%],Landscaping	

TABLE 3-7 Resource Useage Chart

ID	Resource Name	Details				Apr 20, '97			
			T	F	S	S	M	T	W	T
1	Facility Specialist	Work	4.8h	5.37h			5.8h	5.8h	5.8h	5.8h
	Select Architect	Work	4.8h	5.37h			5.8h	5.8h	5.8h	5.8h
	Recruit & Train Managers	Work	0h							
	Select Real Estate Consultant	Work								
	Building Concept	Work								
	Building Design	Work								
	Site Selection	Work								
	Select General Contractor	Work								
	Building Construction	Work								
	Plant Personnel Recruiting	Work								
	Equipment Installation	Work								
	Landscaping	Work								
2	Project Manager	Work	3.6h	2.88h			2.88h	2.88h	2.88h	2.88h
	Select Architect	Work	0.8h	0.9h			0.97h	0.97h	0.97h	0.97h
	Recruit & Train Managers	Work	1.2h	1.2h			1.2h	1.2h	1.2h	1.2h
	Select Real Estate Consultant	Work								
	Pre-Production Plan	Work	1.6h	0.78h			0.72h	0.72h	0.72h	0.72h
	Select General Contractor	Work								

TABLE 3-8 Tracking Gantt with Bar Chart

ID	Task Name	Duration	Start	Finish
1	**HUNTSVILLE PROJECT**	**193.1 days**	**Thu 4/17/97**	**Tue 1/13/98**
2	Select an Architect	2 wks	Thu 4/17/97	Wed 4/30/97
3	Recruit and Train Managers	8 wks	Thu 4/17/97	Wed 6/11/97
4	Select Real Estate Consultant	6 wks	Thu 4/17/97	Wed 5/28/97
5	Pre Production Plan	8.5 wks	Thu 4/17/97	Mon 6/16/97
6	Create Production Plans	4 wks	Thu 5/1/97	Wed 5/28/97
7	Building Concept	2.86 wks	Thu 5/1/97	Wed 5/21/97
8	Building Design	7.39 wks	Thu 5/1/97	Tue 7/1/97
9	Site Selection	4 wks	Mon 5/19/97	Mon 6/16/97
10	Select General Contractor	2 wks	Thu 5/1/97	Wed 5/14/97
11	Permits and Approvals	5 wks	Thu 5/29/97	Wed 7/2/97
12	Building Construction	24 wks	Tue 7/1/97	Tue 12/16/97
13	Plant Personnel Recruiting	8.8 wks	Thu 5/1/97	Wed 7/23/97
14	Equipment Procurement	28 wks	Thu 5/1/97	Wed 11/12/97
15	Raw Material Procurement	8 wks	Thu 5/1/97	Wed 6/25/97
16	Equipment Installation	4 wks	Thu 7/3/97	Wed 7/30/97
17	Product Distribution Plan	2.67 wks	Thu 5/1/97	Tue 5/20/97
18	Landscaping	2.61 wks	Mon 6/16/97	Thu 7/3/97
19	Truck Fleet Procurement	5.71 wks	Thu 11/13/97	Tue 12/23/97
20	Pre-Production	4 wks	Tue 12/16/97	Tue 1/13/98
21	Production Startup	1 wk	Tue 5/20/97	Tue 5/27/97
22	Distribution	1 wk	Thu 7/31/97	Wed 8/6/97

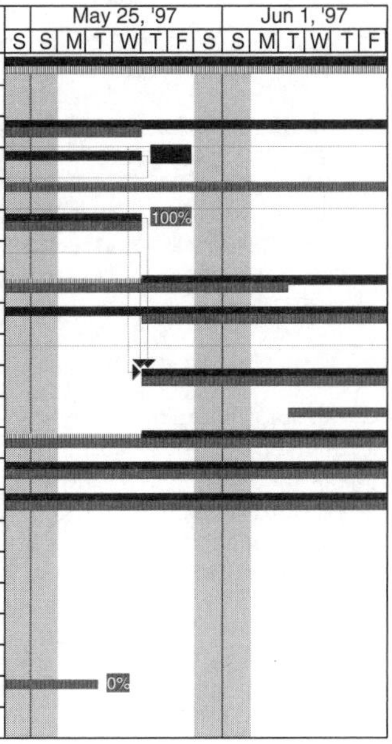

Table 3-8 shows the tracking GANTT with bar chart that shows actual progress (black bar) within the planned duration (grey bar).

Five-step scheduling process

The program manager generates the scheduling process; the department manager serves as a resource on product functionality and department resources and ensures that the technical procedures are in place to complete the work. This process works effectively only with a constant dialogue between program managers, department managers, system engineers, and the team.

The scheduling process for product development involves five steps, culminating in the product deliverable. The general sequence of work is first to define the work from customer requirements; structure the work into an outline or work breakdown structure; define an overall, top-level task structure and work flow; identify tasks, durations, and interdependencies; develop department staffing plans to accomplish the work, estimate the costs; kickoff, monitor, and close-out the program.

Table 3-9 outlines the five functions: a description of the function, and the roles of the program manager, department manager, Director of Product Development, and project team.

Schedule control

Product specification changes that impact schedules are approved by the program manager and departments involved. The program manager initiates two

TABLE 3-9 The Scheduling Process

Scheduling function	Description of function	Program manager role	Department manager role	Director of product development role	Project team role
1. Develop top-level work breakdown in Microsoft Project	Develop top-level program structure consistent with Product Development Process	Lead role, working with systems engineer and department management	Participate in developing top-level structure	Review, comment, approval	Help program manager develop structure, as requested
2. Flesh out schedule, establishing tasks and sub tasks, durations, interdependencies, and constraints	Prepare preliminary schedule	Lead role with inputs from department staff; work to find ways to accelerate work in concurrent tasks when feasible and address risks	Help define how scheduled work can take advantage of prior work, supports concurrency in work schedules	Review, comment, approval	Help department manager find efficient ways to work in parallel
3. Assign resources to schedule, estimating hourly requirements for each task	Establish resource needs to support schedule; assign work from scheduled tasks to staff; confirm resource availability	Work with departments to identify program team and meet other resource requirements	Lead role, department managers are responsible for staffing program with competent, adequately trained personnel and providing adequate resources	Review, comment	Participate with department manager and program manager in fitting assignments into current and projected workload
4. Establish schedule baseline, confirming program "kick-off"	Save schedule as "baseline" in Microsoft Project software, placing file in central directory on network	Take lead to kick-off program, handing out hardcopy of baseline schedule at meeting, along with resource usage table that defines resource commitments	Participate in kick-off meeting, support resource usage plan	Approval of baseline before it is saved as such	Review, comment
5. Monitor performance against baseline, report on performance, variances, and cost to complete, manage change	Enter actual data on % complete and/or cost data from time-sheet project codes; revise start and finish dates as appropriate	Get percent complete and other performance information from project team members; report weekly to Director of Product Development	Review actual and planned, BCWP (Budgeted cost of work performed), and make recommendations	Review actual and planned	Recommend corrective action in case of schedule slippage

kinds of changes—schedule updates based on tracking information, e.g., percent complete, and more fundamental changes from customer inputs, design change notices, and other more substantial changes in the scope of work. Affected department managers and the appropriate program manager must agree to all schedule changes. After the baseline schedule is saved, the director of product development must approve any slips or changes in scheduled milestones on the critical path.

Baselining the schedule

After a preliminary schedule is prepared and analyzed, cross-program resource conflicts resolved, and successful delivery of the product within the schedule determined to be feasible by the program manager, the schedule is baselined. Establishing the baseline schedule is a significant action in the ID program management process, signifying the official kick-off of the work and indicating a strong commitment to the schedule and resource plan. The baseline is the point of departure for the monitoring and tracking process. When a project is baselined, the project schedule is complete. Here are some rules of thumb for baselining:

1. The purpose is to get to a baseline schedule that captures all the work to be done. This includes key documentation and procurement tasks. The baseline schedule does not change unless the basic scope changes. After agreement is reached, the program manager confirms the baseline by saving it and making it available on the network *as the baseline*. There is no uncertainty where the baseline is and how to access it.
2. The baseline schedule is the agreed-upon, scrubbed schedule for the program, and linked to the resource pool. The baseline shows all interdependencies, linkages, and resource requirements; includes all tasks necessary to get the work done, and shows impacts on parallel programs and resources. All procurements and test equipment are covered in the schedule.
3. The baseline schedule is resource-leveled—the schedule can be implemented with current, available resources. Assigned staff are aware of the commitments and have "signed-on" to complete their tasks to meet the schedule milestones.
4. Getting to the schedule baseline involves collaboration between the program manager and all departments and staff involved in planning and implementing the schedule. A baseline meeting is held to arrive at the final agreement on schedule and resources committed before the baseline is saved to the network. The program manager facilitates the meeting, and all department managers attend and come prepared to commit their resources to the final, agreed upon baseline schedule.
5. The final review of the schedule at the baseline meeting involves reviewing all stages and tasks, linkages, and resources assigned, line by line.
6. The baseline schedule is monitored weekly, with actual percent complete data and changes in start and finish dates entered weekly and reported at program review.

Baseline procedures

The program manager uses Microsoft's Planning Wizard (or Tools, Tracking, Save Baseline) to set a baseline schedule. At the same time that a baseline is created, a backup copy of the project file is created as a permanent archive of the original schedule for later reference and comparison to actuals.

Sometimes it is necessary to create a baseline schedule before the complete schedule and task structure is determined, simply to serve as a basis for capturing actual progress. This is because some work, which is clearly on the critical path, e.g., mechanical design or long lead-time procurement, must begin immediately to meet key milestones, sometimes before all the details of a schedule are worked out. To accommodate this, when the final schedule is completed the baseline can be updated by saving an "interim plan." The interim plan saves particular start and finish date changes that are made after a baseline has been saved. Interim changes can be made for the entire project or for selected tasks.

Managing schedules on the network

The basic objectives of network management of program schedules are to enable the program management department to control schedule updates and schedule versions, and provide department managers and staff with an easy way to review and provide input to schedules and schedule assumptions. The following are the steps involved.

1. All schedules will be housed on the server in individual program manager folders (My network places/pm-admin on MAIN 01/folders).
2. The central resource pool file will be named "Integrated Resources—project file name," and will be housed in the schedule folder. Archive versions of schedules will be housed in a separate folder.
3. The program management department will control access to schedule files. The director of product development and the program management department (program managers and program administrator/planner) will have "write" access to the schedules. Department managers, systems engineers, and team staff and other users will have "read" access to program schedules.
4. The program management department is responsible for maintaining and updating program schedules on the network. After the director of product development, the program manager, and department managers agree on a proposed schedule and/or update, the schedule will be linked to the resource file and resource conflicts will be identified and resolved. The program manager will then save the schedule as a baseline schedule. After it is baselined, the schedule will be placed in a project manager-admin directory. The baseline schedule will be the only version of that schedule housed on the network (except for archives) and will serve as the source of "planned versus actual" tracking information.

Resource planning and control

Scheduling is essentially the process of planning for use of personnel and equipment resources. Good program management requires that there be a process to plan for the acquisition of future resources, to allocate current and projected resources to schedules, and to make shifts in resource management as required. The process provides for a central resource pool to identify impacts of project schedules and ensure the efficient utilization of the workforce. The resource pool information on the network is shared with management staff and all team members to allow each team member to evaluate the scheduled work assigned and to provide guidance on task definition, durations, start and finish dates, and interdependencies.

While work actually done on a project is tracked automatically by Project when percent complete data is entered, the program management and finance departments collect actual work done from time sheets to gain a more accurate assessment of actual work. Actual work is tracked from time sheets that collect hours of work against the project account-numbering scheme. The program manager is responsible for establishing the account numbers for charges to the project and for ensuring that time sheets are kept for all work on the project. The project administrator/planner is responsible for working with the finance department to collect data each week and enter it into the appropriate schedules.

Tracking and program review

The director of product development will hold weekly program review meetings to discuss broad program issues, detailed technical, resource, and schedule problems, project team performance, and risk mitigation. Program managers are responsible for preparing presentations for these reviews and identifying key agenda items. Department managers and team members attend program review meetings as appropriate.

The program manager tracks the progress of the program on an on-going basis and updates the schedule on a weekly basis. The program plan will be updated as changes in plans warrant. Program managers hold periodic reviews with the program team in preparation for reporting and to support task assignments and feedback, either as a single meeting with all functions represented, or as a series of meetings with major functional areas represented at each meeting. Program managers report progress for their portions of the weekly report, due to the director of product development COB (close of business) each Thursday.

Schedule update procedures

Using Microsoft Project, program managers can track and update actual performance information once a project is underway, including percent complete for each task, change in task start date, change in task finish date, task duration, task cost, and total work.

1. Review percent complete on a task. At the very minimum, the program manager updates percent complete for each task on which work has been done

during the past week. This data is gathered from the appropriate project team members based on their assessment of the percentage of work actually done compared to the baseline work definition. Program managers are responsible for briefing team members on the importance of accurate assessments of percent complete and how their estimates are used to update project performance. When percent complete is entered, Microsoft Project changes the actual start date to match the scheduled start date and calculates the actual duration and remaining duration.

2. Add a new task to a baseline or interim plan.
3. Enter actual duration of a task.
4. Enter actual start and finish dates for a task.
5. Enter actual work (e.g., hours) completed on each task (from time sheets).
6. Reschedule uncompleted work.

Analyzing variance

The program manager uses tracking data to determine how the actual progress of the scheduled work compares to the original baseline schedule, but more importantly to determine impacts of actual work done on the overall schedule and on critical milestones.

Variances that are tracked include:

1. Tasks that are starting or finishing late. Along with updating the task start and finish dates through the Tools/Tracking command, the program manager identifies impacts of the change on linked tasks.
2. Tasks that require more or less work than scheduled. Changed durations and/or additional resources can be assigned to tasks that are running late; Microsoft Project will shift durations and start and finish dates automatically when resource units are changed.
3. Tasks that are progressing more slowly than planned. Tasks on the critical path that are progressing more slowly than planned must be addressed, either through redefining the task to stay within the schedule or adding resources. Tasks off the critical path provide the program manager with some slack time, or "safety buffer," before they go onto the critical path because of delays.
4. Resources that aren't working hours as scheduled. If there are major variances in actual work versus planned work, this can have implications for several projects. Based on actual hours worked, program managers reassign resources and link to the resource pool to broad impacts and conflicts. Microsoft Project's resource leveling capability can be used to address resource conflicts as well.
5. Earned value. Earned value is an indicator of cost and schedule variance. It is important to track both whether the actual work is on schedule *and* whether

the actual resources expended on the work is consistent with what should have cost to get the work done based on the baseline budget. In other words, earned value tracks whether the work accomplished actually cost what it should have cost, given the original budget. Reports on earned value show schedule variance and cost variance.

6. Corrective action. The real issue in variance analysis and earned value is what corrective action the program manager takes to put a program, which is showing substantial schedule variance, back on schedule. Program managers are responsible for alerting the team to these variances, reporting them in program review, and coming up with corrective actions. Some corrective actions include:

   a. Making sure there is no scope creep; that is, work the team is doing that is not on the system requirements specification
   b. Change task dependencies and linkages
   c. Assign overtime work
   d. Hire or assign additional resources
   e. Decrease amount of work necessary to do a task
   f. Reassign resources
   g. Delay selected tasks
   h. Change working hours

   In support of tracking and program review, the program administrator/planner:

   - Serves as a resource for Microsoft Project procedures and training
   - At the request of a program manager, tracks progress against the schedule and anticipates future schedule problems
   - Flags current and new issues for the week from current schedules
   - Distributes assignments in the central resource pool to project team staff and gathers feedback
   - Identifies conflicts and facilitates resolution

Program close-out and lessons learned

Each program manager meets with the program team at the end of a program to go over the project, identify uncompleted documentation or other tasks, and identify lessons learned. Lessons learned are captured and reported back to the director of product development and the PMO for followup.

The analysis of a new product concept and the setting up of a project to manage its development and marketing are parallel processes which intersect at many points. This Phase 1 activity is aimed at preparing company or agency management to *decide*, to judge the facts from the analysis, draw on their insights and intuition, and on expert advise, and then to make the call on whether a product moves onto the second phase, full development. Thus this process is heavy on analysis and assessment, concept definition, and marketing analysis. This is where the project management process is established to guide the product after it is kicked off, where the tools of project management are integrated into the new product development process to enable innovation and creativity to be balanced with command and control.

Just a word about how the new product development process is handled in this chapter. After describing each phase and the tasks associated with that phase, project management tools and techniques are introduced, e.g., scope of work, schedule, budget, and resource management. This approach provides an early "heads up" to the management regarding organizational issues involved in that phase and applicable important project management tools and techniques. The approach also evidences the importance of setting up a project for command and control.

The presumption here is that it is useful to picture how a sequence of activities would go in the front end of a project *were products developed in a relative vacuum, in sequence, from scratch*. While this is rarely the case, the approach presents a point of departure or framework for looking at the real world of new product development. In the real world, this process will be shaped by the currency of the product concept itself, the culture and processes of the company, the people involved, and the exigencies of the marketplace.

Concept definition

Concept definition is the first step in the new product development process. Concept definition is also necessary for analysis and selection of the project into the company portfolio. Project selection cannot be effective until a product and project go through this process. Concept definition occurs after ideas have been collected and documented, and a portfolio of potential new product concepts has been evaluated and prioritized. The concept definition process addresses how a new product or service idea would work in a market and customer setting. Enough detail is provided to flesh out and evaluate the concept without doing detailed design or engineering.

Initially some kind of documented idea generation process produces a pool of concepts for consideration in developing the company's portfolio of potential new product projects. Resources are allocated to explore the concept as part of the funding of a portfolio of projects, along with *ballpark* deliverables and timelines. The concept is framed in the company's strategic and business plan and related to favorable and unfavorable developing market trends.

Project setup for control

If you talk to people actually responsible for and involved in new product development, their major concern is not how to create new ideas and bring new products through to market test, but rather how to control new product development processes to stop bad products from moving on because there is no way to stop the inevitable process from completing. Control is a big problem in new product development because new ideas and products become associated with sponsors and "champions" and with work commitment. Stopping products at various points is a management challenge despite evidence that the product is not feasible, will not bring value to customers, and will lose money for the company. This is because it takes proactive action to stop work of any kind, particularly in large companies.

Project management is a command and control system and, therefore, a logical, underlying management system for new product development. Unfortunately, many companies see new product development as a system of creativity that should not be obstructed by management and control systems designed to run production-oriented processes. This leads to bad products, and costly, ineffective product development.

New product development must be controlled more effectively than normal operations, but that control must be timed and guided so that control is on resources and time, and not ideas. In other words, the skill in leading new product development is a guiding creative work to various key decision points and providing opportunity for *disinterested* analysis. The term *disinterested* here means the capacity to look at a product and its prospects in analytic terms, but also to join analysis with the judgment and sometimes even intuition of company leadership to decide whether products move to the next level. These decisions should not be driven by cost/benefit or strategic decisions, but clearly should be informed by such input.

We have relied on the traditional project management concept of *project review* in this book to articulate this decision system, sometimes popularly referred to as "gates" or "decision windows." The purpose of project review has always been to assess project progress for the purpose of adjusting direction, resources, or processes given objective information on project progress, or to stop a bad project.

It is important up front to see the difference between the *product* and the *project*. The *product* is the output of the new product development process, e.g., an instrument, a system, a consumer product, a report, etc. The *project* is the effort, energy, time, and resources given to designing, producing, marketing, distributing, and supporting the product and to controlling its development. The project is the whole organizational process from concept through to support and logistics. With this definition, the project becomes more than simply producing the product prototype. The project team is, in fact, chartered to bring the product not only to production, but on into market and support.

Thus the up-front analysis of finances and resources becomes broader than simply budgeting for the cost of producing the first new product prototype. It is as important to ascertain what the cost will be of manufacturing or producing and distributing a product as it is to control its up-front design and development costs. Analysis must include the design-to-cost considerations, e.g., what will the product cost to product and distribute, as well as revenue projections for the product in "revenue" service. Design issues do not only address performance criteria, but also material and cost issues, e.g., how can we produce a prototype for this product that will minimize the cost of production and marketing, by far the largest costs associated with the product.

In project life cycle terms, this is where the concept begins to be a *project*. This is the early project planning phase, the early documentation of the criteria for project selection, the draft scope of work, and broad-gauged schedule and budget information, as well as financial estimates for the project. The scope of work is where the concept is captured in terms of deliverables and work requirements, and although the details of the product are not yet clear in this phase, the draft scope of work should be prepared in this phase as a vehicle that will be updated and carry the project through to completion.

Structure, science, and research

Most companies need structure in their new product development, but structure must be balanced with flexibility and responsiveness. Structure is introduced in the project management process. This means that there is a new product development process identified and products are expected to move through the process in an orderly and controlled way. When management sees opportunity and the right market conditions, products can be "fast tracked" to get them out the door. And when management sees a bad product with no future, it can stop the process without major organizational hassle and distress. Professionals involved in the process see a decision to stop a product that is not feasible as important as the decision to allow it to proceed.

Because of increasingly complex external regulations and the high level of technology in today's new products, the company's capacity to do good science—or, more importantly, to exploit good science results associated with a new product—become extremely important. For instance, Coca-Cola maintains a high-level office in charge of regulatory compliance and science to make sure its new products meet current and anticipated safety and environmental requirements, globally, if appropriate. That office must be capable of doing independent research on new products to keep ahead of external and environment forces.

Some companies look to a research department to generate new products. The research department often generates product ideas simply because research and development is aimed at uncovering new concepts and ideas and transitioning them into consideration for new product development. The weakness of having a separate research effort is the impact it has on the rest of the company; if it is management's intention to build in innovation and creativity in all of its operations, it is harder to do so when research and development is slotted to do all the creative thinking.

Preliminary project plan

A preliminary project plan is developed, updated, and enhanced as the product concept gains clarity and support. A project sponsor, a mid- to high-level company manager, is appointed to shepherd the proposal through and a project leader is designated.

The project plan includes goals and objectives, project deliverables, workforce and skills needed, contractor requirements, and the timeline and budget of the project. The project plan includes:

1. Project Goals and Objectives—A list of project/product goals and objectives as agreed by the program manager and leadership team. The goals and objectives include a high-level definition of the product concept itself.
2. Major Tasks, Milestones and Deliverables—List major tasks, milestones, and deliverables over the lifetime of the project. All development and launch, evaluation activities should be included in the project plan.
3. Duration—Time to completion/days to launch
4. Resource Allocation Needs—Internal and external
5. Cost and Performance Criteria
6. High-Level Project Schedule: Gantt Chart—High-level Gantt schedules will be developed using the outputs listed above. These charts will provide a foundation for project information related to goals, deliverables, time, resources and budget.

Remember that this is only a preliminary document that identifies the tasks, skills, and resources needed to execute the project plan. The resource pool is identified, including a list of vendors, suppliers, and manufacturers that may be needed to execute project plan. Baseline data of the resource plan will be incorporated into the schedule and Gantt chart of the project plan.

This plan includes several key project management elements, including project goals, team charter, scope of work, schedule, budget, risk management plan, and, most importantly, the project team composition. The project plan provides the project team with purpose, direction, timing requirements, and cost constraints, to the extent possible. Because this is a product development plan, many variables in the project plan may be *soft* to begin with, then *firmed up* as the product enters the next gate.

Key requirements are project charter, project scope statement, schedule, budget, configuration management, and change control.

Project charter

Cross-functional teams that work together and communicate outward about the project with internal stakeholders, such as finance, purchasing, and engineering, tend to integrate their projects more effectively than teams that work in isolation. Therefore the major determinant of successful integration is not just technical, but rather social and organizational. The way the team sees the priority of integration at all levels is to *see it explicitly in the charter itself*. Therefore every charter should have a statement such as the following:

> "The project team will integrate project activities at all levels, including with business planning and marketing, finance and budget, functional departments, and customers, to ensure that the new product development and project outcomes reflect

all the stakeholder interests to the extent possible. Project planning should include a comprehensive project schedule that integrates cost, time, and quality factors to create an optimum outcome—cheaper, better, and faster."

The charter addresses:

- Project manager
- Priority of project
- Date
- Owner/sponsor
- Mission
- Scope
- Objectives
- Assumptions
- Constraints
- Schedule and major milestones
- Cost/budget/financial assumptions
- Quality specifications
- Major risks and contingencies
- Project core team
- Subject matter experts
- Contractors

Financial analysis

The purpose of the financial analysis is to assess the financial viability of the product, initial project costs, return on investment over its life cycle, discounted cash flows, life cycle costs, and other financial information that may be useful in the gate decision to develop. The key here is to keep the analysis at a high level, scaled to the project as a whole and providing a macro view of financial return and financial risk.

Task 1—Perform net income and net present value analyses. The net present value analysis calculates the net present value of future net income from the product, at a designated discount rate. Net present value is also used to compare various products to rank them in terms of return on investment.

At a given discount rate cash flow, both costs and revenues are projected over a five-year period. Costs are derived from a first-cut estimate of the product development and production costs, while revenues are derived from estimates of sales and other product-driven income. Net income for each year is multiplied by the discount factor for the selected discount rate, and NPVs are totaled. A positive NPV means that over the period the product is a good investment, in this case, at 12 percent discount rate.

Care must be taken here to estimate costs accurately, which may require breaking down the project into a work breakdown structure, estimating resources and resource costs, developing a schedule, and calculating a total cost estimate. Assumptions behind cash flow projections should be documented and confirmed with marketing at some point before updating the financial analysis later in the process.

Template (sample)

Year	Net Income	NPV Factor (12% DR)	NPV
0	-$2,475,000.00	1	-$2,475,000.00
1	$700,000.00	89.29%	$624,999.90
2	$800,000.00	79.72%	$637,755.20
3	$800,000.00	71.18%	$569,424.00
4	$800,000.00	63.55%	$508,414.40
5	$800,000.00	56.74%	$453,941.60

NPV = $319,535.10

Task 2—Break-even analysis. This task involves calculating the point at which the company breaks even on its product development and production investment. This means taking the cash flows estimated for the NPV analysis and figuring the year in which net income exceeds initial costs. Rate of return can also be calculated using the same data.

Task 3—Identify life cycle cost. This task identifies the life cycle cost of the product (first cut) based on projections of initial product development costs, production costs, distribution costs, marketing costs, and service and maintenance costs. This assessment helps to frame the product in the long-term financial picture of the company.

Task 4—Sensitivity analysis. This task identifies those factors or variables in the product development process that will *drive* costs and revenues, and financial performance in general. The goal is to see how project outcome is affected by various product or project factors, e.g., price of materials, testing validation, customer requirements, and to identify the most important factors. The outcome allows the project team to focus on those factors in design and development that are most important in delivering a successful product.

Project Scope Statement

The project scope statement is the definition of the project—what needs to be accomplished. The *develop preliminary project scope statement process* addresses and documents the characteristics and boundaries of the project and its associated products and services, as well as the methods of acceptance and scope control. A project scope statement includes:

- Project and product objectives
- Product acceptance criteria
- Product or service requirements and characteristics
- Project boundaries
- Project requirements and deliverables
- Project constraints
- Project assumptions
- Initial project organization
- Initial defined risks
- Schedule milestones
- Initial WBS
- Order of magnitude cost estimate
- Project configuration management requirements
- Approval requirements

The preliminary project scope statement is developed from information provided by the initiator or sponsor. The project management team in the scope definition process further refines the preliminary project scope statement into the final project scope statement. The project scope statement content will vary depending upon the application area and complexity of the project and can include some or all of the components identified above. During subsequent phases of multiphase projects, the develop preliminary project scope statement process validates and refines, if required, the project scope defined for that phase.

The scope is the basis for contracting the work with a vendor. Therefore the scope must be detailed enough to provide direction for contract negotiation, costing, and schedule development with a vendor.

Schedule

The schedule is a fully linked schedule in Microsoft Project or equivalent scheduling software, with resources and costs loaded. The schedule is a full work breakdown of all tasks necessary in each stage to accomplish the goals of the project.

Resource plan

A preliminary resource plan document identifies the tasks, skills, and resources needed to execute the project plan. The resource pool is identified including a list of vendors, suppliers and manufacturers that may be needed to execute project plan. Baseline data of the resource plan will be incorporated into the Gantt chart of the project plan.

Budget

The budget is built up from the individual tasks in the schedule, again using Microsoft Project or equivalent software. The budget has a labor element, equipment and capital, and is loaded with overhead and general and administrative costs.

Configuration Management System

The configuration management system is a subsystem of the overall project management information system; however, its main purpose is control of the design of the product for later production and support. It captures the official definition of the product as it moves through changes to its final production configuration. The system includes the process for submitting proposed changes, tracking systems for reviewing and approving proposed changes, defining approval levels for authorizing changes, and providing a method to validate approved changes. In most application areas, the configuration management system includes the change control system. The configuration management system is also a collection of formal documented procedures used to apply technical and administrative direction and surveillance to:

- Identify and document the functional and physical characteristics of a product or component
- Control any changes to such characteristics
- Record and report each change and its implementation status
- Support the audit of the products and components to verify conformance to requirements

Change control system

In practice in product development, there is always a choice of when a product and its components are documented in a configuration management system, thus beginning the change control process. The plan should state when that point is in this product life cycle. The change control system is a collection of formal documented procedures that define how project deliverables and documentation are controlled, changed, and approved. The change control system is a subsystem of the configuration management system. For example, for information technology systems, a change control system can include the specifications (scripts, source code, data definition language, etc.) for each software component.

In practice, the project management plan is a guide for the internal and accounting control of the work. Therefore the plan must include control points, e.g., stage-gateway reviews, that ensure that management authorizes movement from one phase or stage to another. Reporting and monitoring strategies, including the use of earned value to integrate cost, schedule, and quality performance, should be made explicit.

The plan should also address accountability, particularly in view of the recent legislative and regulatory requirements of the Sarbanes-Oxley Act. This requirement is the compliance with internal control and accounting standards and is no longer optional for project managers. In fact, the price of disconnected and inconsistently applied efforts throughout a project and its interfaces, and lack of financial tracking systems that provide for audits, could be businesswide. Compliance with Sarbanes-Oxley therefore is not a choice but a requirement, and the plan should state standards for estimating costs, tracking the costs and relating costs to work performed, and the integrity of the closeout procedure and invoices to customers for work performed.

Key requirements are project charter, project scope statement, detailed plan, configuration management, and change control.

Application to new product development

Our purpose here is to outline a quick and dirty approach to selecting, organizing, and scheduling a new product project from scratch. The following process includes many of the traditional project planning and control steps, simplified and "demystified" for practice use.

Organizing a project from scratch involves six basic steps:

1. Project selection
2. Work breakdown structure
3. Task list, with estimated durations, linkages, and resource assumptions
4. Network diagram
5. Time-based network diagram
6. Gantt chart schedule

ORANGE-AID: New Product Development Case

The project setup process can determine the success of a new product project. Project setup requires a good conceptual grasp of the product concept, a generic project schedule and resource management template for new product development, and an understanding of how to develop a baseline schedule and keep it flexilble throughout development. New product development, although by definition uncertain and risky in its direction and final performance, must be controlled from the beginning of product concept through good setup tools. A new product development project can be kept flexible, but it should not be out of control. Costs and time are controlled through the process. The entire *rollout* process, the way companies plan to design and deliver a new product, should be documented in the front-end through a Gantt chart and resource management scheme grounded in the original work breakdown structure of the product. Table 3-10 shows a typical work breakdown structure and baseline schedule for a new product in Microsoft Project—let's call it *ORANGE-AID*, a new carbonated drink being introduced by a Fortune 100 company that is new to the business and seeking new market opportunity. The company is willing to take on a high risk to penetrate the market, but it has built in a number of go and no-go decisions at the end of each phase—through project reviews—to control advancement of the product through to final marketing.

Notice that the design project concept, research concept, and design packaging, proceed in sequence, moving toward the development of a strategic plan. In this case the strategic plan is not the business plan, but rather the product

TABLE 3-10 New Product Development

ID	❶	Task Name	Duration	Start	Finish	Predeces	Resource Names	2nd Half Jul | Aug | Sep
1		**Orangeaid Product Rollout**	**350 days**	**Mon 4/2/07**	**Fri 8/1/08**			
2		**Design Project Concept**	**27 days**	**Mon 4/2/07**	**Tue 5/8/07**			
3		Brainstorming Session	1 day	Mon 4/2/07	Mon 4/2/07		Marketing Team	m
4		Determine Top 3 Ideas	3 days	Tue 4/3/07	Thu 4/5/07	3	Marketing Team	m
5		Feasability Testing	20 days	Fri 4/6/07	Thu 5/3/07	4	Research and Development	and Developmen
6		Results Review/Analysis	3 days	Fri 5/4/07	Tue 5/8/07	5	Brand Manager	anager
7		Determine Product Concept	0 days	Tue 5/8/07	Tue 5/8/07	6	Product Manager	
8		**Research Project Concept**	**25 days**	**Wed 5/9/07**	**Tue 6/12/07**	**7**		
9		Preliminary Demographic Research	5 days	Wed 5/9/07	Tue 5/15/07		Marketing Specialist	ng Specialist
10		Analyze Demographic Research	5 days	Wed 5/16/07	Tue 5/22/07	9	Marketing Analyst	ting Analyst
11		Perform Conjuct Analysis	15 days	Wed 5/23/07	Tue 6/12/07	10	Assistant Brand Manager	ssistant Brand Ma
12		Present Findings	0 days	Tue 6/12/07	Tue 6/12/07	11	Assistant Brand Manager	/12
13		**Design Packaging**	**28 days**	**Wed 6/13/07**	**Fri 7/20/07**			
14		Present Product Concept to Graphic Draft	1 day	Wed 6/13/07	Wed 6/13/07	12	Marketing Specialist	arketing Specialis
15		Design Packaging Concept	10 days	Thu 6/14/07	Wed 6/27/07	14	Research and Development	Research and De
16		Review designs - First Draft	3 days	Thu 6/28/07	Mon 7/2/07	15	Marketing Team	Marketing Team
17		Make Changes	5 days	Tue 7/3/07	Mon 7/9/07	16	Brand Manager	Brand Manage
18		Review Design Changes - Second Draft	3 days	Tue 7/10/07	Thu 7/12/07	17	Marketing Team	Marketing Tea
19		Finalize Design	5 days	Fri 7/13/07	Thu 7/19/07	18		
20		Present Design to Marketing Team	1 day	Fri 7/20/07	Fri 7/20/07	19		
21		**Determine Strategic Plan**	**40 days**	**Wed 6/13/07**	**Tue 8/7/07**	**11**		

TABLE 3-10 New Product Development (*Continued*)

ID	ⓘ	Task Name	Duration	Start	Finish	Predeces	Resource Names	2nd Half: Jul | Aug | Sep
20		Present Design to Marketing Team	1 day	Fri 7/20/07	Fri 7/20/07	19		
21		**Determine Strategic Plan**	**40 days**	**Wed 6/13/07**	**Tue 8/7/07**	**11**		
22		Product Rollout Strategy	40 days	Wed 6/13/07	Tue 8/7/07	11	Product Manager	Product
23		National Marketing Strategy	40 days	Wed 6/13/07	Tue 8/7/07	11	Assistant Brand Manger	Assistan
24		Preliminary Strategic Rollout Plan	40 days	Wed 6/13/07	Tue 8/7/07	11	Senior Marketing Analyst	Senior M
25		**Review Strategic Plan**	**30 days**	**Wed 8/8/07**	**Tue 9/18/07**			
26		Make Recommendations	5 days	Wed 8/8/07	Tue 8/14/07	21	Brand Manager	Brand
27		Revise Strategic Plan	20 days	Wed 8/15/07	Tue 9/11/07	26	Assistant Brand Manager	As
28		Finalize Design Strategy	5 days	Wed 9/12/07	Tue 9/18/07	27	Research and Development	R
29		**Product Testing**	**110 days**	**Wed 9/19/07**	**Tue 2/19/08**			
30		Determine Test Markets	20 days	Wed 9/19/07	Tue 10/16/07	28		
31		Test Product in Test Markets	60 days	Wed 10/17/07	Tue 1/8/08	30		
32		Focus Group Testing	20 days	Wed 1/9/08	Tue 2/5/08	31		
33		Review Test Results	10 days	Wed 2/6/08	Tue 2/19/08	32	Senior Marketing Ana	
34		Determine Final Strategic Rollout Plan	40 days	Wed 2/20/08	Tue 4/15/08	33	Brand Manager	
35		**Product Launch**	**20 days**	**Wed 4/16/08**	**Tue 5/13/08**	**34**	**Brand Manager**	
36		West Coast Rollout	20 days	Wed 4/16/08	Tue 5/13/08	34		
37		Midwest Rollout	20 days	Wed 4/16/08	Tue 5/13/08	34		
38		East Coast Rollout	20 days	Wed 4/16/08	Tue 5/13/08	34		
39	▦	Review 90 day sales results	10 days	Mon 7/21/08	Fri 8/1/08	38	Marketing Team	

plan—the approach to be used to launch marketing, distribution, and support. Note the several *review* tasks in the project, suggesting continual assessment of product design, performance, costs and financial performance, and business case.

A resource management table (see Table 3-11) is prepared to show resources and standard rates to calculate estimated project cost, as follows.

Then costs are broken down by resources (see Table 3-12).

A resource-based view provides a comprehensive look at when resources are committed, and for what tasks (see Table 3-13).

A task-based view shows tasks schedules and work, and resources (see Table 3-14).

Early/Late Start and Finish Analysis

As tedious as this sounds, sometimes a new product project can be viewed in terms of early and late starts. This is because some tasks are not on the critical path and can be started at different times depending on resource availability and other factors.

TABLE 3-11 Resource Management

ID	ⓘ	Resource Name	Type	Material Label	Initials	Group	Max. Units	Std. Rate	Ovt. Rate	Cost/Use	A
1		Brand Manager	Work		B	Marketing	100%	$41.00/hr	$0.00/hr	$0.00	
2		Assistant Brand Manager	Work		A	Marketing	100%	$26.40/hr	$0.00/hr	$0.00	
3		Marketing Analyst	Work		M	Marketing	100%	$21.25/hr	$0.00/hr	$0.00	
4		Senior Marketing Analyst	Work		S	Marketing	100%	$26.40/hr	$0.00/hr	$0.00	
5		Marketing Specialist	Work		M	Marketing	100%	$18.00/hr	$0.00/hr	$0.00	
6		Product Manager	Work		P		100%	$33.00/hr	$0.00/hr	$0.00	
7		Research Assistant	Work		R		100%	$21.00/hr	$0.00/hr	$0.00	
8		Marketing Team	Work		M		100%	$115.00/hr	$0.00/hr	$0.00	
9		Research and Development	Work		R		100%	$120.00/hr	$0.00/hr	$0.00	

TABLE 3-12 Cost Breakdown

ID	Resource Name	% Comp.	Work	Overtime	Baseline	Variance	Actual	Remaining
1	Brand Manager	0%	584 hrs	0 hrs	0 hrs	584 hrs	0 hrs	584 hrs
2	Assistant Brand Manager	0%	600 hrs	0 hrs	0 hrs	600 hrs	0 hrs	600 hrs
3	Marketing Analyst	0%	40 hrs	0 hrs	0 hrs	40 hrs	0 hrs	40 hrs
4	Senior Marketing Analyst	0%	400 hrs	0 hrs	0 hrs	400 hrs	0 hrs	400 hrs
5	Marketing Specialist	0%	48 hrs	0 hrs	0 hrs	48 hrs	0 hrs	48 hrs
6	Product Manager	0%	320 hrs	0 hrs	0 hrs	320 hrs	0 hrs	320 hrs
7	Research Assistant	0%	0 hrs	0 hrs	0 hrs	0 hrs	0 hrs	0 hrs
8	Marketing Team	0%	160 hrs	0 hrs	0 hrs	160 hrs	0 hrs	160 hrs
9	Research and Development Team	0%	280 hrs	0 hrs	0 hrs	280 hrs	0 hrs	280 hrs

Lets examine the following example. Suppose we were given the following information about our project. We want to establish a basic schedule for the project that provides the following information. We want to know what the basic project duration (see Table 3-15). We also want to know what the critical path for the project. Finally, we want to know what the impact of uncertainty each tasks has on the overall project. We will begin by examining the technique called the Critical Path Method (CPM). This technique will be used in numerous other procedures to establish the project durations, start and stop dates, slack times, and critical paths. The method described here can be performed without having to draw the diagrams or activity-on-arrow or activity-on-node.

TABLE 3-13 Resource-Based View

ID		Resource Name	Work	Details	May 20, '07						
					S	M	T	W	T	F	S
1		Brand Manager	584 hrs	Work							
		Results Review/Analysis	24 hrs	Work							
		Make changes	40 hrs	Work							
		Make recommendations	40 hrs	Work							
		Determine Final Strategic Rollout Plan	320 hrs	Work							
		Product Launch	160 hrs	Work							
2		Assistant Brand Manager	600 hrs	Work				8h	8h	8h	
		Perform conjuct analysis	120 hrs	Work				8h	8h	8h	
		Present findings	0 hrs	Work							
		National Marketing strategy	320 hrs	Work							
		Revise Strategic plan	160 hrs	Work							
3		Marketing Analyst	40 hrs	Work		8h	8h				
		Analyze demographic research	40 hrs	Work		8h	8h				
4		Senior Marketing Analyst	400 hrs	Work							
		Preliminary strategic rollout plan	320 hrs	Work							
		Review test results	80 hrs	Work							
5		Marketing Specialist	48 hrs	Work							
		Preliminary Demographic Resesearch	40 hrs	Work							
		Present product concept to graphic design team	8 hrs	Work							
6		Product Manager	320 hrs	Work							
		Determine Product Concept	0 hrs	Work							

TABLE 3-14 Task-Based View

ID	ⓘ	Task Name	Work	Details	Apr 8, '07 M	T	W	T	F	S
1		**Orangeaid Product Rollout**	**2,272 hrs**	Work	8h	8h	8h	8h	8h	
2		**Design Project Concept**	**216 hrs**	Work	8h	8h	8h	8h	8h	
3		Brainstorming session	8 hrs	Work						
		Marketing Team	8 hrs	Work						
4		Determine top 3 ideas	24 hrs	Work						
		Marketing Team	24 hrs	Work						
5		Feasability testing	160 hrs	Work	8h	8h	8h	8h	8h	
		Research and Development Team	160 hrs	Work	8h	8h	8h	8h	8h	
6		Results Review/Analysis	24 hrs	Work						
		Brand Manager	24 hrs	Work						
7		Determine Product Concept	0 hrs	Work						
		Product Manager	0 hrs	Work						
8		**Research Project Concept**	**200 hrs**	Work						
9		Preliminary Demographic Resesearch	40 hrs	Work						
		Marketing Specialist	40 hrs	Work						
10		Analyze demographic research	40 hrs	Work						
		Marketing Analyst	40 hrs	Work						
11		Perform conjuct analysis	120 hrs	Work						
		Assistant Brand Manager	120 hrs	Work						
12		Present findings	0 hrs	Work						
		Assistant Brand Manager	0 hrs	Work						

The results of our analysis of early and late start are shown in Table 3-16. But to get to this result and see how we got there, we need to establish some definitions and basic rules.

ES (Early Start): This is the earliest we can start a given task.

EF (Early Finish): This is the earliest we can finish the given task given an early start.

LS (Late Start): This is the latest we can start the task without delaying the whole project.

LF(Late Finish): This is the latest we can finish the task given a late start.

Slack: The difference between the late finish and the earliest finish.

TABLE 3-15 Task List

Task	Predecessor	Duration
A	-	5
B	-	3
C	A	8
D	A,B	7
E	-	7
F	C,E,D	4
G	F	5

TABLE 3-16 Final Early and Late Start Analysis

Task	Pred.	Dur.	ES	EF	LS	LF	Slack
A	-	5	0	5	0	5	0
B	-	3	0	3	3	6	3
C	A	8	5	13	5	13	0
D	A,B	7	5	12	6	13	1
E	-	7	0	7	6	13	6
F	C,E,D	4	13	17	13	17	0
G	F	5	17	22	17	22	0
			Proj Dur	22		CP	A,C,F,G

The basic rules are as follows:

We will make two passes through the information. The first starts with the first task's ES and works down each task. The second pass starts at the last task's LF and works toward the top.

Pass 1—Start with first task

1. If task has no predecessor, then ES = 0
2. If task has one predecessor, then ES = Predecessor's EF
3. If task has more than one predecessor, then ES = MAX(EF of all Predecessors)
4. EF = ES + Duration
5. Repeat 1–4 for all tasks
6. Project Duration = Maximum(EF of all Task)

Pass 2—Start with last task

1. If task has no successor, then LF = Project Duration
2. If task has one successor, then LF = LS of successor
3. If task has more than one successor, then LF = Minimum (LS of all Successors)
4. LS = LF–Duration
5. Slack = LF–EF or LS–ES
6. Repeat 1–5 for all Task

From Pass 1 we know the project duration. From Pass 2 we know the critical path.

Any task with a slack of 0 means that we cannot change its start or stop without changing the overall project duration. Those task with slack mean that we can delay the task start by as much as the slack and not change the project duration. Start with a blank table as shown in Table 3-17. We will begin in Task A under ES.

TABLE 3-17 First Cut Analysis

Task	Pred.	Dur.	ES	EF	LS	LF	Slack
A	-	5					
B	-	3					
C	A	8					
D	A,B	7					
E	-	7					
F	C,E,D	4					
G	F	5					
			Proj Dur			CP	

Using our rules we see that Task A has no predecessor (a—under PRED.) Therefore, we will insert a 0 into that cell as shown in Table 3-17 and calculate EF using our formula EF = :ES + Duration. Then move to task B (see Table 3-18). Task B is the same as task A, so repeat the procedure.

Now look at task C, shown in Table 3-19. Task C has one predecessor—task A; therefore, make task C ES = task A EF.

Now we can move down to task D. Task D has two predecessors—A and B. We want to make the ES of task D the maximum of the EF's of these two task. Looking at the data we want to make task D ES equal to the max of 5 or 3, so we take 5, as shown in Table 3-20

Repeat this procedure through all the task until you reach the last task as shown in Table 3-21. The project duration is then the maximum of all of the EF's or 22.

Pass 2

Now we are ready to start Pass 2. We will begin in the lower-right LF in Table 3-22.

We will start with task G. Is there any task that has a G in the predecessor column? In this case, there is not. Therefore the LF for task G equals the

TABLE 3-18 Second Cut Analysis

Task	Pred.	Dur.	ES	EF	LS	LF	Slack
A	-	5	0	5			
B	-	3	0	3			
C	A	8					
D	A,B	7					
E	-	7					
F	C,E,D	4					
G	F	5					
			Proj Dur			CP	

TABLE 3-19 Third Cut Analysis

Task	Pred.	Dur.	ES	EF	LS	LF	Slack
A	-	5	0	5			
B	-	3	0	3			
C	A	8	5	13			
D	A,B	7					
E	-	7					
F	C,E,D	4					
G	F	5					
			Proj Dur			CP	

TABLE 3-20 Fourth Cut Analysis

Task	Pred.	Dur.	ES	EF	LS	LF	Slack
A	-	5	0	5			
B	-	3	0	3			
C	A	8	5	13			
D	A,B	7	5	12			
E	-	7					
F	C,E,D	4					
G	F	5					
			Proj Dur			CP	

TABLE 3-21 Fifth Cut Analysis

Task	Pred.	Dur.	ES	EF	LS	LF	Slack
A	-	5	0	5			
B	-	3	0	3			
C	A	8	5	13			
D	A,B	7	5	12			
E	-	7	0	7			
F	C,E,D	4	13	17			
G	F	5	17	22			
			Proj Dur	22		CP	

TABLE 3-22 Sixth Cut Analysis

Task	Pred.	Dur.	ES	EF	LS	LF	Slack
A	-	5	0	5			
B	-	3	0	3			
C	A	8	5	13			
D	A,B	7	5	12			
E	-	7	0	7			
F	C,E,D	4	13	17			
G	F	5	17	22			
			Proj Dur	22		CP	

TABLE 3-23 Seventh Cut Analysis

Task	Pred.	Dur.	ES	EF	LS	LF	Slack
A	-	5	0	5			
B	-	3	0	3			
C	A	8	5	13			
D	A,B	7	5	12			
E	-	7	0	7			
F	C,E,D	4	13	17			
G	F	5	17	22	17	22	
			Proj Dur	22		CP	

project duration. Now calculate LS using LS = LF–Duration, as shown in Table 3-23.

We will now move up one to task F. Is there any task that has an F in the predecessor? Yes, there is one—task G. Therefore make LF for task F = LS for task G and calculate task F LS as shown in Table 3-24 as well as the slack.

Task E, D, C, and B are the same way. Task E, D, and C each are contained in the predecessor column for task F. Therefore each one's LF is equal to the LS of task F as shown in Table 3-25. Task B is contained only in task D's predecessor; therefore, task B LF = the LS for task D.

Task A is contained in two tasks predecessors—tasks C and D. Therefore, we must take the minimum of the two LS's. Because task C is 5 and task D is 6, we take the 5 as shown in Table 3-26.

We now have enough information about the project to begin scheduling the work to be done, as shown in Table 3-27.

PERT analysis

Now let's add uncertainty to the equation. We will use the beta distribution we have been using with budgeting. The difference is that we are using it to model time instead of dollars. Assume that the durations we have been using are the "most likely" numbers for the beta distribution. We generate the optimistic and pessimistic forecast of task times. These are shown in Table 3-28

TABLE 3-24 Eighth Cut Analysis

Task	Pred.	Dur.	ES	EF	LS	LF	Slack
A	-	5	0	5			
B	-	3	0	3			
C	A	8	5	13			
D	A,B	7	5	12			
E	-	7	0	7			
F	C,E,D	4	13	17	13	17	0
G	F	5	17	22	17	22	0
			Proj Dur	22		CP	

TABLE 3-25 Ninth Cut Analysis

Task	Pred.	Dur.	ES	EF	LS	LF	Slack
A	-	5	0	5			
B	-	3	0	3	3	6	3
C	A	8	5	13	5	13	0
D	A,B	7	5	12	6	13	1
E	-	7	0	7	6	13	6
F	C,E,D	4	13	17	13	17	0
G	F	5	17	22	17	22	0
			Proj Dur	22		CP	

TABLE 3-26 Tenth Cut Analysis

Task	Pred.	Dur.	ES	EF	LS	LF	Slack
A	-	5	0	5	0	5	0
B	-	3	0	3	3	6	3
C	A	8	5	13	5	13	0
D	A,B	7	5	12	6	13	1
E	-	7	0	7	6	13	6
F	C,E,D	4	13	17	13	17	0
G	F	5	17	22	17	22	0
			Proj Dur	22		CP	

TABLE 3-27 Eleventh Cut Analysis

Task	Pred.	Dur.	ES	EF	LS	LF	Slack
A	-	5	0	5	0	5	0
B	-	3	0	3	3	6	3
C	A	8	5	13	5	13	0
D	A,B	7	5	12	6	13	1
E	-	7	0	7	6	13	6
F	C,E,D	4	13	17	13	17	0
G	F	5	17	22	17	22	0
			Proj Dur	22		CP	A,C,F,G

TABLE 3-28 PERT Table

Task	Opt	Most optimistic	Most pessimistic
A	3	5	6
B	2	3	4
C	5	8	10
D	3	7	8
E	7	7	7
F	1	4	5
G	4	5	8

TABLE 3-29 PERT Results

Task	Opt	Most O	Pess	Mean	Variance
A	3	5	6	4.83	0.25
B	2	3	4	3.00	0.11
C	5	8	10	7.83	0.69
D	3	7	8	6.50	0.69
E	7	7	7	7.00	0.00
F	1	4	5	3.67	0.44
G	4	5	8	5.33	0.44

Using the same formulas we used with money, we need to calculate the mean and variance for each task. The formulas are:

$$\bar{x} = \frac{O+4M+P}{6} \qquad s^2 = \left(\frac{P-O}{6}\right)^2$$

The results of the application of these formula to the data are shown in Table 3-29:

At this point, we would normally calculate the project total for the mean and the variance. However, we cannot just add them up as we did with cost. This is because the project duration is a function of the predecessors and the durations given. This is also called the critical path in the CPM process. Let's examine what impact the various estimates make on the project duration. We already know that the most probable scenario has a project duration of 22 with a critical path of A, C, F, and G. Begin by placing the optimistic estimate into the CPM matrix and recalculate the project duration and critical path as shown in Table 3-30.

From this, we can see that the critical path did not change, but the project duration is reduced to only 12. This seems very low and can only occur if every task meets its optimistic estimate. We will look at the probability of

TABLE 3-30 Calculate Project Duration

Task	Pred.	Dur.	ES	EF	LS	LF	Slack
A	-	3	0	3	0	3	0
B	-	2	0	2	1	3	1
C	A	4	3	7	5	9	2
D	A,B	6	3	9	3	9	0
E	-	7	0	7	2	9	2
F	C,E,D	2	9	11	9	11	0
G	F	1	11	12	11	12	0
			Proj Dur	12		CP	A,C,F,G

TABLE 3-31 Results

Task	Pred.	Dur.	ES	EF	LS	LF	Slack
A	-	7	0	7	0	7	0
B	-	6	0	6	4	10	4
C	A	12	7	19	7	19	0
D	A,B	9	7	16	10	19	3
E	-	7	0	7	12	19	12
F	C,E,D	8	19	27	19	27	0
G	F	6	27	33	27	33	0
			Proj Dur	33		CP	A,C,F,G

that occurring in just a minute. Instead, let's look at the pessimistic estimate and its impact of project duration and the critical path. By inserting the pessimistic durations into the CPM matrix we get the results shown in Table 3-31.

Our duration is now extended to 33 with the same critical path. This seems long, doesn't it? Again, we will examine the probability of such an estimate in just a minute. The problem is that we do not expect all of the pessimistic estimates to occur. Instead, if we operate this as an experiment numerous times, we would expect the mean value to occur. This is shown in Table 3-32.

As we did in the budgeting, we can now represent the project duration as a normal distribution centered around the mean project duration. The question is how we calculate the standard deviation to use. The answer is that we will only use the variances of the task that are on the critical path. Examine Table 3-33.

Both of these estimates are unreasonable. To establish a better estimate of the project duration, we will use the project mean and standard deviation to establish a project duration that we are 95 percent sure of meeting. We work this the same way as we did with the budget. This time we will call

TABLE 3-32 Mean Values

Task	Pred.	Dur.	ES	EF	LS	LF	Slack
A	-	5.00	0.00	5.00	0.00	5.00	0.00
B	-	3.33	0.00	3.33	2.50	5.83	2.50
C	A	8.00	5.00	13.00	5.00	13.00	0.00
D	A,B	7.17	5.00	12.17	5.83	13.00	0.83
E	-	7.00	0.00	7.00	6.00	13.00	6.00
F	C,E,D	4.33	13.00	17.33	13.00	17.33	0.00
G	F	4.50	17.33	21.83	17.33	21.83	0.00
			Proj Dur	21.83		CP	A,C,F,G

TABLE 3-33 Final Table

Task	Pred.	Dur.	ES	EF	LS	LF	Slack	Variance
A	-	5.00	0.00	5.00	0.00	5.00	0.00	0.44
B	-	3.33	0.00	3.33	2.50	5.83	2.50	
C	A	8.00	5.00	13.00	5.00	13.00	0.00	1.78
D	A,B	7.17	5.00	12.17	5.83	13.00	0.83	
E	-	7.00	0.00	7.00	6.00	13.00	6.00	
F	C,E,D	4.33	13.00	17.33	13.00	17.33	0.00	1.00
G	F	4.50	17.33	21.83	17.33	21.83	0.00	0.69
			Proj Dur	21.83		**CP**	A,C,F,G	
							Project Variance	3.92
							Project Standard Deviation	1.98

the contingency the project "Buffer." For 95 percent, the buffer is equal to 1.645 * the standard deviation, or 3.26, and the total project duration is the buffer plus the mean project duration, or 21.83 + 3.26 or 25.03.

Decision Trees and Uncertainty

Another tool we will be using in managing new product development is decision tree analysis. This tool is used when you have choices to make in a new product project, e.g., whether to test a new product early in the project despite the possibility that new external product safety regulations might be enacted that would change a key test standard, versus waiting to test the new product later after new regulations are in place but losing valuable time.

We will begin by evaluating a sample problem that is typical of the type of decision we would use a decision tree to evaluate. Consider that following problem.

Decision tree example

Pat is a project manager with a local contractor that has submitted a proposal to install telephone trunk line between Macon central station and Kathlyne, GA. Pat has just received an option on 1,000 acres of right-of-way property at \$100/acre. If Pat purchases the options and the project is not selected, there is a 60 percent chance that Pat can sell the property at what she paid for it; otherwise, she believes there is a 40 percent chance she can sell it at \$90/acre. Pat has another option to wait until the project is awarded to purchase the property. However, there is a 20 percent chance that the property will increase to \$120/acre. Pat feels that there is a 60 percent chance that the company will be awarded that project. The original proposal, based on the \$100/acre netted the company a profit of \$100,000.

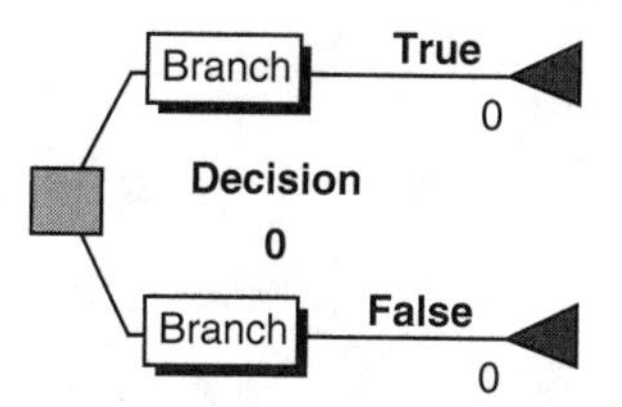

Figure 3-8 Decision component.

Decision tree theory

There are only two components to the decision tree—a decision and an uncertainty. The decision is shown as a box with one arrow for each option available in the decision (see Figure 3-8). The uncertainty is represented by a circle and an arrow for each state of the uncertainty (see Figure 3-9). The arrows for the uncertainty must contain the outcome if that state occurs and the probability of it occurring.

Note that the sum of all probabilities around the uncertainty circle must add to 1.0. Therefore, the states must represent all possible conditions. These components are strung together to give a picture of the decision to be made. With the addition of a method for making a decision using the decision tree called the expected value, we can make our decision and have a method for presenting the results in a consistent manner.

Expected value

Given the uncertainty in the previous section, and a given outcome for each state, we can make the following observations. If the uncertainty was run as an experiment, using the probabilities given on the tree, we can assume that if we performed the experiment numerous times, the results would occur at the probabilities given. Therefore, on average, the result for the uncertainty would approach the weighted average of each outcome at its given probability. For the whole uncertainty this is called the expected value or mean result. Remember this: the expected value is the mean!!!

$$E(v) = \text{Outcome1}(P1) + \text{Outcome2}(P2) + \ldots + \text{OutcomeN}(PN)$$

Consider the uncertainty shown in Figure 3-10.

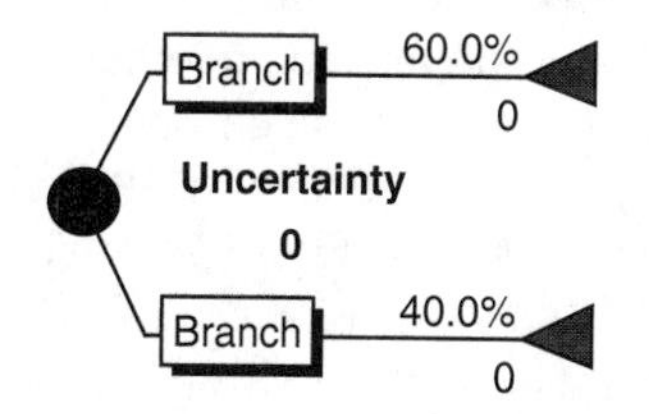

Figure 3-9 Uncertainty component.

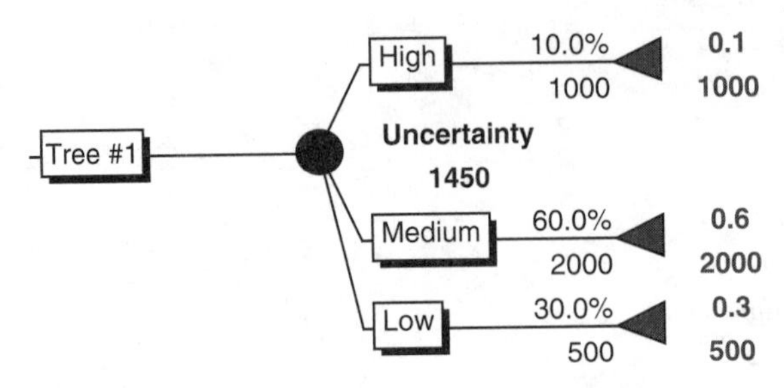

Figure 3-10 Uncertainty.

The expected value of 1450 is calculated as follows:

$$E(\text{Uncertainty}) = 1000(0.10) + 2000(0.60) + 500(0.30) = 1450$$

So, on average, we will get an answer of 1450 if we were to perform the experiment numerous times. How can we use that to get an answer to our decision problem? If we assume that the answers are the actual results of our analysis, and continue to chain such uncertainties together and attach the results to our decision node, we can see that we would want to choose the alternative that "on average" gave us the highest average PW results.

Pat's decision example using decision trees

In building the decision tree for this problem we need to establish what decisions should be made. In this case, the decision is to either buy now or buy later. We then need to establish what the uncertainties are in the problem. In this problem there are two uncertainties. The first is the same no matter which option we choose. That is, whether we win the contract. There is a 60 percent chance that we will when the contract. Therefore, there is a 40 percent chance we will not. The second uncertainty is different depending upon the option being evaluated. That is, if we buy now and lose the contract, there is an uncertainty as to the price we can sell the land for to recoup our expenses. If we do not buy the land and win the contract, then we have to purchase the land. There is an uncertainty as to the price for which we can buy the land at this point. The answer is calculated in the same manner above using expected value. When you have uncertainties that are chained together, the expected value starts at the right side and works to the left. You calculate the expected value around each uncertainty. The resulting expected value replaces the entire uncertainty in the next uncertainty. The result decision tree of Pat's decision is shown as follows.

Using the decision tree we start with the upper decision alternative of Buy Later. At the left side we see that the expected value of the purchase price of land is 100,000(0.8) + 80,000(0.2) or \$96,000. The entire uncertainty can be replaced by the \$96,000. The next uncertainty expected value can be determined as 96,000(0.60) + 0(.4) or \$57,600. Following the same procedure, we can calculate

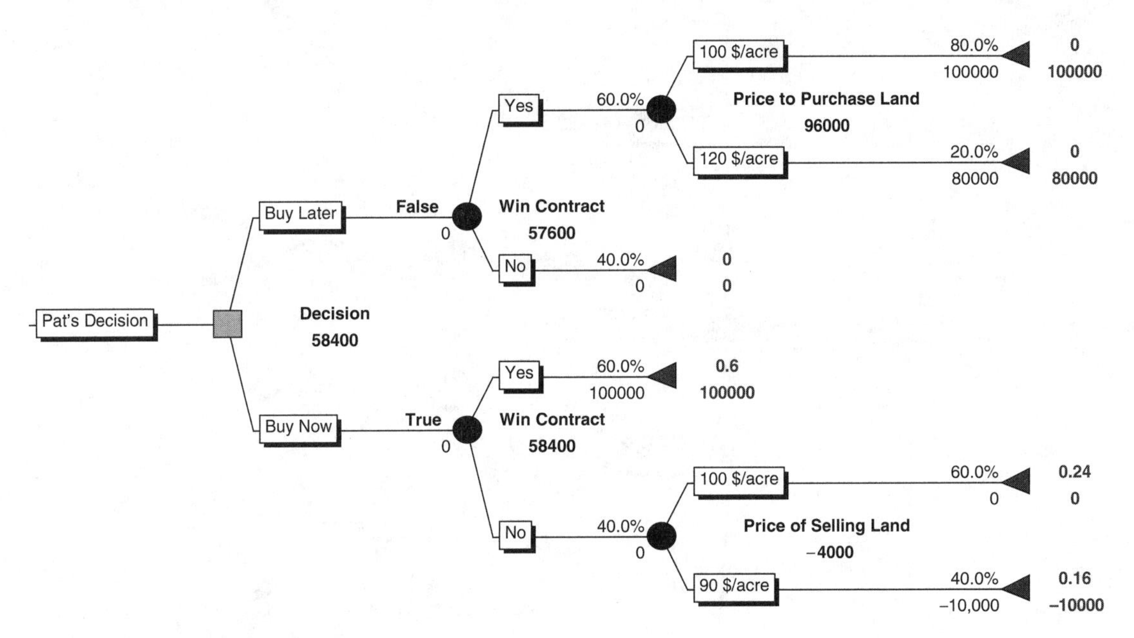

Figure 3-11 Collapsed decision tree.

the lower alternative of Buy Now with a result of $58,400. The tree can now be collapsed to the decision shown in Figure 3-11.

We choose between an average profit of $57,600 for Buy Later or $58,400 for Buy Now. Based on this, we would like to get as much profit as possible, so we would choose to buy now (see Figure 3-12).

We can now expand the concept of expected value to a more generalized concept of a decision theory. In that theory, we can evaluate the decision in terms of a state table or sometimes called a payoff table. The state table shows the various states of the world of the decision for all of the various alternatives. For example, assume that you only have one uncertainty as shown in Figure 3-13.

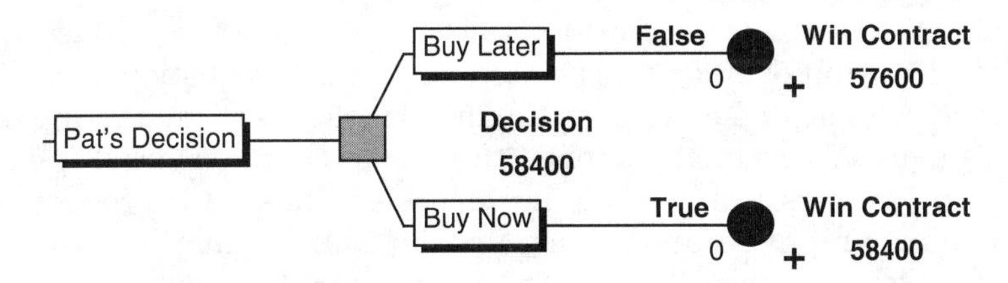

Figure 3-12 Final results.

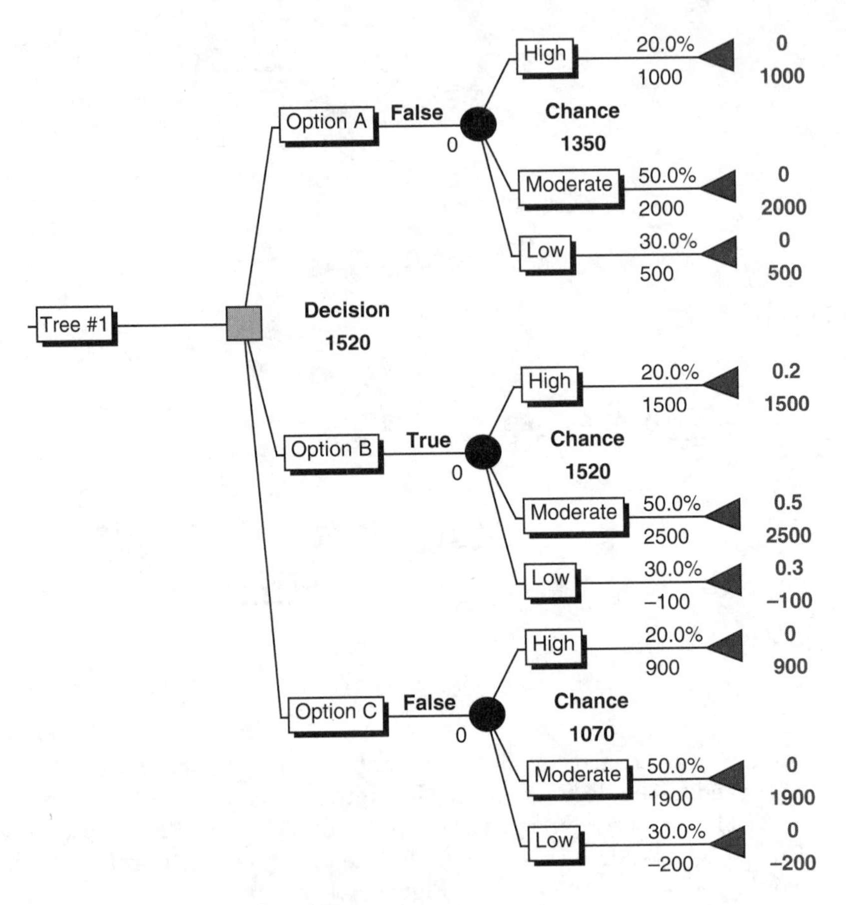

Figure 3-13 Generalized model.

Target cost analysis

The project team should be equipped to perform target cost analysis of the product. This costing assessment addresses the overall, life cycle cost of designing and product and producing and marketing it, as shown in Table 3-34:

In Table 3.34, column 1 is the total cost of product design and development, e.g., the prototype cost.

Column 2 is what it will cost to produce one unit of the product. Column 3 is the total amount of product to be produced in the production plan. Column 4 is the total volume production cost. Column 5 is the cost of marketing the product, and column 5 is the final cost of getting the product to market, e.g., the finals cost of sales. Column 7 is the forecasted sales of the product, column 8 is the product unit price, and column 9 is the final revenue.

Target cost (or design cost) is the maximum cost of producing one unit of the project deliverable, given the target for revenues (plus profit margin) for the

TABLE 3-34 Target Cost: Project Life Cycle (2 Years)

Types of costs	Total cost for product devmnt	Designed unit cost of production	Units to produce	Total production cost (2 × 3)	Marketing unit cost	Final cost of sales (1 + 4 + 5)	Target units to be sold	Target unit price	Revenue + 8% profit	Revenue
Labor	40K	1 dollars			9					
Mater	20K	3 dollar			5					
Indir	25K	1 dollar			1					
Total	85K	5 dollar	50K	250K	15K	350K	50K	7.5 dollars	375K	500,000

project over a given time period. The project must design and produce a deliverable that not only meets schedule and quality goals, but also a deliverable that can be produced in volume at 5 dollars/unit or less, in order to meet revenue goals. Target cost is not simply the cost of designing the deliverable; that is in the prototype cost; target cost is the target for the full design, production, and marketing of the deliverable.

Design to cost is an important concept in new product development. If the project is narrowly defined as the design and production of the product prototype only, then "success" in project management terms may simply meeting the estimated cost of the "front end" of the whole program. However, if the cost of the prototype is added to all other costs associated with getting the product to market, then we know the true amount of revenues we need to gain to cover all development and marketing costs to go beyond break-even.

Chapter 4

Product Concept Definition

The Product Concept Phase

As discussed in Chapter 2, new ideas for new products and services are generated in many different ways in and outside the modern company in the market. Today a more open and collaborative process of accessing innovative ideas and problem solutions is available to companies on the Web in the form of *wiki* platforms. A wiki is a collaborative Website that can be directly edited by anyone with access to it. This virtual market for new product ideas that is generated by a wiki is called an *ideagora*, from the Greek word "agora" (marketplace). An ideagora is a global marketplace on the Web where insights and experiences are shared, debated, and refined. Companies get access to a global pool of talent through this system and can access particular partners or suppliers for specific new product initiatives.

Because there are major issues of intellectual property in this open-ended, *ideation* process, many companies do not want to share their inside ideas on new products and services, and do not trust unknown,"global" partners to act in their corporate interests. But the process is gaining ground simply because it taps into a wide variety of new product concepts effectively and inexpensively.

Entering the Concept Definition Process

Wherever a new product idea is generated in this *ideation* process, the product concept phase begins when an idea or new product innovation is documented and initially reviewed by management. If management commits to supporting the definition and evaluation of the concept for possible funding, as part of the project portfolios, it enters the product concept phase.

After a new product concept has been selected from the portfolio for implementation, it is entered by management into the product concept phase and put through a scheduled concept development phase. This is the first test of the new idea as a *product concept*. The purpose of this phase is to flesh out the product concept so a decision can be made whether or not to proceed to design and development. This decision is made in the project review session at the end of this phase, and the whole phase activity is focused on gathering "actionable" information for a go or no-go decision in project review.

Controlling premature product *lock-in*

Before we discuss the concept definition phase it is important to focus on what is actually *managed* in this phase. In addition to the normal controls on time (schedule), cost and resource, and quality issues, the major challenge to project management is to *avoid locking into a product design before options and impacts are explored.* The purpose of this phase is not to pin down the exact specifications and design for a product, but rather to explore its performance requirements and business promise. Management must work hard, especially in technical and engineering processes, to keep options open and to offset the tendency of engineers and technical staff to lock in on a design too early. When product design begins before a thorough review of customer needs and requirements and alternative solutions is completed, product development often fails in the end.

The way project management avoids this premature lock-in is to frame this phase in terms of options and alternatives, e.g., solutions. Project outcomes should be seen here as performance outcomes, e.g., to meet a customer's need for a mobile communication system, not to design and produce a specific mobile radio product. Product concept is defined in terms of picturing a customer need and exploring alternative solutions.

Concept Definition Phase

The concept phase includes the process shown in Figure 4-1.

Schedule Template

Concept definition is an uncertain process, but it is important to start with a generic new product concept model as the basis for planning and scheduling. In project management terms, this is called a process *work breakdown structure*—a process model that controls the work and assures all the steps in good product concept definition are followed. The process template for a typical new (consumer) product development project might look similar to Table 4-1.

Note that several tasks are going on concurrently, and as the product functional specification is set, another set of tasks proceeds in parallel. The secret to faster, cheaper, and better new products is the management and scheduling of product development as a dynamic, interactive process. The estimates of task durations are made by the project manager using past project experience plus the insights of those who will do the work. The focus in this phase is starting dates,

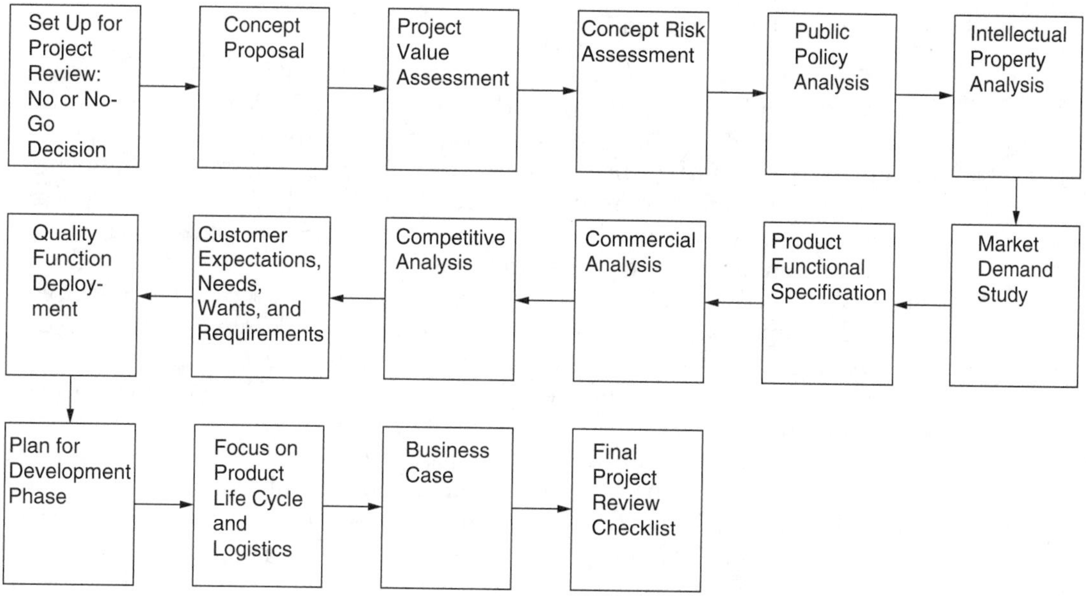

Figure 4-1 Product concept phase.

not task durations. This means, for instance, that the project value assessment (ID 3 in Table 4-1) is estimated to take five days based on past experience, but the task manager is expected to get it done as soon as possible. Durations are not seen as the authorized window for the work; they are seen as outside estimates.

TABLE 4-1 Concept Process Schedule

ID	Task Name	Duration	Start	Finish
1	**Consumer Product One**	**42 days**	**Fri 3/23/07**	**Mon 5/21/07**
2	Set Up for Project Review	4 days	Fri 3/23/07	Wed 3/28/07
3	Project Value Assessment	5 days	Fri 3/23/07	Thu 3/29/07
4	Concept Risk Assessment	10 days	Fri 3/30/07	Thu 4/12/07
5	Public Policy Analysis	5 days	Fri 3/30/07	Thu 4/5/07
6	Intellectual Property Analysis	5 days	Fri 3/30/07	Thu 4/5/07
7	Market Demand Study	20 days	Fri 3/30/07	Thu 4/26/07
8	Product Functional Specification	25 days	Fri 3/30/07	Thu 5/3/07
9	Commercial Analysis	6 days	Fri 5/4/07	Fri 5/11/07
10	Competitive Analysis	7 days	Fri 5/4/07	Mon 5/14/07
11	Customer Needs, Requirement	10 days	Fri 5/4/07	Thu 5/17/07
12	Quality Function Deployment	5 days	Fri 5/4/07	Thu 5/10/07
13	Plan for Development	11 days	Fri 3/30/07	Fri 4/13/07
14	Product Life Cycle and Logistic	9 days	Fri 5/4/07	Wed 5/16/07
15	Final Project Review Checklist	3 days	Thu 5/17/07	Mon 5/21/07
16	Go or No-Go	0 days	Mon 5/21/07	Mon 5/21/07

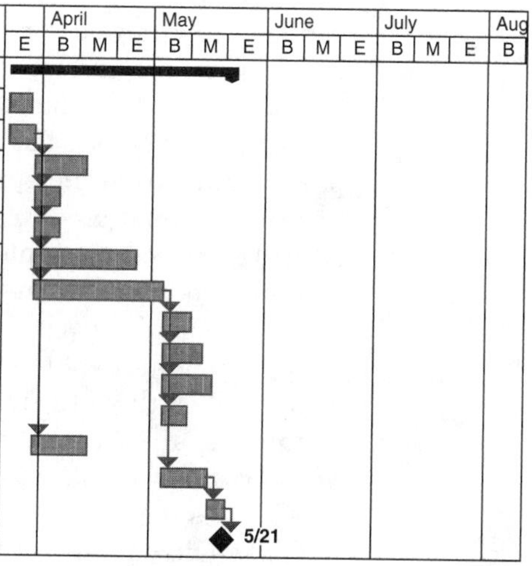

Setup for Project Review: Go or No-Go Decision

The concept phase is important because these front-end activities of analysis and evaluation help to prepare for the go or no-go decision at the end of this phase in project review. Project setup increases the chance that if the product does not have potential, the company can stop the process of development in time to save valuable resources. At this stage, it is key that good data and insight are provided about the product or service under consideration.

The way management commits to the project at this point depends on company policy and procedures, but it is recommended that top management sign off on the scope document and perhaps other outputs of this phase, and that they participate in the concept phase project review to determine whether the product moves on to development. Although members of the new product team should not make this decision, they should be party to it.

The concept phase is also important as a preparatory step for making the financial, technical, marketing, and business arguments in project review for the product. It helps build the case that is made by the project team and sponsor at project review. It is important to remember that project review is designed to *review and pass* on the product. An advocacy process is developed here with the sponsor making the case in project review, and management authorizing advancement—or raising substantive issues that must be addressed before the product goes any farther. Some companies use a scoring system to make the decision, but in the end it is always a judgement call.

Project review: Go or no-go time

Preparing for project review is as a significant step in concept definition as in each phase. Project review in the new product development process serves as the critical milestone for go and no-go decisions, referred to in some circles as the "stage-gate" decision, a term coined by Robert Cooper (author of *Winning at New Products*, Basic Books, Third Edition, 2001). It also serves as the window for advocacy, debate, and analysis of new product concepts—the time for sponsorship to "show up." In part, the project reviews after each phase of the process can be looked at as a social and team building event as well as *decision time*, not just an analytic, data-driven exercise. The social and organizational benefits of this project review process come from the opportunity for advocacy and presentation, for discussion and commitment among company and project management. This is the time for critical reviews of schedules, budgets, quality, and feasibility, and also for strategic alignment of the product regarding where the company intends to go.

In this first phase project review, go and no-go decisions are based on the concept definition, project value analysis, and financial performance projections. They are also based on the intuitive and instinctive responses of the project sponsors.

Agendas are established for the project review, accomplished through email or Web conferencing as well as physical project review meetings that bring the key parties together.

The window for advocacy is the test of commitment and sponsorship; this is the point at which a product becomes an orphan or a family member—when management sponsorship either shows up to advocate or withdraws support for lack of rationale.

Project review is the term applied to the review of the outcomes of each project phase when deciding the next steps for the product concept. The essential question is whether the product should enter the next phase, e.g., that it has been justified in terms of projected costs and benefits.

It is also important to note that in the real world, new products don't proceed cleanly through concepts and phases as described in textbooks on new product development. It is a messy process, filled with unanticipated roadblocks and irrationalities. The concept actually shifts among concept, development, testing, and marketing as it moves toward distribution, or is terminated.

Going from Idea to Concept to Product

There is a progression in getting an idea to concept to product, as illustrated in Table 4-2. Ideas become concepts when they are documented and described in general conceptual terms; concepts then become new products when requirements are identified in the context of the concept, and alternative performance standards and sometimes designs developed; and finally concepts become new product applications when the product is "seen" in a user or customer environment, either through simulation or real customer testing.

New Product Concept Proposal

The first team task after scheduling the project review process from the generic template is the preparation of a *product concept document*. Think of the concept document as a proposal that includes features and functional requirements—customer value. The proposal should also include technical and logistics challenges in marketing the product. The proposal is formulated from available information and structured as a proposal for evaluation. The proposal addresses the product's performance requirements and features in very broad terms, its potential value to customer, its value to the business, and competitive issues. The proposal is typically a short, four- to five-page document, written as an

TABLE 4-2 Idea to Concept

Idea	Concept	Product	Application	Comment
Customer not served well by coffee maker	Develop a new, computer-driven coffee maker	Chip-based coffee maker integrated with coffee bean/grounds supplier	Hotel and home-based use of computer-driven coffee maker	Competition working on it; need four-month time to market
Soft drink vending machine systems archaic	Design new soft drink machine built into user environment	Integrate vending machine into new building design	New building design incorporates vending concepts	Requires partnering with architectural industry

executive summary with appendices and references for further assessment. However, the proposal should not take more than 15 minutes to read at this point; it is a highlight document. Later it may be presented in project review as a PowerPoint presentation in summary form.

The proposal is a preliminary scope of work, of sorts, addressing the plan for making this a product, thus bringing value to the business and its growth plans.

Need, Form, and Technology

The proposal addresses need, form, and technology. *Need* addresses market and customer demand, and whether the product will actually meet an expressed need of the product. *Form* shapes the product in graphics terms so its dimensions—its *look and feel*—can be reviewed. *Technology* defines the technical systems and interfaces that occur inside as well as outside the product. The concept can be seen in Figure 4-2.

Project Value Assessment

Project value assessment is an estimate of the *stream* of value, e.g., benefits the product will bring to customers and to the business. This takes into account various consumer price ranges as well as what revenues, opportunities, and benefits it will produce for the business. All costs and logistics issues associated with the product, both pro and con, are identified. This *valuation* activity tests the product's cost/benefit early to get the bottom-line issues on the table in preparation for the first project review to determine *go or no-go*. Again, concept definition, like all phases in the process, is aimed at preparing the business for the next project review decision that either enables the product to proceed to the next phase, or terminates it.

Valuation requires some understanding of customer needs, requirements, and wants, and the company should see clearly the differences among these issues. *Needs* refers to a customer's functional demand, e.g., it is evident that a pilot of a business aircraft needs a digitized and reliable altitude instrument. *Requirements* cover the actual functional performance standards of the instrument and its implications for design to meet the need. *Wants* are customer desires, evidenced in market research or market tests, demands that may not match needs or requirements. What does the customer need to enhance the

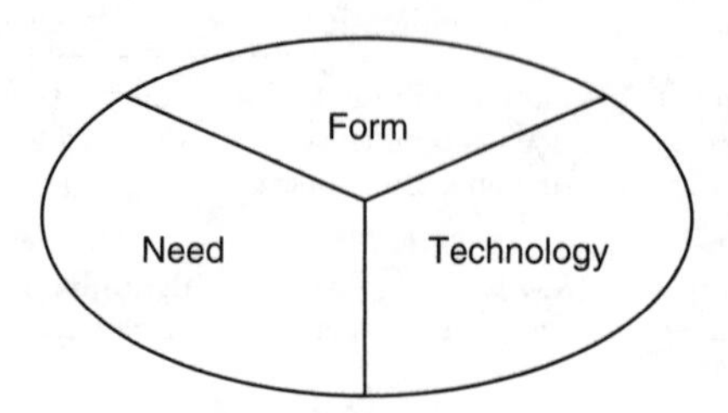

Figure 4-2 Product model.

customer value of the product, what are the requirements of a system or product to meet that need, and what do customers actually want? Sometimes the answers to these questions generate conflicting objectives and goals for the product, which are worked out in an alternative analysis later. The focus here is the general performance of a product to meet a need, not its detailed design or dimension.

This analysis gives dimension to the uniqueness of the product concept in meeting a current or yet unanticipated need by the customer. What differentiates the product from others? What features could it have that will satisfy the cost, convenience, efficiency, effectiveness, profitability, and regulatory needs of a customer or sector?

In new product development, we are primarily looking for incremental value or breakthrough value. Incremental value adds benefits to an already functioning product, but in a distinct way. Breakthrough value induces demand to meet a current or new need that is neglected in the marketplace. Why would the customer be willing to pay a particular price for the product; what process of thinking and rationalizing would the customer go through?

The analysis of value also includes a preliminary overview of the financials. The goal is to provide preliminary insight into the market, including size and potential market share, and into sales price, sales forecasts, expected margin, cost to produce, and revenues. In general, this deliverable should answer a major question—is the projected margin high enough, given the company hurdle rate, and does the idea/concept have potential and sustainable competitive differentiation in the marketplace?

The analysis should build on data generated earlier in portfolio development and project selection and will include:

- Description of the market and market share—What does the market look like, and what share can the company expect to gain through this new product initiative?
- Logistics—What are the distribution, support, and logistic issues associated with the product that could present problems down the product development and marketing path?
- Sales forecasts—What kinds of sales and pricing can be expected under specific conditions?
- Estimated cost to develop and produce—What will be the project costs to get the product into the market quickly?
- Profitability and preliminary margin—What will be the product selling price and when will revenues cover costs?

Estimating Product Value in New Systems or Process Concepts

Product value is in the product concept and system, not simply the product itself. In other words, new products and services fit into a user or customer environment or *user system*. It is not simply the product function that creates value; it is the product function in the context of a customer setting. So you must know the customer setting.

Consider new products to include new systems and processes. A new product for bagging groceries creates the opportunity for improving the whole process of bagging groceries in the customer's place of business. It is not just the new sack; it is the way the new sack helps customers get groceries safely to their cars and the confidence with which the grocery baggers use the sack. It is a mistake to limit thinking about new products to consumer product issues alone. New products include internal and support products and systems, e.g., new computer networks and new manufacturing processes, that add value to the business. Value is calculated in terms of cost avoidance, productivity, and management effectiveness.

Concept Risk Assessment

Concept risk assessment is a preliminary assessment of all risks associated with the concept/product. This includes feasibility, test failure, technical performance, market, regulatory, development, and resources. Each risk area receives an evaluation of high, medium, or low. Activities judged to be medium or high require explanation. A risk matrix is created at this point, formatted as shown in Table 4-3 using a public safety broadband concept as an example.

In completing the risk matrix at this point, we are looking at the project as a whole, but this activity may also highlight particular project tasks that will

TABLE 4-3 Risk Matrix: New Public Safety Broadband Concept

Nature of risk	Risk defined	Probability	Impact	Severity	Contingency plan
Feasibility	New broadband concept is found to be unfeasible, given current public policy and regulations	50%	Showstopper	Major	Partner with regulatory agencies and get subsidy from public sources to float concept
Technical performance	New public safely platform does not work	25%	Medium	Minor	Simulate platform options and test in working community environment
Resources	Outside funding is not found	50%	Medium	Major	Find venture capitalist
Test failure	System does not perform to specifications	25%	Low	Medium	Run early simulated tests and change standards if unattainable
Competition	Competition comes up with concept and markets it	50%	Medium	Medium	Work out partnership with competing interests

create risk—and opportunity. Because overcoming risk in a new product creates opportunity, risk assessment is focused on what the company can achieve in the new product to create advantage in the marketplace.

The following defines each analysis as shown in Table 4-3.

- Nature of risk—This activity categorizes the risk in terms of general source.
- Risk defined—This activity defines the risk in terms of what the risk event is and how it could occur.
- Probability—This is a subjective qualitative assessment, ranking risk in terms of likelihood of occurrence, say in terms of 25, 50, or 75 percent.
- Impact severity—This is a ranking activity, placing the risk in a category of High (showstopper, will terminate the project), Medium, or Low.
- Contingency plan—This is the corrective response to the risk, an action plan to either prevent or offset the risk. This activity is sometimes included in the schedule of the project later if the probability and impact of the risk are high. The contingency is not simply "doing this task better"; rather, it is a new approach to overcoming the risk, perhaps through a new process or technique.

External analysis: Public policy analysis

A public policy analysis is the external view of the product, looking at the economic, regulatory, and social factors that will impact product success. If the product concept is subject to governmental or regulatory requirements in any country in which it will be marketed, an analysis is made of the potential safety, environmental, economic, or local impacts. The aim is to identify anticipated regulatory and safety constraints by agency and country, if applicable. This environmental scan is likely to capture other outside forces, including potential current events that could influence the success of the product in a given market setting. For instance, a political change in a given country may influence the potential success of a given product or service. This process should not be seen as a defensive one simply to protect the company; often, these outside standards provide new ideas on how to improve product performance in a given customer setting.

Intellectual property analysis

If the new product has potential intellectual property value, e.g., it can be patented or copyrighted, an analysis is performed on the risks and processes associated with intellectual property control and potential loss. This analysis requires a full understanding of current and projected intellectual property law and regulatory structure, and thus should be performed by IP specialists. This analysis will include a look at opportunities to take advantage of patent, license, and copyright opportunities, and other ways to preserve and protect intellectual property. A risk assessment is completed on risks associated with failure to control intellectual property.

Market demand and other impacts

This is a rough assessment of market demand. The concept is looked at in terms of the dimension of potential market demand as well as in terms of market characteristics and financial performance for business growth given the estimated demand. What is the demand likely to be for the product, and will it generate adequate revenues in the marketplace to cover development costs and produce profitability and market share goals? If so, over what period of time? If the new product is an internal system or process improvement, the activity becomes an assessment of whether the improvement is really needed, and whether it promises to produce continuous improvement in business operations and effectiveness. Other impacts of the product or system on corporate or agency operations are also reviewed. For instance, a product may replace another or impact on the performance or market share of another. These impacts need to be injected into the decision process.

This is the beginning of the process for developing a market launch case and rationale. Most data addressed here are preliminary and do not require in-depth market analysis, but there needs to be enough of a case for the product to suggest moving on. On the other hand, if the product or system is a breakthrough improvement, the issue is how much the product will induce new demand; that is, how will the product change the current demand picture simply because it will create new value for customer, user, or client?

Here is where the "voice of the customer" is sought out; where potential users, clients, and/or customers, as well as stakeholders, are identified and dimensioned using focus groups, simulations, and other tools.

Product Functional Specifications

How will the product work and what are its functional specifications? The concept phase fleshes out technical specifications for the product in high-level terms. In this phase the product is defined in terms of function and performance, not necessarily in terms of design. In other words, the focus is on how the product will serve customers' needs in the user setting. Specific design issues are left open for the development phase, although models can be developed to clarify the look and feel of the product.

The product's functional specification is outlined in high-level terms. Functional specification describes how the product will perform and what design specifications *might* enable it to perform at its highest level of efficiency. Not all the design data is available at this point, but enough information is gathered to allow management to gain a vision of the product in an operating situation.

Table 4-4 shows an example of a high-level product.

The definition of functional specifications for a new product is a tricky process in this initial concept development phase because the tendency is to go into too much detail (premature lock-in) on how the product will work and how it should be designed before analyzing its potential value to the customer and its business value to the company at a high level.

TABLE 4-4 High-Level Product Specification

Desired performance attribute	General functional specification	Customer's priority	Comparison to industry or competitive standard	Comments
Helps cell phone users use phone while doing other things	Must create place for cell phone in various mobile environments	High	Current cell phone holders	Current holders do not perform well over time

The management challenge in product functional specifications is to find a balance between detailed and graphic representations of the product, which are often time consuming and expensive to produce, and an inexpensive and workable qualitiative model or prototype of the product that can be tested with customers and users. Engineers will sometimes go too far in detailing a product before its feasibility and customer value are established, spending valuable time defining tolerances, risk analyses, design options, and networks of a product that may not survive the test of customer and business value.

Commercialization Analysis

Here is where the product is looked at in terms of how it will be commercialized. Commercialization embodies the successful volume production, distribution, marketing, sales, and support system required to be a winner in the commercial sense. A strategy for commercialization will require a marketing plan and a partnering agreement with other companies in the supply and delivery chain, defining direct and indirect costs and barriers to market success, and identifying the capacity of the company to bring the product to customers and support it over time.

Here is where logistics costs and opportunities are identified, including distribution channels, trends in economic and social developments that might impact on logistics, e.g., availability of a labor base, and competitive forces that might inhibit free flow of the product *pull* to the customer.

Competitive Analysis

Competitive analysis is the search for competition, competitive differentiation, and value unique to the company in this product concept. Differentiation is the value of the product and the company's capacity to deliver and support it, as compared to the competition. It is relational, based on the opposition. It requires some speculation on how the competition is planning to meet the customer need, and where the competition is in the new product development process.

This analysis relates directly to the company's strategic plan and *risks, threats, and opportunities* assessment. The analysis also looks at what is liable to happen after a product is made public; that is, how the market will react to the product in terms of competition, barriers, and risks. Technological developments are reviewed here because new product development often spurs technology development and the generation of new companies—even new industries.

Sometimes this means looking for partnering opportunities early to offset potential competitive forces. It may make sense to sit down with the competition to work out a partnership in a growing or new market.

Table 4-5 shows what a competitive analysis might look like.

Finding drivers of competition

What are the drivers or key forces of competition, and what are the key sensitivities in the market that open up opportunities for competition? *Drivers* of future competition include all the critical success factors and forces that combine to create competition for the target product market. For a new plastic grocery sack, the driver might be capacity to distribute the sacks just-in-time. Any firm might be able to make the sack, but few firms have the distribution and support capacity to get it to stores when they need it. The value in identifying such drivers is that they can be monitored and, in some cases, offset by effective product development and marketing approaches. Drivers include:

- Technology development. New products create technology challenges that many companies cannot handle, thus a major force in competition is the relative technological strength of other companies in the industry.
- Visibility of future markets. Future markets may not be visible to competitors, depending on the effectiveness of their marketing activity.
- Timing. Some competitors will not be a position to take advantage of new markets, even if visible, because of inconsistencies and timing of their own competitiveness.

TABLE 4-5 Competition Analysis

Competition	Pros and cons	Potential response	Potential contingency	Comments
XWZ Company	Has a product in development like ours, but its performance is suspect over time	Competitor is liable to get out of this market if challenged	Partner or merge	

- Financing. New product development requires funding, often from venture capitalists, thus the availability of financing, e.g., interest rates, determines entry for many companies.
- Scale. The scale of the markets—and the costs of entry—may be too large for many competitors to enter.

Working Out Customer/Client/User Expectations, Needs, Wants, and Requirements

This is process of pinning down the potential customer base, e.g., a marketing function. The phase must define the "pull" of the customer or client for the product or system, as opposed to a process of "pushing" a product at a customer to induce demand. The pull concept is a function of different forces from the push concept. Customers *expect* a product to perform in a way sometimes different from their needs; in other words, customer expectations are driven by past experience plus customers' vision of success. Customers *need* a product because of a functional gap, but they may not see that need. *Needs* is a functional term, one that results from a functional analysis of the customer's environment. *Wants* are customers' emotional attachment to a given product or system, derived from their wish list of how the world ought to operate. And requirements are functional attributes, e.g., features, that a customer may be looking for.

The way a product is connected to customer needs and expectations is key to this first phase—how will the product under review integrate with customer expectations, needs, wants, and requirements?

Again, customers are analyzed in terms of:

- Expectations—the users' "sense" of what they would really like to see in the product category, what they would embrace if cost was not a factor.
- Needs—an analysis of what the user needs from a marketing perspective, our judgment of what the user needs, e.g., performance requirements, given their expectations.
- Wants—a listing of what the customer/user *says* they want, which might be different from what they expect, need, or require.
- Requirements—product performance specifications and later, design requirements, more technical product criteria.

This analysis identifies the *customer voice* by determining what customers/consumers want in a superior, differentiated product or system enhancement. A user needs and wants study is conducted to produce information on how the new product matches with customer expectations, needs, wants, and requirements. Table 4-6 presents the approach for each study. Its basic output is a vision for the future, based on customer feedback and articulated wants and needs. Studies are conducted regionally to yield customer insights in the context of local and regional settings and cultural characteristics. A summary analysis

TABLE 4-6 User Needs and Wants Study Template (Sample)

Steps	Questions	Analysis	Implications for market growth	Comments
Customer interview group	Define focus group characteristics.	Review focus group data for customer characteristics and transferability.	Must be a representative group or results are misleading	
Current customer setting and situation	How is problem solved now; what would be a better solution?	Look at what has been done to correct these problems in the past.	Fundamental breakthrough needed or incremental?	The facilitator must orient the discussion in the direction of the new product benefits
Perception of current offerings	How do current products rate on a scale of 1–10?	Establish consistent scale and scoring approach across all focus groups.	Is it service or equipment performance?	
Desired improvements	What do customers want; what should improve?	Analyze underlying issues in desired improvements.	Why?	
Vision for the future	What is their concept of how their needs might be met in the future; what factors did they consider?	Why are they thinking that way?		

will be developed capturing all the results of local and regional studies to allow for market-wide assessment.

Quality Function Deployment (QFD)

QFD is applied in the concept design phase to link customer requirements with product specifications, e.g., design and specification. It is a translation of *what* the customer wants to *how* the new product is going to meet that want. QFD is useful both as an agenda and a framework for discussion as well as a methodology. QFD suggests a way of thinking, an approach to asking questions of the customer and brokering answers to those questions to figure out how the customer's needs are going to be reflected in internal processes, product design, and production capabilities. QFD is a process, not simply a quantitative template for analysis, which assures that the project team is in dialogue with the customer and can assure that it can employ the appropriate systems and processes to produce what the customer wants. The QFD exercise should create standards or criteria for evaluating later the extent to which a product will be acceptable to the customer, e.g., that it conforms.

TABLE 4-7 QFD Template (Sample)

What the customer wants	How the product will meet those wants	Comments and challenges
Defect-free dispensing of a product	Embedded software for self-repair or remote repair	Need to resolve gaps between expectations, needs, wants, and requirements
Light weight	Aluminum construction	Review impact capacity and aesthetics.
Easy to use	Simple, user-friendly design	User-driven design, tested in customer/user setting
Reliable	Six sigma development and manufacturing	Install six sigma culture and systems
Replacement parts	Modular design	Design challenge
"Green" requirements	Recyclable parts	Review design-generated energy requirements

QFD involves two summary tasks; (1) obtain customer requirements in writing; and (2) link customer requirements to product specifications. As these steps are documented, the project manager is assured that the team can verify or trace the scheduled work directly back to the customer requirements. This is shown in Table 4-7.

Plan for the Development Phase

Looking ahead to project review and the next phase—full development and marketing—some planning should occur now for development. Again, the concept phase must produce information necessary to plan for full design and development. Looking ahead, this process includes functional and equipment requirements down the road, supply chain and partnering issues, outside government impacts, logistics, and packaging and distribution needs.

But the most important preparation action is the definition of the development *process*—what technical designing, testing, and prototyping will the product go through to justify it for marketing?

Focus on product life cycle

There is a tendency for project managers to see only the short horizon of a new product, and not the long game. This is because project management has sometimes looked only at cost, time, and quality in terms of producing a prototype, rather than the long haul involved in production, marketing, and revenue enhancement. This activity to outline the entire life cycle costs and benefits of the new product will offset this tendency to limit the project. Analysis includes total costs of development, production, distribution, logistics, marketing, and support over the time period the product will survive in the marketplace.

This raises a key management issue in new product development. Is the project manager responsible just for producing the new product prototype for production, or for the whole process of bringing the product to market and producing business value? Who is accountable for the whole process? The best organizational approach to this problem is to appoint a program manager (or a product manager) for the product who is accountable for the entire process. Specific project managers report to the program manager for concept and design, development, production, and marketing.

Equipment and logistics plan

Because the costs of logistics and distribution are major elements of total product life cycle costs and are often left out of concept definition, special attention should be paid to preliminary estimates of logistics costs and support needs for the product. It may be that the product reduces logistics and support costs, and that this value may be *the* value of the product. A logistics risk analysis should be conducted here, at a high level, simply to identify the risks and opportunities in product support, logistics, and maintenance.

Business Case

Will the new product help to grow the business? Drawing on the results of an earlier portfolio development and project value assessment, the business case activity requires a review of the combined impact of market, financial, alignment, and risk issues. The business case is made in final project review for the concept phase based on how the product is likely to further business plans and goals, and how the product will enhance market share. A project manager or project sponsor usually makes the case. Market analysis takes a look at the potential market for the new product, trends in similar marketing efforts, commercialization issues, and competitive forces. Financial performance is reviewed in terms of cash flows and assumptions, discounted by net present value calculations at the business hurdle rate. Alignment is looked at in terms of whether the product fits into the company portfolio and capacity. Risk is assessed using a risk matrix, including a definition of risk, and its impact, severity, probabilities, and contingency actions to prevent or offset risk.

Final Project Review

Final project review at the end of the concept phase raises the following questions aimed at go or no-go.

1. Does the company have a project review committee or council that represents company management? Can this committee or council organize and conduct a project review for this product? These organizational questions go to the company's capability to get management attention on this product so that when and if it goes to development, it has a company sponsor and company backing.

2. Does the product/systems concept make sense from a business perspective? Does the concept create *opportunity* for the business, and will the payoff be worth the investment and risk? This is the business case issue, generating a discussion of what the concept phase has demonstrated in terms of the financial (net present value at the company hurdle rate), alignment (weighted scoring model index), commercialization, feasibility, risk (risk matrix), and marketing potential of the product concept.
3. Do the initial analyses in portfolio development hold up after this phase in terms of qualifying this concept for company financial backing and support? This requires the initial portfolio analyses be updated so management is assured that the original concept still holds water. This kind of review is necessary because the concept now enters development and production, the most costly and time-consuming aspects of new product development.
4. Does the product/systems concept make technical and functional sense? Is the concept technically feasible, and can the functional specifications be met in development and production?
5. Does the product/systems concept make process sense? What does the product do to current key business processes, e.g., design, development, production, configuration management, distribution, logistics, marketing, support, procurement, and human resources? Can the product be sustained within the current capacity of the business, or will it generate new process needs and opportunities?
6. Does the product/systems concept make sense from an ethical, safety, or regulatory perspective? In other words, will the product/system raise ethical issues, e.g., a new drug concept that may be currently illegal or unethical, and will the product be vulnerable to current regulatory, safety, or public policy trends?
7. Does the product/systems concept make sense to potential customers, users, and clients? Does the product have the potential to meet real need, expectation, and requirements of real customers in the market or users in the real work setting?
8. Does the product/system pose risks that cannot be overcome with current contingency planning? Although the risk analysis in the concept phase is preliminary at best, it has uncovered risks, impacts, probabilities, severity ratings, and has generated contingency plans. Does the product/system pose risks that the business cannot afford, given its capacity and financial strength? Are the contingency plans developed for each risk identified feasible and *costed out*, and are they included in the current project schedule to assure that they can be triggered, if necessary?
9. Does the product have the support of the workforce? In other words, do the people who have been associated with the product/system to date support its development, or are there real issues and bottlenecks that they see in

the product that have not been brought out in the *official* concept phase reporting? Does the company have the capacity to produce the product?

10. Can the product be supported in the current business supply chain, and are there partners available to provide necessary support to the development and production of the product/system? This is the partnering question, challenging the company to ensure that supply and vendor chains are there for this kind of new product.
11. Are there special reliability issues inherent in the product/system that might cause barriers or bottlenecks in development and/or production? How will the internal product testing accurately replicate the user setting for the product, and what are the risks that new reliability standards are developed during production and distribution?
12. Does the project plan include a schedule and cost estimate for completing the project within a timeframe that the company can handle? Can this product be delivered to market in time to beat the competition?
13. Have all the assumptions inherent in the product/system been identified and evaluated? What assumptions have been made in assuming that this product can be marketed successfully?
14. Does the company have the organization, managerial, and technical *maturity* to market this product? This is an issue relating to the company's organizational, development, and management capacity to make decisions, and to the technology performance of the company in terms of the challenges of this particular product.

Chapter

5

Full Product Development and Marketing

Development and Marketing

The development process includes designing and building the product into a prototype; creating a defined and configured product design that is tested for performance against performance requirements; and then planning for production, marketing, and distribution. In effect, full development prepares the product *fully* for marketing and distribution. During development, the product design is documented through a configuration management process. Configuration management involves a continual updating of the exact product component breakdown as it evolves through design and development in order to assure that the final product is defined for production, supply/inventory, and logistics. A specialized configuration management software program is used to preserve product components and to document suppliers who may be required to produce components later in production.

The reason development and marketing are addressed in one chapter is that too often they are planned and implemented in isolation from each other. For organizational and cultural reasons, marketing is often not part of product design and development. Project managers in new product development need to consider a seamless process of development and marketing with staff from each functional division working together throughout the process.

Project Setup and Management

Project setup here can be defined as setting up the project to control the process and to support the go or no-go project review at the end of development. This involves confirming product requirements, project scope of work, schedule and budget, risk management plan, and project team composition. The project management challenge is the coordination of people and work efforts from different company divisions—in design, development, configuration management, and manufacturing/production. All workers from these different departments have a direct interest in the success of this stage, and in the final product design for production and assembly. Frequent meetings between these roles and functions among department members should be scheduled to allow for a free exchange of information and a status assessment *assessing the status of* the product. Communication among these players in new product development will pay off down the road as the product nears final design and production transition, as well as in project review.

Figure 5-1 shows a detailed new product development process flow chart. Many of these functions will proceed in parallel, and some will be minimal, depending on the nature of the product.

It is important from a project management standpoint to schedule and control the development process. The focus should be on making task *starting times* since because durations and end dates will be very uncertain in the development process. The end game for management is control by monitoring progress, and making sure all work that is planned for parallel activity is coordinated. All task resources should be identified, and a core team established to work the process and pull in needed outside contractor assistance.

The schedule template for a typical new consumer product might look similar to Table 5-1.

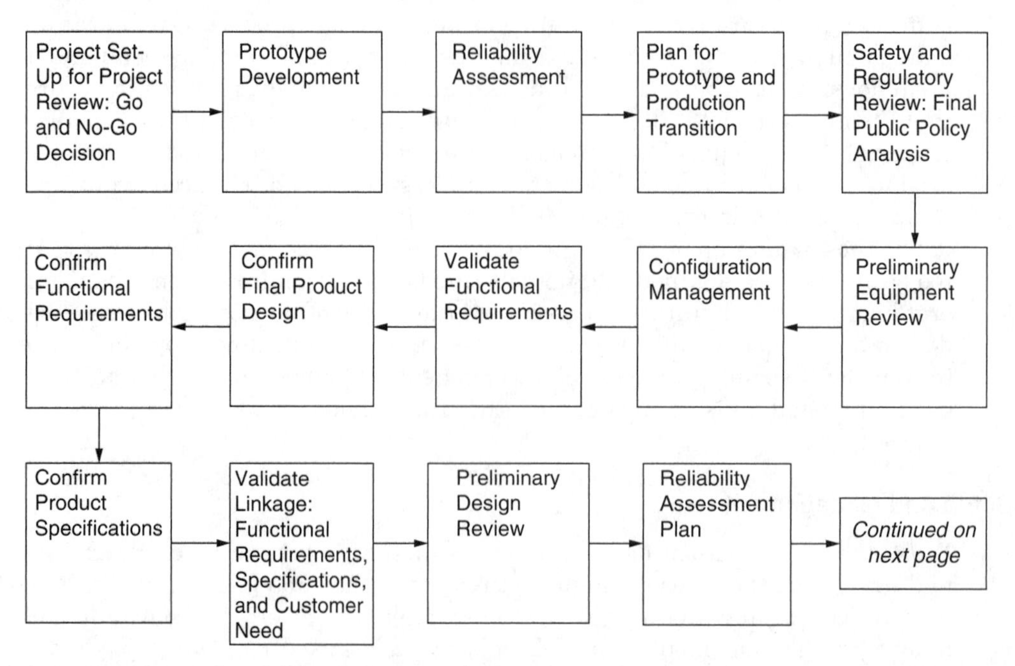

Figure 5-1 New product development process flow.

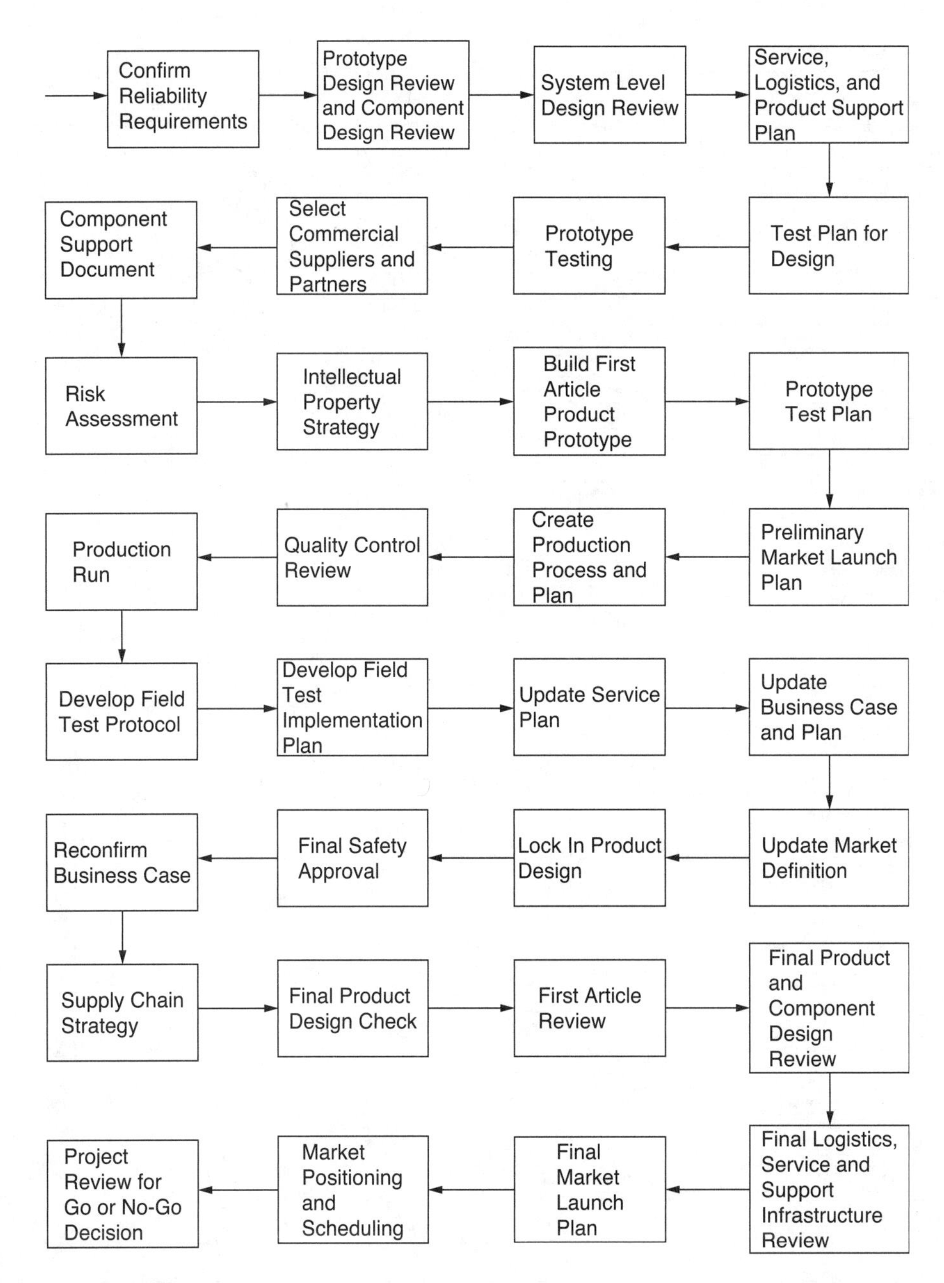

Figure 5-1 *(Continued)*.

TABLE 5-1 Sample Schedule Template

ID	ⓘ	Task Name	Duration	Start	Finish
1		**Consumer Product One Development**	**61 days**	**Fri 3/23/07**	**Fri 6/15/07**
2		Project Set Up for Project Review	5 days	Fri 3/23/07	Thu 3/29/07
3		Prototype Development	10 days	Fri 3/23/07	Thu 4/5/07
4		Reliability Assessment	10 days	Fri 4/6/07	Thu 4/19/07
5		Plan for Prototype and Production Transition	4 days	Fri 4/6/07	Wed 4/11/07
6		Safety and Regulatory Review: Public Policy	15 days	Fri 4/6/07	Thu 4/26/07
7		Preliminary Equipment Review	4 days	Fri 4/6/07	Wed 4/11/07
8		Configuration Management	4 days	Fri 4/6/07	Wed 4/11/07
9		Validate Functional Requirements	5 days	Thu 4/12/07	Wed 4/18/07
10		Confirm Final Product Design	2 days	Thu 4/19/07	Fri 4/20/07
11		Confirm Functional Requirements	5 days	Thu 4/12/07	Wed 4/18/07
12		Confirm Product Specifications	5 days	Thu 4/12/07	Wed 4/18/07
13		Validate Specifications, Requirements, and Need	5 days	Thu 4/12/07	Wed 4/18/07
14		Preliminary Design Review	2 days	Thu 4/12/07	Fri 4/13/07
15		Reliability Assessment	9 days	Thu 4/12/07	Tue 4/24/07
16		Confirm Reliabiilty Requirements	4 days	Fri 4/20/07	Wed 4/25/07
17		Prototype Design Review	2 days	Thu 4/26/07	Fri 4/27/07
18		System Level Design Review	8 days	Mon 4/30/07	Wed 5/9/07
19		Service, Logistics, and Product Support Plan	10 days	Thu 5/10/07	Wed 5/23/07
20		Test Plan for Design	2 days	Thu 5/24/07	Fri 5/25/07
21		Prototype Testing	5 days	Mon 5/28/07	Fri 6/1/07
22		Select Commercial Suppliers and Partners	10 days	Mon 6/4/07	Fri 6/15/07
23		Component Support Document	3 days	Mon 6/4/07	Wed 6/6/07
24		Update Risk Assessment	2 days	Mon 6/4/07	Tue 6/5/07
25		Final Intellectual Property Assessment	5 days	Mon 6/4/07	Fri 6/8/07
26		Build First Article Product Prototype	3 days	Mon 6/4/07	Wed 6/6/07
27		Preliminary Market Launch	10 days	Mon 6/4/07	Fri 6/15/07
28		Create Production Process and Plan	5 days	Mon 6/4/07	Fri 6/8/07
29		Quality Control Plan	2 days	Mon 6/4/07	Tue 6/5/07
30		Production Run	4 days	Mon 6/4/07	Thu 6/7/07
31		Develop Field Test Protocol	10 days	Mon 6/4/07	Fri 6/15/07
32		Develop Field Test Implementation Plan	6 days	Mon 6/4/07	Mon 6/11/07
33		Update Service Plan	3 days	Mon 6/4/07	Wed 6/6/07
34		Update Business Case and Plan	4 days	Mon 6/4/07	Thu 6/7/07
35		Update Market Definition	3 days	Mon 6/4/07	Wed 6/6/07
36		Lock in Product Design	2 days	Mon 6/4/07	Tue 6/5/07
37		Supply Chain Strategy	4 days	Mon 6/4/07	Thu 6/7/07
38		First Article Review	2 days	Mon 6/4/07	Tue 6/5/07
39		Final Product and Component Design Review	2 days	Mon 6/4/07	Tue 6/5/07
40		Final Logistics, Service, and Support Infrastructure	6 days	Mon 6/4/07	Mon 6/11/07
41		Final Market Launch Plan	3 days	Mon 6/4/07	Wed 6/6/07
42		Final Project Review	2 days	Mon 6/4/07	Tue 6/5/07
43		Go or No Go	0 days	Tue 6/5/07	Tue 6/5/07

A product can be pulled or pushed through to production. In development, the product should be pulled by the need to conduct a project review at the end of the phase to make the go or no-go decision. The product should not be pushed through. Pushing the product in this case suggests that work is conducted with no end point in mind except to finish tasks. Ideally, the product is *pulled* through development toward information needed in project review, configuration management, and production requirements. Everyone in the process understands that development produces information for the key go or no-go decision. Everyone knows the product may not make it because of technological or business reasons.

The danger here is that configuration management will lock into a product prematurely, before its final development is completed, simply through the forces that press the product to market. The challenge to project management here is to keep the process open but controlled, and focused on data and information. Configuration management tends to press development to create product details, to move quickly to a production model, and to preserve the supply and contractor chain associated with each component.

The management function during development is more hands-on than in concept definition. This requires project management and team members to be attentive to the passage of uncontrolled time and excessive consumption of resources beyond the project development budget—*playing* with the product instead of developing it.

Development pulls new products through two essential phases: preliminary development and final development. Sometimes these phases are called *alpha* and *beta* testing, or prototype development and prototype testing in a user acceptance mode.

Prototype development

The role of prototype development is to move from functional requirements for the new product to a *forward* development stage in which documents (CAD drawings, schematics, pictures, models, and so on) are generated, and outputs are planned that support preliminary component and equipment design as well as other development work. The information generated in this activity will be used to demonstrate that the design is capable of meeting the quality, performance, reliability, and cost specifications of the project. This is where the product begins to take shape and look like its production and marketing design. This activity produces a hard copy or electronic design that can be manipulated and altered as project and design reviews suggest. Test data will be gathered on the physical or engineering prototype, and the design is signed off by a concept designer to assure that there is continuity between early performance requirements, early concept design, and the more mature development prototype.

The final output of prototype development is a workable design with tested performance. Also included are preliminary feasibility and risk assessments, and functional requirement verification test plans. Other miscellaneous design documents to support the review process are also generated in this activity. All plans will be tested in the later tasks of full development.

Development means transitioning the concept into a model of the final product, *hard* or *soft*. It is important to note here that new products will go through this development phase whether they are "hard" products, such as engineering, construction, or production products, or "soft" products, such as software or information systems, online reports or Web products, or marketing and sales promotion outputs. For a physical consumer or systems product, the outputs of development prototype include CAD models, schematics, part lists, wiring diagrams, and sketches. Output should be adequate so that the entire design can be reviewed and scrutinized by peers. Every part—subassembly and assembly—should be documented well enough to convey preliminary thoughts on design and assembly. Calculations and other supporting data should all be generated from these models and sketches.

The concept of peer review is often used here, opening the product to review by an inside, or sometimes outsourced group of peers. This process works against *locked in* views of the product that are created when a company cannot see any data that do not confirm product feasibility and performance. This is the *paradigm* issue again. During this phase, figures, drawings, and text will be dated and clearly labeled so that the peer review can easily find and understand drawings. All of this data is to be presented in a peer design review.

The Failure Mode and Effects Assessment (FMEA) identifies the hypothetical failure modes of the product design, and projects the internal and external effects of such a failure (severity and probability of occurrence). It is a risk assessment at the product and operational levels, and is useful even in "soft" products such as software. The FMEA is important because it allows the designer to rank the failure modes, and facilitates discussion around the worst failure modes of the product. This gives the developer a process to identify which failure modes need to be prevented or reduced in severity (or probability). This is not a full risk assessment, which looks at a wide range of failures, both technical and managerial, but a supplement to it for the product itself.

Risk assessment involves the dimensioning of risk under alternative operating conditions and performance requirements, with a risk matrix including risk definition, probability, severity, impact on cost, schedule, quality, and contingency (mitigation) plan. The designer will need to assess all possible risks in the design, and that includes risk associated with: part supply, design, product performance, operating conditions, schedule, and cost. Risk may also come from an obscure area, e.g., logistics or distribution of the product, or packaging, so care must be given by the designer to consider all possible areas of risk.

Reliability

Product reliability is a big issue in development. Reliability is tested at the highest level to confirm technical specifications and produce performance information on customer, user, and functional requirements. This question here is:

"Can the product components and the product as a whole meet reliability targets of the design in a user setting?" For instance, for a digital avionics instrument, the test might be whether the low voltage system can run properly at the voltage stated in requirements. The test plan may indicate that the design targets will be verified using analysis, testing, or by similarity. Similarity means that the component, say a resistor or a building component, is the equivalent of another component that has already passed muster, in an earlier product development process, and that the test results have been documented. Risk assessment will uncover critical subassemblies or untested components that require special attention. Failures of critical parts are highlighted in project reports and documented to allow retesting and possible redesign. It is the role of the functional engineer or technical specialist to identify untested product components and to flag them for testing, but the configuration management system is the final confirmation that all components have been tested.

Build and production transition plan

It is not too early to plan for production of the product, so this preliminary activity identifies how the product will move from development into the production or manufacturing process. Often called the production transition plan, the build plan addresses build issues and generates a *build* strategy. Included in this process are designers and developers, assemblers, manufacturing engineers, production schedulers, and the project manager.

Safety and regulatory review

It is also not too early to look at the safety aspects of the product in terms of outside public regulations and industry safety standards. This is an early look at regulations that will control the product in various markets and requirements for approvals, paperwork, and so on, that will allow the product to be marketed.

Preliminary equipment and component review

Remember that development is aimed at *informing* project review at the end of this stage. The purpose is to help management make the decision to proceed to the next stage. This preliminary equipment design review assesses the design for technical competence and functional requirement adherence. This is where the product is tested against requirements using tracking software that links performance results to requirements, one by one. The output of this preliminary equipment design review (and potential design rework that is required) is a solid product design and set of specifications that serve as the basis for future feasibility, reliability, and the FMEA (Failure Mode Effects Analysis) assessment.

The review is actually a meeting that all relevant participants attend. Project management needs to make sure key technical and management people attend and contribute to this review. This is sometimes a challenge because technicians and new product developers in general tend to avoid such meetings, which they often believe do not directly involve their work. The design review package is sent out to all participants one week ahead of time so that everyone can adequately prepare for the design review meeting. All preliminary equipment design outputs will be approved by project leaders who conduct these downstream tests and assessments.

Top-level drawings, models, and schematics are produced here to identify the product prototype and its assemblies, subassemblies, and components. The documentation here needs to be relevant documentation that includes an official description and design. Supporting documents needed for testing include:

1. Product configuration and numbered parts list from configuration management system
2. Mechanical, electrical, and software design specifications
3. Final Bill of Material at subsystem level
4. Preliminary product design FMEA (Failure Mode Effects Analysis)
5. Updated product technical and performance specification.
6. Electronic access to all the design content, test results, and so on, for safe and secure storage

Configuration management

The configuration management function is extremely important in new product development because it is this function that documents design and component configurations as they are changed and finalized. It is important that there be continuity and compatibility between the software used to preserve design and the software used to document designs in configuration management. However, from a management standpoint, configuration management cannot be allowed to inhibit good design and development. Configuration management should be managed by a functional department associated with development and managed as part of the project.

Validate functional requirements

The validation of functional requirements ensures that the product design will perform to the product specifications. The product is taken through a *build and test* cycle to validate its performance. Design and development plans are updated as the design evolves during the alpha process. Functional requirements are finalized from concept definition and product specifications in this process. In the process of developing specifications, incomplete, ambiguous, and/or conflicting requirements are resolved with both the concept definition and potential customers. As requirements are defined, individual subsystems may be released into detailed design after design reviews. To achieve validation of functional requirements, requirements must be clearly defined with no ambiguity and eventually documented in a configuration management system.

Design validation (how the product will be designed to function properly) ensures that the product will meet all of the requirements imposed by the product specifications. This includes the product system as a whole, hardware, software, electrical and other components, and failure modes. Validation is typically performed to an internally generated requirements document to ensure that the product meets the standards required for thoroughly testing the production product. Design validation data is documented for use in project review and go or no-go decision making.

To make sure there is a continuing linkage back to customer need, functional design validation includes "touching" customers as well. Through focus groups, simulations, and meetings, customers validate that the product functional requirements and specifications will meet their needs. This avoids the risk of suboptimizing around a limited product viewpoint, and experiencing failure in the test cycle.

Confirmation of Final Product Design

Development of the final product design requires a final update and documentation for market execution. This task is critical to product success because it confirms the final design to be produced and marketed. The task documents all results for testing as well as any changes in product configuration, design, and parts lists from earlier development. Traceability is critical at this point; a final design check is compared to product specifications and customer needs to ensure that the product design can be traced from specifications directly back to customer needs.

Final product requirements, specifications and drawings used to produce the prototype, and production products are evaluated through a critical, final design review. Review includes detailed design outputs, written or electronically formatted drawings, acceptance test requirements, design documents, product models, and other analysis documents. Updating these outputs enables the product to be manufactured, and its performance validated against requirements once after beta testing is completed. Three steps are involved

1. Final Product Models Prepared—Final versions of product models are prepared to facilitate market execution.
2. Final Analysis Documents Prepared—These updates may include interface requirements, applicable design standards, trade studies, layout drawings, product test procedures, and configuration management documents.

3. Final Bill of Materials—The final bill of materials identifies all of the components of the product and gives them preliminary part numbers for eventual documentation into a configuration management file.

Confirm Functional Requirements

This task involves confirming product specifications and functional requirements so that they can be validated with customer needs. This task produces a validated design requirement that conforms to the functional requirement template. Table 5-2 shows what a functional requirement might look like at a high level.

Confirmation of Product Specifications

The confirmation of product specifications involves a review and assessment of how current specifications may have changed as a result of design and prototype testing as well as other outcomes of the product development process. Particular attention must be given not only to documented changes—changes that were fully documented in the product configuration and reflected in the bill of material—but also *de facto* changes that may not have been documented. In other words, this is the point at which *any* changes are documented for field testing.

A special project management challenge here is to assure that no undocumented changes in product design occurred in contractor activity. Intellectual property issues can surface here because product design has potential value to contractors.

The risk involved is that outdated field test units will be built and go to test. This can happen if some specification issues remain open and are not resolved in this confirmation process. This task *locks in* the product configuration for field testing.

TABLE 5-2 Functional Requirement Template

New product performance function	Functional requirement	Comments
Produces Digital GPS Data on Mobile Phone	Receives GPS data	
	Shows digital data on various phone screens	
	GPS data change during movement	

TABLE 5-3 Customer Need/Functional Requirement Template

Customer need	Corresponding functional requirement, e.g., fitness for use	Comments
Ease of use	Digital screen easy to clean and maintain	
Capacity	Can handle full digital data on screen	
Cost	Low cost access to GPS data	
Maintenance	Easy maintenance	
Noise	No noise	
Aesthetics	Pleasing color and changeable fonts	

Conforming requirements to customer need

When preparing the development prototype, this task again confirms and validates that the features of the product meet customer needs. This is accomplished through focus groups, meetings, and simulations in which customers interact with the product in local settings. Table 5-3 shows the template for capturing data.

Preliminary Design Review

The purpose of a preliminary design review is to identify product and equipment design issues and potential problem areas before they become major reliability risks, and before the prototype is built and tested. Preliminary equipment design reviews concentrate on the design of the entire product as a system. The preliminary review raises issues that become the agenda for a full design review later in this stage, and for project review.

At this stage, the design of product components and equipment are still evolving. Therefore, the design process is not ready to turn out a configured product ready for production. But this review helps to *status* the design process and surface challenges.

The design review meeting brings together all the stakeholders in the product design. The agenda is heavily oriented toward engineering issues, but other problems, e.g., schedule and customer requirements, typically come up. The program manager prepares the agenda. A lead engineering facilitator leads the meeting, and appropriate subject matter experts participate as needed. Drawings, documents, and graphics are projected for group review and discussion, and items are recorded for follow-up and attention during gate review.

The issues reviewed in the meeting include:

1. Confirm specifications
   - Status of customer needs document
   - Status of product specifications document
   - General status of equipment design activity

2. Identify Design Issues
   - Performance requirements
   - Design approach
   - Review of equipment and layout drawings
   - Applicable design standards and tolerances
   - Design trade studies
   - Applicable design models
   - Design to cost goals (if any)
   - Environmental and green issues
   - Preliminary Bill of material
   - Producibility and manufacturing issues
   - Test requirements
   - Reliability requirements (preliminary to full Reliability Assessment)
   - Robust design goals and issues
   - Critical parts review
   - Test Readiness review items
   - Safety and regulatory issues, e.g., electrical grounding
   - Unanswered design questions, e.g., material specifications, potential supplier design problems, cost

   The outputs include:

1. Prepare Design Issues Document—This task requires a design engineer to produce a design issues document that guides the preliminary design review meeting and follow-up action. This document identifies design issues, such as:
   - Size and function of critical parts
   - Tolerances
   - New technologies
   - Form and features
   - User issues
   - Design options and alternatives, e.g., materials, etc.
2. Conduct Review Meeting—The preliminary review meeting agenda includes:
   - Review of design documents, drawings, etc.
   - Design definitions
   - Review of design and functionality of product
   - Review of design issues document, e.g., potential design risks
   - Bill of materials and parts list
   - Supplier issues
   - Generation of follow-up actions and responsibilities
3. Design Review Matrix Template—A design review matrix is prepared and used to guide preliminary review and follow-up. Table 5-4 shows a design review matrix.

TABLE 5-4 Design Review Matrix

Design topic	Document or reference	Issues	Follow-up
Design concept	Design engineer document		
Progress of design	Design engineer assessment		
Technical adequacy of design approach	Review meeting outcome		
Compatibility of design with internal and external hardware, test equipment, and facilities	Inventory of test facilities, hardware suppliers document		
Physical interfaces and integration issues	Drawings		
Design risks	Risk assessment		
Manufacturing risks	Assessment by manufacturing engineer		
Scheduling issues	Program manager assessment of schedule variance		

Reliability Planning

The key feature in the success of most new products is reliability, the capacity of the product to consistently perform according to requirements. Customer satisfaction is based on reliability, not just first-time performance. But reliability is a major cost factor in development, thus it is important to use the right tests for reliability and get them right the first time.

The reliability plan is the approach to defining acceptable reliability and testing the product in development. Reliability criteria are driven by customer requirements, another reason to make sure customer requirements are correct to begin with.

Reliability requirements can be very technical and quantitative, stating how the product will perform under specific conditions and stress. These requirements can also address field support standards, e.g., one service call per year. Sometimes technical standards such as "mean time between failures" can be used. Sometimes a reliability model is developed to simulate performance under various conditions. For instance, in the case of the plastic grocery bag, the extent of flexibility in the bag when loaded, e.g., how much it stretches, is stated in terms of maximum amount of stretch.

Setting reliability objectives

Setting reliability objectives is a tricky business, and the process has management and cost implications. Since customer requirements drive reliability standards, these requirements should be clear. However, for a brand new product it may be difficult to state those requirements without field testing. This is not a case of getting the best reliability you can get in development. This approach would be expensive and sometimes dysfunctional, and yet without careful

TABLE 5-5 Reliability Objectives and Requirements Template

Product or component	Conditions of use	Reliability objective (narrative)	Reliability targets (measure)	Comment
Product	Temperature range	Describe range for design	MTBF, number of failures in range, operating time	Customer is instructed to operate within specified temp ranges; product is designed to operate within those limits
	Vibration			
	Water			
Subsystem A	Temperature			
	Vibration			
	Water			
Component A	Temperature			

management attention new product teams tend to overdesign and overdevelop the product. The point is to relate what the customer needs and will pay for to a given range of reliability, and then to design to that standard.

Table 5-5 shows a template for reliability objectives and requirements.

Prepare reliability plan

The reliability plan is the approach to testing the product for reliability and follow-up contingencies. The plan will address:

- Key product reliability objectives
- Linkage of product design and components to objectives
- All subsystems and components that are reliability critical, e.g., that would have major impact
- Risks associated with testing failures
- Methods and test protocols
- Supplier and contractor testing requirements, e.g., to make sure all suppliers incorporate proper testing methodology and can be counted on to test future components

Table 5-6 shows what a plan format might look like.

Confirmation of Reliability Requirements

The FMEA (Failure Mode Effects Analysis) document is key in confirming reliability objectives and requirements for testing. The FMEA is performed by an objective engineer or technician who is not directly associated with the new

TABLE 5-6 Sample Plan Format

Product or component	Reliability objecive	Related risk	Test protocol	Conditions of use	Industry standard
Product	2 service calls per year	Excessive downtime	Test product failure rate; equip reqts	Temperatures below 0 degrees in outside setting	TBD
Subsystem	MTBF (Mean Time Between Failures) = .152 failures per service action given conditions of use	TBD	TBD	TBD	TBD

product team. This ensures objectivity in the analysis. As requirements change during design, the FMEA is updated.

Here the product is reviewed for performance objectives, components, and functional requirements that are sensitive to reliability. For instance, in the development of an avionics instrument, if the optics of an LCD screen in the instrument are critical to a pilot's understanding and use of the projected digital data, optics reliability becomes a critical reliability function. Testing is focused on the high reliability functions of the product, in this case optics. In sum, the confirmation process *locks in* product reliability requirements for field testing.

Contractor and partner arrangementsmust also reflect the company's reliability goals and objectives, so this activity involves a full review of contract provisions to ensure that suppliers and contractors have the capacity to meet reliability needs and will meet them as part of the contract.

Pre-prototype design review

This is the final design review before the product is entered into prototype development and full development testing. This review starts with the results of the preliminary design review and associated follow-up actions, as well as with updated technical specifications, design models, and previous FMEA assessments. These reviews are scheduled tasks on the new product project schedule and should be attended by key team members and contractors. Management needs to focus on design review so that necessary communication occurs between sometimes disparate team member functions. For instance, if the materials engineer and the FMEA engineer are not linked, a key material requirement could change without being reflected in the testing protocol.

Detailed product and component design review

Because this is the final review before testing, the review should establish and confirm the final design of the product and its performance against specifications. In other words, the product configuration under review at this point should satisfy all performance requirements.

Again, the design review is a meeting that brings together all the stakeholders in the product design. The agenda is heavily oriented to physical and engineering issues, but other problems, e.g., schedule and customer requirements, typically come up. The program manager prepares the agenda. A lead engineering facilitator leads the meeting and appropriate subject matter experts participate as needed. Drawings, documents, and graphics are projected for group review and discussion, and items are recorded for follow-up and attention during project review.

System-level design review

System level design review focuses on the whole system within which the product operates. For instance, in developing and testing the plastic grocery sack, the whole store and home environment of the customer is reviewed so that reliability requirements are traced from "bagging" to ultimate disposal in the customer home and recycling. This task requires a design engineer to produce a design issues document that would guide the preliminary design review meeting and follow-up action. This document identifies design issues, such as:

- Size and function of critical parts, e.g., how reliable is a push button on an instrument under various conditions
- Tolerances, making sure mechanical engineering tolerances and interfaces are reviewed
- New technologies and testing new components, e.g., a new low voltage resistor
- User issues
- Design options and alternatives, e.g., materials

The preliminary review meeting agenda includes:

- Review of design documents, drawings, etc.
- Design definitions
- Review of design and functionality of product
- Review of design issues document, e.g., potential design risks
- Bill of materials and parts list
- Supplier issues
- Generation of follow-up actions and responsibilities

Prepare test protocols and facilities

While testing protocols must be accurate and link directly to requirements, it is often logistics and proper execution of testing that are the key management challenges in new product testing. Testing facilities and equipment must be scheduled and committed to the scheduling testing dates, technicians must be trained to perform the testing, and the protocols must be understandable.

Testing requires defined procedures and protocols so that test results can be documented. This task involves preparing test protocols, e.g., testing procedures and equipment. The *user*s for these testing protocols are the test technicians, either in-house or supplier/partner personnel, who will actually conduct the tests and document results. Industry standards are often used here to assure conformance with accepted requirements. Regulatory agencies in product safety often issue testing standards to assist in effective reliability standards, particularly if the regulatory agency itself sets those standards. After product testing is completed, the root causes of any failures are analyzed, and a new design is prepared when necessary. Then the test is repeated.

Acceptance test procedure is a term used to describe how the new product test demonstrates product performance against each requirement. ATP anticipates the user acceptance of the product. The test protocol also addresses the testing report format, including the types of analyses to be conducted, the results of all analyses and tests, a summary of applicable problem reports, and how each has been resolved. The test protocol also defines when a new product is *test ready*, e.g., when a product in development is to be tested.

Service, logistics, and maintenance plan

This step is a key development function. This is the plan for servicing and maintaining the new product in the field. The plan includes a program for preventive maintenance and service calls based on the projected reliability standards for the product. The plan also addresses the availability of trained field support personnel to support a new product during its entire life cycle.

Final test plan

The final test plan covers all pretest, test, and test follow-up activities, including performance requirements, reliability objectives, test protocols, and FMEA analyses. The final test plan is scheduled as a key project task and milestone in support of project review.

Special project management issue: Test space and equipment

As indicated earlier, the project management issue here is not only to control the testing process, but to provide for testing facilities and space when they are called up in the schedule. This is not easy in a multiproject environment when other projects are using valuable and possibly scarce test facilities. Project managers have to pay special attention to how testing space, equipment, and technical resources are scheduled by other projects to make sure the windows for testing in the product development schedule can be used to test when scheduled. This can be a major problem in development.

Prototype development and testing process

The development and production of a new product prototype involve close collaboration between engineering or R&D and manufacturing, because the new design must be manufactured to allow testing. Some companies produce prototypes in their normal manufacturing process, but the danger here is that operators and manufacturing processes need to be changed for the new product. Because of this, it is advisable to have a separate production process in the engineering or R&D operation to product prototypes that reflects the new designs. This process will help not only to develop good prototypes for testing, but will also develop new production and assembly processes that will be necessary when the product is produced for distribution and marketing. This process should be supervised by a manufacturing engineer who bridges the gap between design and manufacturing.

Conduct Prototype Test

The development and testing of the product prototype present major technical and management challenges. From the technical standpoint, testing requires the use of a relevant testing protocol that is not always accurate, a testing capacity, and space and equipment that is often scarce and subject to many demands. From the management standpoint, the problem is the supervision of test technicians and engineers who often do not agree on the protocol or on the interpretation of results, and never on the documentation requirements.

The testing protocols serve as the basis for all prototype testing and reporting. The following steps show the prototype testing procedure.

1. **Produce prototype model**—The first task in the development of the prototype is to ensure that the product design has been engineered into a physical model of the product that will serve as the basis for production of the prototype units. All product, subassembly, and component specifications, tolerances, and so on must be released to configuration management for documentation, including all supplier and other contractor information on parts. A parts list is then generated.
2. **Produce parts list**—The parts list must be reconciled with the product breakdown structure (PBS) and configuration management system. The PBS is a hierarchy of the product components, decomposed down, e.g., the beginning of the bill of material. The configuration management system is the software that carries the *official* product description for the purpose of later supporting manufacture and parts inventory.

3. **Produce prototype**—The prototype should be manufactured in a production facility like the one that will do the production of the final product. This will create conditions for gathering information on production issues, parts and inventory problems, and assembly problems during manufacture of the prototype(s).
4. **Test prototype**—Actual testing is conducted by a supplier or in-house test technician staff, following prototype test plan procedures and protocols. Reports and documentation of all test operations and results are produced by the testing engineer and test technicians.

Here, again is a project management dilemma. Do you have the prototype produced by the engineering department, or is it produced in the regular manufacturing process? In new product development projects, it is important not to rely on manufacturing to produce the prototype because key lessons learned could be missed unless the process is under the control of a manufacturing engineer. On the other hand, many companies do not have a manufacturing capacity in engineering, thus making it necessary to rely on production.

Select commercial partner

Commercialization means the successful marketing and sales of the product consistent with business plans. The selection of the commercial supplier is key here because contractors are typically used in this process. The process includes stating the scope of work for selecting a commercialization partner/supplier, identifying potential suppliers, generating criteria for selection, and making the partner selection. A partner is typically used to ensure market implementation and field support, e.g., an understanding of market conditions and sales force needs.

Selection of supplier partner

The selection of the new product development and test partner/supplier is a key decision because it establishes the basis for a long-term contractor partnership that will be required to design and test the product and then move the product into commercialization. The partnership must work seamlessly; the supplier must understand company goals, objectives, and processes; work in a close relationship with the program manager and team; and serve as an extension of the project team. Thus the selection of the partner must reflect the values and goals of the company and the team.

Steps in partner selection

The following steps should be followed when selecting a partner. A qualifying partners template is shown in Table 5-7.

1. **Qualify potential contractors**—This process is based on the development of criteria for qualifying potential development partners, reviewing their

TABLE 5-7 Qualifying Partners Template

Criteria	Question	Research	Comment
Capacity	Does the company have the workforce and physical capacity to serve as a partner?	Review dimensions of workforce, skills mix, etc., and office and research space	
Proven experience	Does the company have proven experience in design in new product development?	Review contracts and previous work	
Technical competence	Does the company have the technical know-how to do reliability and validation testing?	Review previous work	
Product development processes	Does the company have key staff that understands new product and prototype development processes and can work within them?	Review previous work	
Quality	Does the company have a quality improvement process?	Review company's R&D program	
Business processes	Does the company have adequate business processes, including bench and testing systems, testing procedures, and protocols?	Review product development processes	
Management	Does the company have reputable top and middle management, and a core of project management talent?	Review resumes of key management	
Contract independence	Is the company too dependent on outsourcing staff to be able to learn from its development work?	Review mix of in-house and contract engineering and other personnel	

performance in past supplier relationships, and determining which contractors should be invited to prepare proposals.

2. **Request for proposal (RFP)**—The request for proposal is a contractual document developed with the contracting office, and includes scope of work, concept definition, and technical specifications.
3. **Screen proposals**—Proposals are screened by a committee and include all key company departments. The proposal uses two basic criteria: quality and cost.
4. **Selection of partner**—The selection of the key partner is made by the program manager in consultation with key staff and team members.

Prepare Product Component Support Document

A new product support document is created to support the configuration management and bill of material documents. Its purpose is to go into more detail on component sources because the component supplier information can be lost if

TABLE 5-8 Sample Template for Component Document

Product component	Technical requirements	Equipment and parts	Drawings and part numbers	Manufacturing and assembly manuals	Component sources
Electrical	Reference technical specifications	Reference Part definitions	Reference product drawings and part numbers, e.g., specification, equipment, drawing, serial, manufacturer, and change order	Reference manufacturing engineering manual	Reference suppliers, sources, and parts numbers

left to the supplier. In other words, to preserve key information on components, suppliers cannot be trusted to have and hold information that may be needed in the future simply because contracts were written to provide that information. This document preserves information in-house to support production and inventory, as well as logistics and product support. Table 5-8 shows a sample template for the component breakdown.

Risk Assessment

A full risk assessment is conducted in the development phase. The process begins with a full review of each system, subsystem, and component, and asks the questions, "what can go wrong; what are the risk issues, probability, severity, and impact; and what contingency plan must be created to either prevent or respond to the risk?"

Changes are likely in the product design and plans since the concept stage and the first risk assessment were conducted. Therefore, the risk assessment plan needs to be updated at this point in time. Updates will include new anticipated task/risk events, impacts and implications of risks, views of probabilities, impact assessments, and contingency plans. A broader view is taken of the business risk of the product, as well; for example, "how will the product affect the business brand and image?"

The task involves a full review and analysis of the original risk assessment, interviews with all parties to the design and development process including appropriate suppliers, and a listing of new issues and risk events. An updated risk assessment is then prepared, highlighting changes and new issues. The final product should be distributed to key staff and suppliers, especially those who will participate in beta testing. Each risk is documented in a risk matrix as shown in Table 5-9.

Risk assessment includes the development of contingency plans to address identified risks. While risks are identified by the project team, contingencies

TABLE 5-9 Sample Risk Matrix Template

Product and/or component name	Risk event	Impact on schedule, cost, and quality	Probability of occurrence scale 0–100%	Severity of risk scale of 1–10	Worst-case cost of risk	Total risk	Contingency action recommended	Linkages to other component risks
LCD screen for aircraft altimeter	Digital connections can be incompatible with designed electronic components	Could delay whole project schedule, increase cost, and damage partnership with foreign LCD supplier	Medium (25%)	High, critical impact on quality of altimeter	$1,000,000 Redesign time of six months and/or tooling modifications could cost upwards of 750,000. Six-month delay will have lost opportunity cost of $250,000	$4.5MM	Send engineer to foreign manufacturer to develop testing protocol to prevent incompatibility	Could influence how LCD is assembled and packaged, as well as electrical components, including resistors and screens

are prepared by specific team members assigned by the project manager to address particular risks. Contingencies are not focused simply on preventing or mitigating risks by doing a better job. Contingencies are new tasks to offset risks, tasks that are often integrated into baseline schedules to ensure they get done if risk events do occur. These risk contingencies are also used to develop expected, pessimistic, and optimistic scenarios for risk-based scheduling using the Microsoft Project PERT tool.

Intellectual Property Strategy

There are two conflicting forces acting in the development of an intellectual property strategy. One is the need to protect against a competitor copying the product or its components. A close reading of applicable IP regulations is necessary here to avoid losing the value of the product to a competitor. The other force is the increasing need to collaborate in a global process of finding and applying new product ideas and opening the process to unknown partners. The is the so-called *wiki* process addressed earlier. Wiki uses the Internet as a new product idea and concept generator.

The secret to reconciling these forces is to use the wiki approach to get and contribute ideas, but when a new product concept is documented and ready to go to concept definition, to close out global collaboration and prepare for an intellectual property strategy.

Develop Preliminary Market Launch Plan

While it is early in the process to start locking in on marketing launch, it is not too early to be doing preliminary market launch planning. This process involves looking at launch obstacles and bottlenecks and preparing for a sales force, distribution and logistics, and promotion.

Identify potential market launch issues. This task involves listing and describing anticipated market launch issues. These issues include:

- Where will the product be launched?
- What sales and marketing resources will be needed?
- How will customers and users be *prepared* for launch?
- What are the anticipated logistics and costs?

Market launch is a formidable project management challenge because there are so many unknowns and potential developments that can sidetrack a good launch plan. These include customers who change their minds on the product, unanticipated regulatory or *green* (environmental) requirements, and disconnections between sales and production.

Field Support to Market Launch

What is often missing from market launch plans is provision for hands-on customer support at every point in the marketing and sales process. New product marketing often requires training and development of customers and users in product use and support, beyond what is anticipated. Murphy's Law prevails in market launch; whatever can go wrong, does. This training is provided by a seasoned professional in the field over an initial trial period, and can be costly and time consuming.

Create Production Process and Plan

Establishing the production process for the design involves the development of a manufacturing/production process to build test units. This task will identify whether the production is done by an internal or external supplier, or through a partnership approach. The specifications for the production process will be established, from raw material input through final packaging. Detailed plans for production and equipment/tooling acquisition will be included. This includes making sure the final bill of materials is accurate.

Create preliminary production plan

The production plan will include input-output analyses in terms of raw material inventory, assembly processes, tooling, and equipment. A production control process will be described to ensure that the units proceed through a defined set of operations. Modules as components are integrated into the final first article units. Table 5-10 shows the production plan template.

Quality control review

Now is the time to create a preliminary production quality control process. Quality control will be addressed in terms of component and inventory control, process control, quality assurance, and inspection processes. Component and inventory control will be established through a stabilized bill of material and configuration control. Inventory will be ensured through an acquisition process

TABLE 5-10 Production Plan Template

Product or component	Raw material needs	Assembly process	Tooling and equipment	Personnel
Circuit board	Printed circuit board from X supplier meeting Y specifications	Automated followed by hand solder	Automated circuit board installation; requires board installation tool	Requires trained assemblers familiar with tooling equipment

triggered in time for production to proceed on schedule and guided by the final bill of material. Quality assurance will be addressed through a description of each production process step, including assembly, testing, and integration standards, as well as product inspection.

The quality control review is necessary at this point to ensure that defects and quality problems in manufacturing are avoided now, in development. Table 5-11 shows a quality control template.

Produce test units

Again, the actual production of the test units involves updating the preliminary production process for test units to a final production plan and then implementing the plan and monitoring the production line as the units are produced. Final test units produced are tested under laboratory conditions to ensure that they exactly duplicate the final design requirements, and that the configuration is consistent with the final bill of material.

Finalize the production plan. The production plan is made final for the production line and signed off by the production control manager and manufacturing engineer. Inventory is checked and personnel proficiency in assembly and tooling control is assured by the production control manager.

Conduct production and built process. This includes product schedule, process definition, assembly operations, personnel, and inventory. Table 5-12 shows a sample conduct production template.

Develop field test protocol

This task involves the development of a procedure for testing the product in the field. The "protocol" is a series of tests and data collection in a user or customer setting. Both technical performance data and customer views will be collected. The plan will address success criteria, logistics issues, service and support requirements, training needs, marketing and brand considerations, and post-test equipment disposition plans.

TABLE 5-11 Quality Control Template

Component	Specification	Control technique	Inspection timing	Anticipated defect problems
Mechanical housing	Dimensions, tolerances	Checks, inspections, statistical process control, and variance analysis	Every step in process	Tolerances violated in supplier production of component; check incoming raw material

TABLE 5-12 Conduct Production Template

Production schedule	Process definition	Assembly operations	Personnel	Inventory
Schedule can be implemented; no conflicting scheduled production	Process is defined in terms of assembly requirements, necessary manuals, and assembly aids; work plans prepared where necessary	Actual assembly is monitored by manufacturing engineer; dry runs conducted when necessary	Assembly personnel and equipment operators are certified by manufacturing engineer	Inventory for build volume is checked and assured by production control manager

The plan specifies how customer views will be documented and reported, and ensures that customer expectations as well as needs are obtained. This is especially important because product requirements may no longer be consistent with "real" customer and user settings.

Prepare technical protocol. The technical protocol includes all prescribed tests under conditions of use. These prescribed tests are prepared by appropriate design engineers and technicians, and documented in a test manual that can be used for training and guiding actual field tests.

Prepare customer interview protocol. The customer interview protocol involves development of customer interview surveys, open-ended focus groups, and other prescribed techniques to solicit customer and consumer views, expectations, and needs generated during field testing.

Develop field test implementation plan

This task is the actual field test implementation, on-site in selected field test locations and markets. Field test is managed by marketing and sales people. The task is supported by the test plan as the basic guidance document, but lots of flexibility is maintained during actual field testing to allow variances, new techniques, and contingencies depending on the dynamics and challenges in actual test settings.

Field testing success criteria, service and support requirements, training needs, marketing considerations, and post-test equipment disposition plans are written into a field test operations guide that prescribes how the field test will be conducted, step by step.

A field test schedule is prepared to guide field testing, with day-by-day operations specified at selected field sites. The schedule includes individual tests under conditions of use, with durations, resources, and equipment.

Prepare field test schedule. The field test schedule includes all field test operations with timing, resources, and site arrangements specified.

Conduct and document field test results. The field test team conducts field test operations, monitors results, and documents data and information collected in a beta field test report.

Update service and logistics plan

This task involves updating the plan developed earlier for servicing and maintaining the product over its entire life cycle. The plan will include a program for preventive maintenance and normal repair or service calls based on the projected reliability standards for the product. The plan also includes a listing of anticipated service requirements based on design and development factors, such as life cycles for components, and supplier recommendations for servicing specialized parts and materials. A listing of required replacement parts is prepared to guide inventory management to support maintenance.

A training and development program is developed to ensure that adequate numbers of maintenance contractors are available to implement the service and maintenance plan worldwide. Updating the plan will take account of changes in product design, specifications, reliability, market launch strategy, remote service technology, and other factors that impact service and maintenance costs, resource needs, and scheduling. This update activity reviews the technology involved in maintenance, as well assures that the appropriate technology is available for servicing the product.

Update Business Plan

While the earlier update of the business plan in product design involved a quick review of changes in the business plan, this update is more comprehensive. This review of the business plan anticipates the approaching market launch and thus must assure that the business plan reflects all changes up to this point in product development. Special attention is paid in this task to financial performance involving the updating of forecasts of cash flow and net present value of the product over its life cycle.

Issues reviewed for change in the business plan include market definition, competitive assessment, marketing positioning, customer-internal linkage, brand/trademark linkage, commercialization plan, regulatory/agency plan, financials, intellectual property plan, financial performance, and strategy for supplier. All of these task outputs are again reviewed to ensure that any changes or different perspectives on the product are reflected in ensure business planning. These updates might specifically include new forecasts of product pricing and financial performance, new customer and market information indicating how the product might induce new demand, new commercialization issues, e.g., the use of the Internet, and new supplier and contract factors. Supplier and contract issues might include a fresh look at the product from the standpoint of creating a new business brand or image. This update reviews the financial viability of the product at this point in development, including initial project costs, return on investment over its life cycle, discounted cash flows and net income, life cycle costs, and other financial information that may be useful in the gate decision to proceed to launch. Financial risk is reviewed to update any new information on financial impacts, e.g., project costs and net present value, from changes in product development.

Update market definition

Market definition is updated to ensure that the market profile has not changed in design, and that market launch is based on the effective market definition. Market definition update addresses changes in:

- Market history
- Customer base
- Market size
- Market location
- Conditions of use
- Pricing factors
- Monitoring market launch
- How product is currently positioned
- Market placement
- Market volumes
- Risks in brand impacts
- Contingencies

Locking in product design

This task *locks in* product design for market launch. The freezing, documentation, and release of the final product design require a comprehensive review of design changes to ensure that the design does not change during production and market launch. Along with freezing product design, production tooling is also locked in because production tooling must be aligned with product design.

Freezing product design assumes that there has been a documented *change control and configuration management* system in place that has recorded changes and documented them through a formal configuration management system. If such a formal system is not in place, this task must perform the control and documentation function before market launch begins. Documentation of the product design accomplishes the following objectives:

- Ensures that all product design change orders have been clearly stated and agreed to by stakeholders
- Ensures that all product components have been documented in a final bill of materials, with product numbering and coding, supplier information, specifications, assembly and production guides, and testing results

This task can uncover open issues that must be resolved before the design is frozen. For instance, changes in design may be necessary as a result of testing that revealed revised needs, early design errors, omissions, and customer or engineering afterthoughts.

This task brings closure to these issues, and facilitates agreement among all key parties and stakeholders on the final design.

Final Regulatory Approval

Final regulatory and safety approval involves confirming what regulatory requirements are applicable in getting the product to market. The supplier must be an integral part of this decision. Any remaining major regulatory issues are raised at this point so they can be addressed and resolved with the appropriate regulatory agency. For instance, if an avionics instrument is being developed and the Federal Aviation Administration is considering a new regulation that might affect the new product, project and company management have to make decisions that protect them from noncompliance after the product is launched.

Reconfirmation of the equipment performance requires a full review of equipment specifications both for the product and for packaging and distribution. Equipment specifications are reviewed for changes, including the following. A final agency approval template is shown in Table 5-13.

TABLE 5-13 Final Agency Approval Template

Market execution region/location	Applicable agency	Applicable regulation or requirement	Compliance confirmation documented	Comments and corrective actions
Middle East-Israel	Consumer Product Commission	Commission must test product for consumer safety	Required proposal	
Europe-France				
Pacific Islands				
Caribbean				
Asia-China				

Final production transition and scheduling

This step involves preparation for final production and inventory operations that produce the product for distribution. Manufacturing engineers work to produce a process that maximizes quality and reduces defects in product. Full-scale production is planned around volume, shipping, and packaging requirements. What raw materials and logistics must be addressed? What user manuals and instructions are applicable?

Reconfirm Final Business Case

This is one more pass through the business case to prepare for project review. This task finalizes the business plan and business case to reflect any changes from beta testing results, or from analysis of business or brand impacts from the last business plan update. This final review of the business plan anticipates approaching market execution and thus must ensure that the business plan reflects all changes up to this point in product development. Special attention is paid in this task to financial performance, updating forecasts of cash flow and net present value of the product over its life cycle.

Issues reviewed for *freezing the business plan* include market definition, competitive assessment, marketing positioning, customer-internal linkage, brand/trademark linkage, commercialization plan, regulatory/agency plan, financials, intellectual property plan, financial performance, and strategy for supplier. All of these task outputs are once again reviewed to ensure that any changes or different perspectives on the product are reflected in updated business planning. These updates might include new forecasts of product pricing and financial performance, new customer and market information on how the product might induce new demand, new commercialization issues, e.g., through the use of the Internet, and new supplier and contract factors. Supplier and contract issues might include a new look at the product from the standpoint of creating a new business brand or image. This update reviews the financial viability of the product at this point in development, including initial project costs, return on investment over its life cycle, discounted cash flows and net income, life cycle costs, and other financial information that may be useful in the gate decision to proceed to market launch. Financial product and component risk is reviewed from the risk assessment to update any new information on financial impacts, e.g., life cycle product costs and net present value, from changes in product development.

Supply chain strategy

The purpose of this high level plan is to outline initial issues in the supply chain. This strategy covers all suppliers involved in production, distribution, and marketing. Potential suppliers are identified, and qualifications and reliability confirmed, especially with respect to company quality control and configuration management requirements. The plan also addresses initial component sourcing, supply terms and conditions, supplier chain issues, supplier selection processes. Potential supplier quality assurance plans are reviewed and training needs identified.

Update market assessment

The market assessment is updated by addressing how the competitive environment has changed, what channel opportunities are available, and what market forces and factors may have changed. Changes in market conditions and economic and social factors can upset the best marketing plans. These changes can include interest rate changes, political and social upheavals in foreign countries, unanticipated government regulations and requirements, safety and public health accidents and events, and changes in business planning.

First Article Review

First article review is a manufacturing engineering function, usually carried out in partnership with the supplier, confirming that the first manufactured product meets product design specifications. A full review of the first article is conducted before volume market launch units are produced. The manufacturing engineer and the manufacturing operations manager manage production and review of the first manufactured article. The manufacturing process itself is also reviewed to ensure manufacturing process integrity in preparation for volume production.

Prepare manufacturing operations plan

The manufacturing operations plan includes a process flow diagram of the manufacturing process, how final product design will be translated into parts and inventory (configuration management), inventory, key assembly stations and equipment, applicable industrial standards and controls, the assembly team, a training program for assemblers, and a production schedule for the first article.

Produce first article

The first article is produced in a controlled production line with manufacturing engineer oversight. Each step is documented with quality control and assurance issues, assembly problems, and other documented information about production.

Final financial performance analysis

The financial performance data from earlier analyses are updated to again ensure that the product's financial contribution to the company bottom line is dimensioned. Rate of return, break even point, net present value, cash flow projections, market and pricing information are revisited. Sensitivity analysis is performed to ensure that all the key factors that contribute to financial performance are identified.

Final Logistics Plan for Market Launch

This task involves updating the plan developed earlier for servicing and maintaining the product over its entire life cycle, plus support systems for each execution program. The plan will include a program for preventive maintenance and normal repair or service calls based on the projected reliability standards for the product. The plan includes a listing of anticipated service requirements based on design and development factors such as life cycles for components and supplier recommendations for servicing specialized parts and materials. A listing of required replacement parts is prepared to guide inventory management to support maintenance.

A training and development program is developed to ensure that adequate numbers of maintenance contractors are available to implement the service and maintenance plan worldwide.

Because infrastructure includes resources and support systems across the board, updating the plan takes account of changes in product design, specifications, reliability, market launch strategy, remote service technology, and other factors that impact that infrastructure. This update activity reviews the technology involved in maintenance, as well ensuring that the appropriate technology and infrastructure are available for servicing the product.

Prepare listing of infrastructure and support needs

The listing of infrastructure needs is categorized and then content provided based on market execution locations and plans. A generic listing involves service and support requirements and ensuring that all requirements are ready for market execution. Infrastructure includes all the equipment, resources, teams, product market units, agency approvals, and contacts, and company, sponsor, and local support systems necessary to execute in market. This listing can be seen in Table 5-14.

TABLE 5-14 Infrastructure and Support Needs Plan Template

Function	Resources	Equipment	Teams and key personnel	Company and agency support
Packaging	Materials	Tooling	Trained personnel	Staffing
Distribution				
Installation				
Promotion and advertising				
Agency approvals				

Prepare checklist for each market location

A checklist is prepared for each market location, tailored to that region and anticipated infrastructure needs unique to the area. For instance, in a desert location, special needs for equipment protection from dust and sand would be included.

Market Launch Plan

A final market launch plan is prepared that finalizes the locational, channel and distribution, logistics, and pricing strategies. Reviews are made of sales force requirements, training, promotion, incentives, and roll out. The following issues are reviewed to ensure market success.

1. Have marketing goals been determined for the product? Have all marketing, advertising, geographic, channel, distribution, logistics, and pricing goals and objectives been determined and confirmed by management?
2. Has a product manager been designated to manage the product into the marketplace? Has the product manager been part of concept definition and full development so that there is complete understanding of the history of the new product?
3. What kind of sales force is available for the product, and how will the sales force be trained and deployed? Does the sales force have experience or are they "winging it?"
4. Has product value and superiority been confirmed, including product features and functionality that differentiate the company in the marketplace?
5. Has the competition been assessed, and has the process avoided "showing your hand" too early in marketing and distribution?
6. Has the customer been involved in the product design and development, either directly or through focus groups and surveys that tested concepts on real users?
7. Have product support and maintenance issues been resolved?
8. Have contingency plans been developed to assure that risks can be offset?
9. Is the right team in the field to accomplish effective marketing, e.g., experienced sales and marketing people who know the customers and know the product value?

Market positioning

This task is designed to do a final forecast of what it will take to position the product for the target market. This task includes gaining a preliminary understanding of how the product or system will be launched in target markets. Sales force needs, customer/consumer service support, and resource requirements are outlined. The potential distribution channels are identified, as well as potential challenges to reach the target market. Opportunities to leverage existing channels of distribution and channel control issues are considered. Brand issues are resolved so that it is clear how the product fits into current brand structure. Potential marketing, communication, and promotional requirements are addressed, including internal communication needs, trade incentives, packaging and graphics, marketing collateral, and timing. Launch costs and resources are explored, as well as training needs.

Manage product marketing

Marketing must be managed and controlled. A project manager addresses the product marketing process from a cost, time, and quality standpoint, with resources under control. This perspective on marketing will emphasize management, resource, scheduling, control, and measurement issues involved in marketing a product. A new product development and marketing manager must not only understand the basics of new product marketing, but also *make decisions* during the process. These decisions often involve data and information that members of the project team may not consider. This is because a new product development and marketing team typically focuses on the product and the marketing issues and not on time and cost factors.

Service and product

Remember that we are addressing new product marketing from the standpoint of product *and* service. A new product involves a physical consumer or commercial good or asset, but a new service involves a new service, such as a new Web-based maintenance support system for cell phones, or a new adult education or advanced degree program delivered through interactive, satellite radio.

Market launch planning

A market launch program requires a disciplined schedule and budget to avoid long and costly marketing campaigns that do not pay off in successful financial performance. In other words, market launch tends to be a costly process if not managed and controlled.

A different project team

Market launch should be conducted, managed, and driven by marketing considerations, not product development factors. In other words, the flow of this process moves from design and development to marketing. Therefore, it is logical that this process will be directed by a team dominated by marketing, sales, logistics, and distribution staff but managed by the project manager. This means that there is a natural handoff, or interface, from development to marketing at this point as a product moves successfully through project review and is authorized for market launch.

A Marketing Launch Plan

The actual plan is developed by the new product development team, which now involves marketing, sales, and logistics staff in addition to a residual staff from early design, development, and configuration management activity. Marketing objectives are framed in terms of market period, product distribution, sales, customer satisfaction, and information goals. Market segments and demographics are included to confirm earlier market research in the development process. Product distribution goals are described in terms of how and when products will be produced and moved to marketing locations. This will require a distribution schedule prepared by logistics or distribution staff, with specific start and end dates for the market launch period.

Key Role of Experienced People

As discussed in Chapter 1, the introduction of a new product into grocery stores in the 1980s—the plastic sack—demonstrates what is really important in market launch people. The key issue in this case was the dedication and commitment of field personnel and strong headquarters leadership to *sell* the new concept to stores and convince them to commit. Their experience demonstrates the importance of the energetic leadership of a true entrepreneur and a sales force with "sleeves rolled up" to do whatever needed to be done to prove the benefits of the new system to stores. The story also illustrates the need to anticipate the key processes through which a product is dispensed and used when designing and launching the product.

Market Scheduling

Table 5-15 shows the schedule for a typical market launch program with linked tasks, resource assignments, and duration estimates.

Each of these tasks represents time and costs factors in controlling product launch. Note that the schedule includes 22 days for load and unload. This estimate may have come from logistics staff based on anticipated risks in locating load and unload locations and possible accidents and risk event factors. The "safety" factor here might be 50 percent, that is if the logistics staff had been asked to estimate duration on the basis of 50 percent probability, e.g., what is the most optimistic estimate that you might make with 50 percent probability. The answer might have been 20 days. Savings in time here would be 12 days if the optimistic estimate is achieved. It is this kind of ratcheting down of estimates that occurs here because each participant in the process will protect themselves with padded estimates based on anticipated risk. This is also why a risk matrix is a valuable component in market launch.

TABLE 5-15 Market Launch Program

ID	Task Name	Duration	Start	Finish	2nd Quarter
1	**New Product Project Distribution**	**65.5 days**	**Wed 3/16/05**	**Wed 6/15/05**	
2	**Plan**	**11 days**	**Wed 3/16/05**	**Wed 3/30/05**	
3	**Scope**	**11 days**	**Wed 3/16/05**	**Wed 3/30/05**	
4	**Contract**	**6 days**	**Wed 3/16/05**	**Wed 3/23/05**	
5	Write contract template	2 days	Wed 3/16/05	Thu 3/17/05	
6	**Negotiate contract**	**4 days**	**Fri 3/18/05**	**Wed 3/23/05**	
7	Identify contractor	4 days	Fri 3/18/05	Wed 3/23/05	HR[25%]
8	Schedule negotiation	3 days	Fri 3/18/05	Tue 3/22/05	Bruce[50%]
9	**Technical Specificaton**	**6 days**	**Wed 3/23/05**	**Wed 3/30/05**	
10	Packaging Requirement	4 days	Wed 3/23/05	Mon 3/28/05	
11	Distribution Requirement	2 days	Tue 3/29/05	Wed 3/30/05	
12	**Implement Distribution**	**54.5 days**	**Thu 3/31/05**	**Wed 6/15/05**	
13	Logistics	3 days	Thu 3/31/05	Mon 4/4/05	
14	Transfer	15 days	Wed 4/13/05	Tue 5/3/05	Lateisha[75%]
15	Load and Unload	22 days	Wed 5/4/05	Thu 6/2/05	Bill
16	Point of Sale Set Up	10 days	Wed 5/4/05	Wed 6/15/05	Kim

Table 5-16 shows the work table for tasks, a table which can be used to track costs of a new product project. "Work" indicates how many person hours are needed to complete a task, given current costs to date. The work table includes baseline information so that the project manager can compare progress to the original estimate, the baselines. You can see large variances in this report,

TABLE 5-16 Work Table for Tasks

Resource name	Cost	Baseline cost	Variance	Actual cost	Remaining	Details	W
Unassigned	$0.00	$0.00	$0.00	$0.00	$0.00	Work	
Write contra	$0.00	$0.00	$0.00	$0.00	$0.00	Work	
Packaging R	$0.00	$0.00	$0.00	$0.00	$0.00	Work	
Distribution	$0.00	$0.00	$0.00	$0.00	$0.00	Work	
Logistics	$0.00	$0.00	$0.00	$0.00	$0.00	Work	
Joe	$26,000.00	$8,000.00	$18,000.00	$450.00	$25,550.00	Work	8h
Plan	$4,200.00	$4,000.00	$200.00	$200.00	$4,000.00	Work	
Implement D	$21,800.00	$4,000.00	$17,800.00	$250.00	$21,550.00	Work	8h
PM	$5,880.00	$5,600.00	$280.00	$280.00	$5,600.00	Work	
Plan	$5,880.00	$5,600.00	$280.00	$280.00	$5,600.00	Work	
Kim	**$36,120.00**	**$16,800.00**	**$19,320.00**	**$1,400.00**	**$34,720.00**	Work	16h
Implement D	$30,520.00	$11,200.00	$19,320.00	$560.00	$29,960.00	Work	8h
Point of Sale	$5,600.00	$5,600.00	$0.00	$840.00	$4,760.00	Work	8h
Barry	$30,520.00	$11,200.00	$19,320.00	$560.00	$29,960.00	Work	8h
Implement D	$30,520.00	$11,200.00	$19,320.00	$560.00	$29,960.00	Work	8h
Lateisha	$6,300.00	$6,300.00	$0.00	$630.00	$5,670.00	Work	
Transfer	$6,300.00	$6,300.00	$0.00	$630.00	$5,670.00	Work	
Bill	$12,320.00	$39,200.00	($26,880.00)	$0.00	$12,320.00	Work	8h
Load and Un	$12,320.00	$39,200.00	($26,880.00)	$0.00	$12,320.00	Work	8h
Bruce	$840.00	$14,000.00	($13,160.00)	$0.00	$840.00	Work	
Schedule ne	$840.00	$0.00	$840.00	$0.00	$840.00	Work	
Tom	$0.00	$4,000.00	($4,000.00)	$0.00	$0.00	Work	
equip1	$0.00	$0.00	$0.00	$0.00	$0.00	Work	
equip2	$0.00	$0.00	$0.00	$0.00	$0.00	Work	
HR	$560.00	$0.00	$560.00	$0.00	$560.00	Work	
Identify cont	$560.00	$0.00	$560.00	$0.00	$560.00	Work	

suggesting major problems in actual work and costs incurred, versus the original baseline plan.

Risk-Based Scheduling

At this point we need to identify the high-risk tasks uncovered in development and adjust task durations if necessary to reflect new information. Our purpose is to use risk assessment and analysis information to calculate a *risk-based* project schedule in Microsoft Project. The risk-based schedule is calculated from your original project schedule, but uses your weighted estimates of three possible task durations (expected, pessimistic, and optimistic) to come up with a new project schedule. The new schedule is calculated for individual tasks and "rolled up" to the whole project.

Procedure

The risk-based schedule is usually a better schedule estimate than your original one because it reflects your best estimates of what could go wrong (risk) and what could go right (controlling risk). Here is the procedure:

1. Prepare your regular project schedule using Microsoft Project. Use your best estimates of task durations and linkages. This project schedule does not reflect any risk assessment.
2. Prepare a risk matrix. Using the work breakdown structure (WBS) and the project schedule, rank *all* project tasks in terms of risk, designating them high, medium, or low.
   a. A high-risk ranking shows a high probability (>50 percent) of the risk actually occurring, and that the risk will have a relatively severe impact on schedule, cost, and/or quality.
   b. A medium-risk ranking implies less probability (<50 percent) of happening and less schedule impact
   c. A low-risk ranking implies very low probability (<10 percent) that the task will occur and low impact.
3. Select the five highest task risks (or more if you have more tasks that present risks that you want to reflect in your risk-based schedule).
4. Calculate the risk-based schedule. Your objective now is to calculate a risk-based schedule by taking each of the five highest-risk tasks and calculating a risk-based duration for each. Using Microsoft Project, follow these steps:
   a. Pull the PERT Analysis toolbar up from the View pull-down menu
   b. Highlight one of the high-risk tasks on the Gantt chart
   c. Go to the PERT Entry Form and enter your duration estimates for that task for three scenarios—expected (use the duration in your original schedule), pessimistic (worst case impact if risk occurs), and optimistic (best case, all risks controlled with no impacts)

d. Use the PERT Weight button to set the weights for each scenario (weights reflect the probability that a given risk and impact will happen). Microsoft Project uses a total weight scale of 6 points; your job is to divide the 6 points up among three scenarios—expected, pessimistic, and optimistic. Note that the Microsoft Project "default" is 4 for expected (based on the high probability that the actual duration will fall somewhere between the two extremes) and 1 each for pessimistic and optimistic. You may want to change those weights based on your estimate of the relative probability that a given scenario is going to happen.
e. After you have entered weights, go to the PERT Calculation button and calculate the risk-based duration for that task based on your inputs.
f. Click the PERT Entry Sheet to see the newly calculated risk-based duration for the task compared to the three scenario durations (expected, pessimistic, and optimistic).
g. Repeat this procedure for the remaining high-risk tasks.

The resulting "rolled up" schedule is now a risk-based schedule, reflecting a new project duration.

A Note on Microsoft Project PERT and Risk Matrix Terminology

Microsoft Project uses the terms *pessimistic, expected,* and *optimistic.* "Expected" usually means the duration you originally estimated without concern for risk, although it may not. "Optimistic" means the risk and impact are low, and you think you might be able to "beat" the expected—you are optimistic about it. "Pessimistic" means that the risk and impacts are high,

TABLE 5-17 Risk Matrix Comparison

Risk ranking in risk matrix	High (risk severity is high and probability is high that it will happen)	Medium (risk is moderate and impact not so severe)	Low (risk is low and impact low even if it occurs)
Microsoft Project Terminology	Pessimistic (duration reflects concern based on probability that risk will occur and will have major adverse impacts and slip the schedule)	Expected (original estimate of duration without considering risk, unless there is a reason to change it)	Optimistic (duration reflects low risk and therefore "hope" that the risks can be controlled by contingency plans and the task can be completed quicker than expected)

and you don't think you will be able to make the expected duration—you are pessimistic about the expected duration based on risk. In the Risk Matrix you use the terms *high*, *medium*, and *low* for risk rankings. In general, "high" is over 50 percent probability and high severity; "medium" is less than 50 percent probability and moderate severity; and "low" is less than 10 percent probability and low impact.

Table 5-17 compares the terms from the risk matrix and your PERT analysis.

Chapter

6

New Product Development in Consumer Products and Electronic Instrumentation

Special Challenges in Electronic and Computer-Based New Product Development

The management of new product development involving electronic and computer-oriented consumer and system products is a special case because of the challenges created by high technology and complex systems. This chapter explores the process of producing new products in this field and especially the management of project risk and the big picture. By "the big picture," we mean the high level, external factors that often lead to new product project failure despite the fact that the product itself performs as designed in development. These factors are often external to the project team—in the company organization and culture, or in the global marketplace. They are often missed by the technical and engineering personnel in the project team because of the tendency to fix narrowly on the product itself and its design and function, and not on the the company's success in getting it successfully to market.

Technical new product development in this field can go wrong because of several factors:

- Developing and producing a new product that functions as required but cannot be produced, distributed, marketed, and supported successfully. This is the function of program management as opposed to project management. Program management sees the product through all of its cycles from requirements to sales and support, while project management may tend to take a narrow view within the bounds of cost, schedule, and quality.
- Lack of top management support, e.g., if the new product does not have a top manager or sponsor who can *steward* the product, it may not get the attention and resources it needs to succeed.

- Organizational mismanagement and lack of good administration, e.g., because new product development involves innovative and creative people who often do not value routine procedures, administration, time, and cost, program management can lose control
- Misalignment with business plans, e.g., offline from the company's real strengths, i.e., sometimes the ideation process produces product concepts that are simply too far removed from the company's *core competence*, and therefore a whole new system of thinking and new workforce resources are necessary to implement, often unsuccessfully
- Keeping marketing out of the process until too late, e.g., excluding marketing and sales issues and commitments from the product design and development process, or vica versa, marketing goes out too far to commit to a product unproven and untested
- Letting engineers run projects without training in management, e.g project *management by accident*, the process of appointing project leaders because of their technical skills rather than their management, leadership, and communication skills
- Too many new product projects in the pipeline at once; excessive multitasking because of an overly aggressive management posture

Missing the forest for the trees

The strong tendency of technical and technology professionals, including engineers, is to focus on the details of a system or product without seeing the big picture. In fact, these professionals often don't want to see the big picture because their focus is producing a product that meets higher and higher product requirements. Experience in this industry suggests that project engineers and technicians get caught up in task and technical details; indeed, that is their job. But the focus on internal product components and performance sometimes leads to overdesigning products to meet requirements beyond customer or system need. This is especially true in today's economy because technology development has produced higher and higher performing components that may be inconsistent with slower developing client needs.

Top management support

We have seen the results when finely tuned new products, well planned and scheduled, fail because of lack of management support and sponsorship. This means that although the project may be supported and needed by the customer, the project organization itself is not committed to it. When inevitable problems surface in new product projects, top management must be prepared to intervene and assist in problem solving, especially if resources and customer satisfaction are involved.

Organizational mismanagement

New product development efforts are notoriously late and undermanaged, often the victim of bad organization alignment, excessive multitasking, and poor division of labor. Although a project may proceed on time and within budget, it may be scuttled by internal conflict and coordination problems created by weak upper management and inflexible organizational structure. Project managers must protect their projects from this development by making sure there is a top management sponsor for the project, and that organizational accountability for results is placed in the proper hands. Further and maybe most important, the organization must have a project management system in place to guide the process, to account for time and cost, and to make key go or no-go decisions about the product along the way.

Misalignment with business plans

Many a new product is out of sorts with the business plan and competency simply because the project was sold without consideration of internal capacity to perform. This development can lead to failure, despite the best project management and new product performance. To offset this tendency, project managers establish close ties with corporate plans and strategies, and ensure their projects are linked into planning and budgeting processes.

Keeping marketing out

Marketing should *pull* new product development products through the process to the market launch, based on their access to real data on market and customer demand. If marketing is involved, the reality of new product success will be reflected in team decisions; without marketing, projects can get "handed off" that have no potential for successful marketing and sales. On the flip side, marketing and/or sales makes commitments to customers in the field that cannot be achieved in product design and development unless marketing and development interface constantly during the process.

Project management by accident

Engineers are not trained to manage, and they often do not want to manage. They often become project managers by accident, e.g., by default. Project management training can help to empower engineers and specialists to see the big picture. Project managers can see the broader issues of organizational management, product marketing, sales, and company alignment, and are trained to communicate with each other and management on project progress. Without such training, engineers can lose track of the key factors in product success.

Focus on task durations

Eliyahu M. Goldratt has written a seminal work on critical chain theory, a new approach to production and project planning and scheduling. In his book

(*Critical Chain: A Business Novel,* The North River Press 1997) he states that in projects where products and outputs are uncertain, e.g., new products, the focus should be on starting times, not finish times, and that most individual task duration estimates should be taken lightly. Original duration estimates should be cut in half to avoid "safety" issues created by overstating tasks to protect from failure, and all work should proceed on start dates. Work should progress as fast as it can, and managed to completed work.

Too many projects in the pipeline

Multitasking creates time delays, costly mistakes, and quality problems in outputs because of the inability of team members to do many jobs at once. Goldratt says that new product projects should be sequenced and completed one at a time, not in parallel with each other. Whatever the approach to deciding when to introduce new product projects into an already busy workforce, a company needs to develop a way to assess the capacity of the workforce to take on new work. The most effective way to guage the readiness of a team for new product work is to ask them if they can handle it. The best indicator of an overworked team is a high incidence of errors in routine tasks.

Project Risk Management

Risk management is at the heart of good new product development. Good project risk management in new product development requires substantial organizational capacity to handle technical, risk-related information and data, and to calculate schedule, quality, and cost impacts of various risk events. As projects become more and more complex, risk management requires an effective project management software program—and supporting network systems to allow for exchange of data and analysis. Microsoft Project provides a good base for project planning and control, and for scheduling and costing out risk mitigation actions. But the organization needs a network system with a workable directory so that project teams and stakeholders can access and communicate timely risk data.

In addition to its usefulness in documenting contingency tasks as a part of the baselined schedule, MS Project's PERT analysis allows for estimating alternative scenarios and impacts on schedule and cost. The PERT analysis provides a template for placing weights on three scenarios: expected, optimistic, and pessimistic, as well as estimating durations.

The product development process and risk

New product development means risk simply because creating a new product generates uncertainty at every turn. The new product development process typically produces designs and prototype products for manufacture and sale. Product requirements are translated into engineering design specifications, drawings, and so on, through the product development process. The process is oriented to develop, control, verify, and validate the design as it progresses to completion. There is risk in design, prototype development and testing, configuration management, and manufacture. But the most critical risks are in the management of the process, producing costly work that does not produce good information for the go or no-go decision.

The most critical upfront technical risks—requirements and design risks—are triggered by the potential for failure in understanding customer requirements and in failing to validate the design. In other words, the requirements will be wrong thus the design is irrelevant, or the design will not align with requirements.

Risk management in product development: Embedded verification and validation

Design involves two major risks: a product prototype will not align with the requirements/specifications for the product, and even if there is alignment with specifications, the product will not work in a user setting. Product development involves these key risk concepts that underlie the project risk management process, verification, and validation. Verification is the process of ensuring that a product design meets the specification for the product, e.g., that the design will do what the specifications would have it do. Validation is testing the product in a user setting to ensure that the product meets customer needs. Each of these risk strategies should be integrated into the product development process to yield useful data on both risks. Thus product development is a good example of the integration of risk into the project process.

Stages in Product Development in Electronic Instrumentation

Here are the stages in instrumentation development:

Stage 1—Requirements definition

Stage 2—Detailed design

Stage 3—Prototype development (design verification)

Stage 4—Design validation

Stage 5—Production transition

As stated earlier, project reviews are scheduled at the end of each stage to control risk and make go or no-go decisions. Each of the stages has distinct entry and exit criteria and risk exposures; however, each stage will overlap in time,. For example, the detailed design activity can and will begin on certain parts of the product before all of the requirements are known and defined.

Design and development plans are prepared and updated as the design evolves. These design and development plans include risks, risk matrix information, and contingencies. This way, risk is *directly integrated* into planning, not a separate process. These plans guide the design and risk management process, serve as the basis for defining all design tasks, and establish program schedules that include contingencies for each design activity.

Steps in Product Development

The five steps of the product development process are described in more detail in the following sections.

Step 1: Requirements definition

The purpose of product design is to define:

- The product setting—how the product interfaces with a system
- Functional requirements—performance standards
- Product architecture—design factors
- Hardware requirements—the platform and system configuration
- Software requirements—software and network
- Design validation requirements—how the product performance will be ensured
- Test equipment requirements—how the product will be tested

Each program and/or project develops product requirement specifications based on performance requirements that stem from customer needs, expectations, and wants. Here is where conflicts are resolved when it is apparent that the product cannot meet all requirements. This involves the risk assessment process, identifying and resolving potential requirements, conflicts, and risks at each step. Product requirement specifications are then converted to product and subsystem design and performance specifications. The specifications are reviewed to ensure adequacy and to satisfy applicable regulatory and statutory requirements.

Software such as *Requisite Pro* is used to document requirements. This is necessary to establish the basis for tracking and tracing requirements to design and testing later in the process.

The concept of *release* is important in new product development; it is the point at which the design process task manager or engineer verifies that a product meets requirements and can proceed to the next step. As requirements are defined, individual subsystems may be released into detailed design. Product design culminates with clear definition of the requirements.

Step 2: Detailed design

There are inherent risks in handoffs from one stage to another, thus the importance of design reviews before entering the next stage. The purpose of detailed design is to convert the product requirements into graphic design and/or hard copy drawings used to produce the product. Detailed design for each subsystem is started when the requirements for the subsystem are understood. The risk here is that the design will commence before requirements are understood, or that requirements fundamentally change during design.

Detailed design also produces acceptance test requirements, analysis documents, support requirements and interface control documents. Detailed design works to enable product manufacturing, and validates performance against design input requirements. Design reviews ensure that all requirements are addressed, and that all the documentation is available for project review at the end of the process.

Step 3: Prototype development

The purpose of prototype development is to produce a model of the *real* product. Prototype development begins with the release of completed parts, subassemblies, and assemblies to configuration management, and includes the following:

- Procurement of material for the prototype product under development
- Manufacture of prototype product(s)
- Test and integration (hardware and software) of the prototype products
- Special development tests (as required)
- Development of the Acceptance Test Procedure (ATP)

Configuration management is a risk management action in itself; the function addresses the highly probable risk that in changing designs toward development of a manufactured product, product components are not adequately described in detail and changes in design are not controlled by a separate risk management function. The ultimate risk is that the final product works, but is not documented cannot be described for purposes of manufacturing and assembly.

Procurement of prototype material and manufacture of prototype products are performed by the purchasing and manufacturing departments. The product development organization provides definition of the material to be ordered and the product to be manufactured through drawings generated as part of the detail design process. The risk here is that suppliers are not found who can produce the required components at the quality level specified and who do not document the component properly.

Prototype development and design verification ensure that the design fully meets the design input requirements. Of course, if the input product requirement is wrong, the prototype will be wrong as well, hence the importance of getting the input requirement right. Production design is the design that will be used by production to create inventory and produce the market version of the product. Verification is accomplished by testing and/or analyses, and through the design review process (where appropriate) wherein design concepts/approaches together with the actual end item, product, subsystem, and elemental level designs are progressively evaluated and approved.

Step 4: Design validation

The risk here is that the product will not work to meet user needs, even though the design has been verified. The purpose of design validation is to ensure that the product *in operation* meets all of the requirements imposed by the applicable specifications. Hardware and software validation of a production product are governed by the appropriate industry standards. The specific requirements for a given product are contained within the systems requirement documents. For test equipment, validation is performed to an internally generated document to ensure that the equipment meets the standards required for thoroughly testing the production product.

Step 5: Production transition

The risk in transition from development to manufacturing is that the product is not defined and/or cannot be assembled and manufactured. The purpose of production transition is to facilitate a smooth transition from the development of the product to production scheduling. The production transition stage ensures that:

- All documentation required to produce the product in a production environment is released and is up to date
- All test equipment, cables, test software, and fixtures are available and documentation is released
- Knowledge of any special manufacturing, test, or inspection criteria is defined
- Training on all aspects of the production product and test equipment is complete, and production personnel have a working knowledge of the equipment and test methods used to test/troubleshoot the equipment

The production transition stage concludes with a successful Production Readiness Review.

Risks in Organizational and Technical Interfaces

There is major risk in the human interactions that occur in a product development process. The structure and membership of the program teams must be clearly defined and documented to identify individual and team responsibilities, task assignments, and technical competencies. Project management must ensure timely and complete transmittal of information among team elements, as well as to and from outside functional organizations supporting the team. Handoffs from functional experts to each other and to the project manager must be performed effectively. Program managers, department managers, and system engineers review documented information, and remedial action is taken to correct any data discrepancies as well as any deficiencies in design.

Design changes

All product configuration changes are documented, reviewed, and approved by authorized personnel prior to implementation through a defined document change and control process. Engineering change notices will be reviewed to identify the need for first part inspection.

Design review and risk

There is inherent risk in design review because of the potential for missing design flaws that will impact on performance and testing, and therefore on schedule and cost. Design risk is addressed in design reviews; risk is why design reviews are completed.

Risk reviews

Interim reviews during development of electronic instrument products help to prepare for go or no-go decisions made in project review. Sometimes individual technical risk reviews are scheduled to keep the product on target, as follows:

Preliminary Design Risk Review	(PDRR)
Critical Design Risk Review	(CDRR)
Production Readiness Risk Review	(PRRR)
System Design Risk Review	(SDRR)
Test Readiness Risk Review	(TRRR)
Task Level Requirements Risk Review	(TLRRR)
Task Level Design Risk Review	(TLDRR)

Table 6-1 defines the various levels as well as reviews to be conducted on each level.

TABLE 6-1 Reviews in Development Process

ID	Task Name	Duration	Milestone
1	**STAGES IN INSTRUMENT DEV**	**89 days**	
2	**Requirements Definition**	**30 days**	
3	Requirements Development	30 days	
4	System Design Risk Review	0 days	9/3
5	**Detailed Design**	**30 days**	
6	Design Details	30 days	
7	Prelminary Design Risk Revie	0 days	10/12
8	Critical Design Risk Review	0 days	10/12
9	Task Level Risk Review	0 days	10/12
10	**Prototype Development**	30 days	
11	**Design Validation**	**30 days**	
12	Validate Design	30 days	
13	Test Readiness Risk Review	0 days	11/23
14	**Product Transition**	**10 days**	
15	Prepare Plan for Production	10 days	
16	Production Readiness Risk Re	0 days	12/7

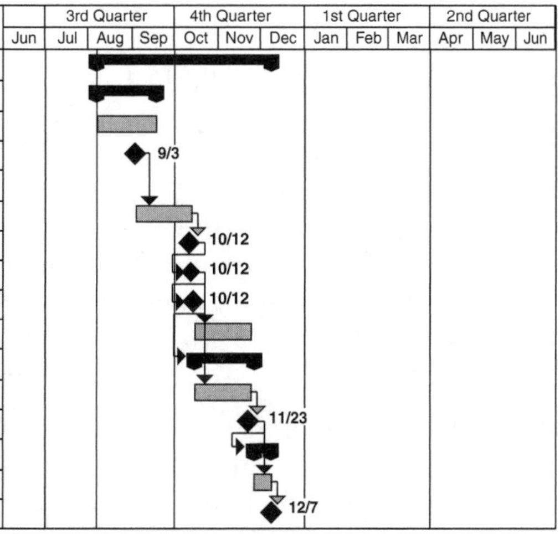

Timeline: 3rd Quarter (Jun, Jul, Aug, Sep) · 4th Quarter (Oct, Nov, Dec) · 1st Quarter (Jan, Feb, Mar) · 2nd Quarter (Apr, May, Jun)

Preliminary design risk review (PDRR)

The PDRR is a project-level review of the design concept for deliverable configuration items. The purpose of a PDR is to

- Evaluate the progress and technical adequacy of the design approach and the risk exposure
- Evaluate the compatibility of the design approach with internal and external hardware, test equipment, or facilities
- Establish the compatibility of physical and functional interfaces
- Evaluate risk associated with production and manufacturing processes

Additionally, a PDRR may be utilized to solidify schedules, finalize work responsibilities, and ensure compatibility of the design approach between corporate divisions.

Critical design risk review (CDRR)

The purpose of a CDRR is to

- Determine that the detailed design theoretically satisfies performance requirements
- Establish detailed design compatibility with internal and external hardware
- Document performance characteristics not addressed by the system specification
- Assess risk areas and develop risk mitigation plans
- Assess the producibility of the detailed design

Additionally, a CDRR may be utilized to address these issues between corporate divisions.

Production readiness risk review (PRRR)

This review is intended to identify specific issues that must be resolved prior to executing a production go-ahead decision. Multiple reviews may be required to satisfactorily resolve all issues. The PRRR will include identification of high-risk/low-yield manufacturing processes and manufacturing development efforts required by the detailed design approach. The PRRR will also address incorporation of design producibility improvements, production planning, acquisition of long-lead items, and facility allocation.

System design risk review (SDRR)

This review is conducted to assess allocated requirements and assess their optimization, completeness, and inherent risks. This will include a review of basic manufacturing considerations and detailed design stage engineering plans. This review should be conducted when system definition has progressed to the point where subassembly configuration items are identified. System design reviews may be utilized to document assumptions made in the flow-down of requirements to subassembly levels.

Test readiness risk review (TRRR)

A TRRR will be conducted prior to out-of-house product performance testing. The intent of this review is to ensure appropriate plans are in place before proceeding. This review will occur during the prototype development and design validation stages of product development.

Task-level requirements risk review (TLRRR)

The objective of this review is to ascertain the adequacy of task-level requirement definition before proceeding with detailed subassembly design. This review will be conducted after a significant portion of the system design and task-level requirements have been established.

Task-level design risk review (TLDRR)

This review will be conducted for each subassembly when detailed design is essentially complete. The purpose of this review is to

- Determine that the design under review satisfies applicable engineering and performance requirements
- Document performance characteristics not specified within the TLRR
- Determine task compatibility with other assembly items
- Assess risk areas on a technical, cost, and schedule basis

General Responsibilities

Program management is responsible for assessing the necessity of project- or phase-level reviews on a case-by-case basis, and for ensuring that risk management is integrated into the process. If necessary, project- or phase-level reviews are included in project- or phase-level plans and schedules.

Program management is responsible for ensuring that engineering, manufacturing, quality assurance, and procurement personnel participate in reviews as appropriate. Program management also coordinates the attendance of customers, subcontractors, and vendors as required.

Program management is responsible for establishing the time, place, and agenda for each project-level review. Additionally, program management is responsible for facilitating the review process by

- Scheduling reviews in coordination with the project schedule and the availability of necessary information
- Publishing the review agenda

- Reserving conference rooms
- Designating someone to record, publish, and distribute meeting minutes and action items

System-level reviews

System engineering is responsible for ensuring that engineering, manufacturing, and quality assurance personnel participate in system-level reviews as appropriate. System engineering is responsible for establishing the time, place, and agenda for each system-level review. Additionally, system engineering is responsible for facilitating the review process by

- Scheduling reviews in coordination with the project schedule and the availability of necessary information
- Publishing the review agenda
- Reserving conference rooms
- Recording, publishing, and distributing meeting minutes and action items

Task-level reviews

Task level reviews address project work at the task level, e.g., individual team member tasks such as a testing procedure, equipment application and software tool.

Function of task-level reviews

The various reviews discussed here are critically important in managing new product development because of what might be termed the *tyranny of small decisions*, e.g., the process through which each team member tends to make important design and development decisions on a new product without checking on the overall impact on product performance and project success. To offset this process inherent in new product development, reviews are conducted constantly to drill into such decisions and uncover impacts and contingencies.

Preliminary design risk review (PDRR)

General: PDRRs are conducted for each item or component in the product configuration under development. For instance, a team member may have discovered a new resistor concept in the Wiki (Internet search for new concepts) process and incorporated the design into the product. But the new resistor turns out to be more expensive than the budgeted resistor for this product, and has not been tested under the appropriate user conditions of this product. PDRRs are normally conducted during the detailed design stage of product development, after a design approach has been formulated to uncover risks. More reviews may be scheduled to address follow-up and unanticipated test results.

Review Items: Here are some of the targets for review:

A. Schedule; is the project on schedule?
B. Design requirement synthesis
   1. Interface requirements; are all component interfaces OK?
   2. Performance requirements; does the product/component perform to requirements?
   3. Applicable design standards; what are the design requirements?
   4. Configuration and document numbering; are all components documented in the configuration management software and appropriately numbered?
   5. Standardization; are there standardization opportunities?
   6. Design trade studies; what will be the product cost and financial impacts?
   7. Design to cost goals; is the product within design and production costs?
   8. Design for growth capabilities; can the product be enhanced later?
   9. Layout drawings and preliminary drawings; are all design documents in proper order?
C. Environmental requirements
   1. Similarity assessment
   2. Proposed verification plan; is there a plan to verify performance?
D. Producibility and manufacturing
   1. Preliminary manufacturing process; can the product be manufactured?
   2. Design concepts that require advancement of current capabilities; will current performance meet all concept requirements?
   3. Identification of tooling requirements; will manufacturing tooling be impacted?
E. Testability requirements
   1. Preliminary production test procedure; is the production test protocol prepared?
   2. Specific subassembly requirements to support production test; are there test articles available?
F. Reliability/maintainability/availability (R/M/A) requirements
   1. Reliability issues with similar designs; have there been reliability results that are relevant to this product?
   2. Preliminary assessment of high-risk, long-lead items; what are the high-risk items, either from a technical or procurement standpoint?
G. Electromagnetic compatibility requirements
   1. Electrical grounding scheme; if electrical systems are involved, is there adequate grounding?
   2. Signal isolation scheme; are signals tested?
H. Technical risk
   1. Project level cost/schedule implications; what are the high risks identified in the risk matrix?
   2. Preliminary risk mitigation plan; are there contingencies for each risk—incorporated into the project schedule and resources planned?

After the Review: The review document is prepared and communicated to all project stakeholders in a company network system.

Critical design risk review (CDRR)

General: A CDRR makes sure design meets requirements.
Review Items: Specific areas to be reviewed are as follows:

A. Engineering documents (layout files in appropriate software and drawings)
B. Design details
   1. Electrical design
      a. Functional/performance requirements compliance
      b. Power requirements and estimates
      c. Detailed schematics
      d. Testability
      e. Bill of material review
   2. Mechanical design
      a. Assembly
      b. Environmental specification compliance
      c. Electrical and mechanical interface compatibility
      d. Producibility and manufacturing
      e. Tooling provisions
      f. Bill of material reviews
   3. Reliability and Maintainability
      a. Temperature and other user conditions
      b. Component selection
      c. Mean time between failure estimate

System design risk review (SDRR)

General: The SDRR addresses the overall system in which the product performs.
Review items: Items to be reviewed at the system level:

A. System engineering management activities
   1. Functional analysis; how does the *whole system* in which the product is embedded perform?
   2. Requirements allocation; have all requirements been traced to a component?
   3. Environmental conditions; what are the external requirements?
   4. Subassembly synthesis; do subassemblies link OK?
   5. Standardization; are there standardization opportunities to reduct cost?
   6. System growth capability; can the system be extended later?
   7. Program risk analysis; what are the system risks?
   8. Producibility analysis and manufacturing; can the system be produced?
B. Results of significant trade studies:
   1. Cost: What are the costs and supplier issues?
   2. Cost versus performance; is the performance level attained worth the cost?

   3. Design versus manufacturing consideration; is the design fit for production?
   4. Common versus unique support equipment; can commonly available equipment be found?
   5. Size and weight; are size and weight requirements met?

C. Updated design requirements; have all design changes been incorporated?

D. Updated requirements for manufacturing methods and processes; are new production systems reflected in design?

E. Updated requirements for operations/maintenance personnel and training; can people be trained to support the product?

Test readiness risk review (TRRR)

General: The TRRR makes sure the product is ready for testing.

Review items: Items to be reviewed are:

A. Test procedures; are all test protocols prepared and approved?

B. Support equipment requirements; will appropriate equipment be available when necessary in project schedule?

C. Physical configuration of the unit under test; is the product configuration preserved in configuration management software?

Task-level requirements review (TLRR)

General: The TLRR is aimed at highly detailed tasks:

Review items: Items to be reviewed are:

A. Preliminary design schedule; is the schedule being met?

B. Design requirement synthesis
   1. Interface requirements; are key interfaces being addressed by team members?
   2. Performance requirements; is each component checked?
   3. Applicable design standards; what design standards apply?
   4. Configuration and document numbering; is each component numbered?
   5. Standardization; are there standardization opportunities?
   6. Layout drawings and preliminary drawings; are all drawings adequately done and documented by each team member?

C. Environmental requirements
   1. Comparison to current levels; does each component meet environmental requirements?
   2. Proposed verification plan; is there a verification plan for each team member?

D. Producibility and manufacturing
   1. Preliminary manufacturing process; process design improvements; has each team member verified production requirements?
   2. Identification of tooling requirements; has each team member verified tooling needs?

E. Testability requirements.
   1. Preliminary production test procedure; can the component be tested?
   2. Specific subassembly requirements to support production test; have appropriate team members working on interfaced components collaborated on testing?
F. Quality assurance provisions
   1. Design provisions for inspection; what should be inspected and should inspection be 100%?
   2. Identification of critical design features; what are the risk-based, critical design issues?
G. Reliability/maintainability/availability (R/M/A) requirements
   1. Reliability issues with similar designs; are there reliability issues that need to be addressed at the component level by individual team members or subteams?
   2. Preliminary assessment of high-risk, long-lead items; has each team member checked procurement availability?

New Product Software Development Risk

Software development in new product development and risk go hand in hand. This is because there are so many "points of risk" in the software design process, so many detailed decision, steps, and coding actions that can go wrong. This relationship between software and risk can be illustrated by reference to the theories and concepts of W. Ross Ashby (*Introduction to Cybernetics*, Chapman & Hall, 1956, ISBN 0-416-68300-2). His "law of requisite variety," a core concept of systems theory, indicates that the job of system design and control becomes more and more difficult as the complexity of the system and its interrelated parts increases. Risk can be said to be a function of complexity and control—the less a complex system can be controlled from one point, the more risk it generates. Software development is a good example of this concept.

As in Ashby's law, software development presents unique risks and challenges associated not only with the design and development process, but also with the complexity of the product and the potential gap between functionality of the product and customer or user needs, requirements, and expectations. Add to this system the natural creativity of software design and development, and you have a process inherently risky. This is why software design and development often includes embedded quality assurance and control to reduce risk at every key "point of risk."

The software development engineer is typically responsible for developing software, creating documentation to meet or exceed a requirement, and ensuring that each software development process step is accomplished. A so-called "certification" engineer is also responsible for reviewing and verifying software development products to ensure that requirements are testable and certifiable. The certification engineer is also responsible for developing and implementing cases and procedures that test code and requirements to specified levels, for capturing the test procedures and results in accordance with company procedures, and for ensuring that each verification process step is accomplished. In addition, a software quality assurance engineer is often responsible for monitoring the software life cycle process to ensure that it meets or exceeds the intent of this document.

Software design involves multiple handoffs and team interfaces, creating many risks. Control of software design is accomplished by regular reviews and by ensuring that software designers are fully integrated into the team.

Chapter

7

Quality, Six Sigma, and New Product Development

Quality and Process Improvement

The purpose of this chapter is to link new product development to quality and process improvement. New products create new processes and new business risks, and therefore new product development initiatives open up opportunities for continuous improvement and associated *Six Sigma projects*.

Six Sigma is a measure of quality that drives an organization to achieve near perfection through a *management-by-fact and data-driven* process that defines a defect as anything outside customer specifications. A defect in Six Sigma terms is six standard deviations between the mean and nearest specification limit. In new product development this process is often called DMADC (define, measure, analyze, design, verify), an improvement program used to develop new products and new processes at Six Sigma levels. The objective is to target customer requirements—nothing more and nothing less. The company goal is to avoid wasteful and expensive rework and processes, and produce a product as close to specification as possible the first time.

One application of this quality concept is called integrated test management. New product processes require complete test coverage and tracking systems to ensure that issues associated with the product that are discovered in testing are evaluated and resolved. This requires traceability, the capacity to link every performance specification with a test and data point. The point is to ensure that new products meet stated requirements and functional standards set by customer need.

The focus of quality in new product development is to get the customer requirement right because it drives everything else. Quality is not the highest performance you can achieve; it is what the customer "requirements." Looking at new products this way, you see new product development from the customer's perspective and recognize that the risk is not only that the product will not meet customer requirements, but also that you will get the customer requirements wrong to begin with. As the major stakeholder and *the* project sponsor, the customer/client *pays the price* at the end of a project if these risks are not well managed.

Customer-Driven Risk Management

A customer-driven, new product project team is a team that responds to customers and manages customer satisfaction as a regular team function. Risk management is an overall obligation of the customer-driven project team. The team continually assesses risk at the project level as well as in each task of the project, not only in terms of time and cost, but also in terms of the technical feasibility. Again, the customer-driven lead team establishes the system for risk management. This risk management system influences the use of the other project management tools and techniques.

No new product project is without risk; that is the probability that a given process, task, subtask, work package, or level of effort cannot be accomplished as planned. Risk is not a question of time; it is often a question of feasibility. It pays in the development of the WBS to assign risk factors to each element of the WBS, separate from the assignments of schedules and milestones for the work. Risk factors can be assigned based on uncertainty, technological feasibility, availability of resources, or competition. Later, the elements with high risk factors are given close attention by the project manager, whether or not they are on the critical path itself. Special attention is given to customers, users, and clients.

One of the most effective ways to deal with risk is to develop customer-driven contingency plans, or parallel courses of action that come into play if the task cannot be accomplished as the customer sees it. In other words, if Six Sigma cannot be accomplished, alternatives are available. Contingency planning is based on "satisficing," as Herbert Simon called it (Hebert A. Simon: AI Pioneer Andresen S. *IEEE Intelligent Systems & Their Applications* 2001). Satisficing is finding solution that is most acceptable to all parties to the process.

Customer-driven contingency plans are carried out by the project manager, based on the level of risk assigned to a particular element of the project work breakdown structure. When testing a new technique or approach, for example, a manager might estimate a 15 percent chance of successfully completing task 1 and satisfying the user because few new techniques work well the first time they are attempted. Similarly, there may be only a 10 percent chance of successfully completing task 2 because of the organization's previous experience with a similar problem. Hence, there is only a 1.5 percent probability of even reaching task 3 if its performance depends upon successful completion of the preceding tasks. After task 3 is completed, however, it may be implemented again and again, depending on how many problems have to be solved.

Illustration of New Product Risk Management—The Defense Risk Program

The Department of Defense (DoD) has for many years provided standard templates to contractors for reducing risk in system development. The DoD uses templates directed at the identification and establishment of critical engineering processes and their control methods. For each of the critical engineering processes a critical path template is provided. The template addresses the following areas for each critical engineering process:

- Area of risk
- Outline for reducing risk
- Timeline

Six Sigma quality template

Six Sigma quality is defined as an organized process of continuous improvement by private defense contractors and DoD activities aimed at developing, producing, and deploying superior material. The primary threat to reaching and sustaining this superiority is failure to manage with a purpose of constantly increasing intrinsic quality, economic value, and military worth of defense systems and equipments. Note the focus on management first, not technical risk. The Armed Forces and defense industrial entities may not attain a lasting competitive military posture and long-term competitive business stature without a total commitment to quality at the highest levels. Six Sigma is applicable to all functions concerned with the acquisition of defense material, supplies, facilities, and services. Being satisfied with suboptimum, short-term goals and objectives has adverse impacts on cost, schedule, and force effectiveness. A short-term approach leads to deterioration in the efficacy of specific products, the firms that produce them, and the industrial base overall. Major risk also is entailed with the inability to grasp and respond to the overriding importance attached to quality by the "customer" or user activities.

DoD outline for quality

- The organization has a "corporate-level" policy statement attaching the highest priority to the principles of total quality and Six Sigma. This policy statement defines quality in terms relevant to the individual enterprise or activity and its products or outputs.

- The corporate policy statement is supported by a quality implementation plan that sets enduring and long-range objectives lists, criteria for applying quality tools to new and on-going projects and programs, provides direction and guidance, and assigns responsibilities. Every employee at each level plays a functional role in implementing the plan.
- All personnel are given training in quality principles, practices, tools, and techniques. Importance is placed on self-initiated quality and Six Sigma effort.
- The quality effort begun in the conceptual phase of the acquisition cycle is vitally concerned with establishing a rapport between the producer and the user or customer and a recognition of the latter's stated performance requirements, mission profiles, system characteristics, and environmental factors. Those statements are translated into measurable design, manufacturing, and support parameters that are verified during demonstration and validation. Early new product activity is outlined in the Design Reference Mission Profile template and Design Requirements template. The Trade Studies template is used to identify potential characteristics which would accelerate design maturity while making the design more compatible with and less sensitive to variations in manufacturing and operational conditions.
- Design phase quality activity is described in the Design Process template. Key features enumerated include: design integration of life cycle factors concerned with production, operation, and support; availability of needed manufacturing technology; proof of manufacturing process; formation of design and design review teams with various functional area representation; and use of producibility engineering and planning to arrive at and transition a producible design to the shop floor without degradation in quality and performance. The Design Analysis template and Design Reviews template provide guidance in identifying and reducing the risk entailed in controlling critical design characteristics. Both hardware and software are emphasized (reference the Software Design template and Software Test template). A high-quality design includes features to enhance conducting necessary test and inspection functions (reference the Design for Testing template).
- An integrated test plan of contractor development, qualification, and production acceptance testing and a test and evaluation master plan (TEMP) covering government-related testing are essential. The plans detail sufficient testing to prove conclusively the design, its operational suitability, and its potential for required growth and future utility. Test planning also makes efficient use of test articles, test facilities, and other resources. Failure reporting, field feedback, and problem disposition are vital mechanisms to obtaining a quality product.
- Manufacturing planning bears the same relationship to production success as test planning bears to a successful test program (reference the Manufacturing Plan template). The overall acquisition strategy includes a manufacturing strategy and a transition plan covering all production-related activities. Equal care and emphasis are placed on proof of manufacture as well as on proving

the design itself. The Quality Manufacturing template highlights production planning, tooling, manufacturing methods, facilities, equipment, and personnel. Extreme importance is attached to subcontractor and vendor selection and qualification, including flow down in the use of quality principles. Special test equipment, computer-aided manufacturing, and other advanced equipments and statistical-based methods are used to check and control the manufacturing process.

Timeline

The *define, measure, analyze, design, verify* process is used throughout the product life cycle. Defense contractors and government activities concentrate on designing and building quality into their products at the outset. Successful new product developers are not content with the status quo or acceptable level of quality approach. Those developers respond to problems affecting product quality by changing the design and/or the process, not by increasing inspection levels. Reduction in variability of the detail design and the manufacturing process is a central concept, and is beneficial to lower cost as well as higher quality. Defect prevention is viewed as key to defect control. Astute contractors are constantly on the alert to identify and exploit new and proven managerial, engineering, and manufacturing disciplines and associated techniques, and are recognized as such.

DoD manuals stress the following, more traditional Total Quality Management (TQM) principles:

- Total commitment to quality
- Continuous improvement
- Involvement of many functions
- Long-term improvement effort
- Customer focus

TQM principles include company actions that:

- Produce a policy statement (vision/mission)
- Pursue a TQM environment
- Stress a TQM implementation plan
- Foster ownership
- Advocate training
- Include quality as an element of design
- Encourage measurements
- Include everything and everyone
- Nurture supplier and customer relationships
- Encourage cooperation and teamwork

New Product Portfolio Management

DoD policy first articulated the importance of project selection and portfolio management in the 1980s and has since refined the concept. The initial and most disruptive risk involved in program management lies in the inability to choose the right portfolio of new product projects in the first place. Fit and consistency with strategic plans, cash flow analysis and rate of return, and company competency ("can we make this product?"), lack of business analysis on markets, technical feasibility, and legal risk are key factors contributing to successful risk management.

Fit with company strategy. Alignment with strategy is a key risk challenge because projects that might otherwise be attractive might not be part of the company's plans for growth and competence. The weighted scoring model is one tool for assuring fit with strategy, but the risk here is in the inability to measure whether in fact a candidate project is going to help implement a company strategic goal.

Cash flow. Because profitability and rate of return are key to company growth, a project must be planned out over the long term to ensure net income growth. This means that candidate projects must be "fleshed out." The risk here is that cash flows are misestimated, and that key decisions and assumptions behind the decision are not made explicit.

Consistency with company competency. If the company does not have experience in a given area, it does not matter whether the project is consistent with strategy and cash flows are promising. Company capacity to perform is directly related to past experience and capacity; "sticking to the knitting" is still a major principle of success, and therefore a useful risk management concept.

Market analysis. If the market and future demand for a given project outcome or product/service are not well researched, the risk is that an efficient project may meet schedule, cost, and quality objectives, but produce a product that is not marketable. Therefore, the focus in risk planning must be in the adequacy of early business market analysis and research.

Technical feasibility. Because technical feasibility is key to new systems, if a project involves new, unproven technology, there is inherent risk in the project. Technical feasibility can be offset by embedded risk measures as described earlier in the product development process.

Legal risk. Legal and regulatory risk is attributable to changing government regulations and legal issues involved in liabilities for a given product. The risk here is that the company fails to understand and anticipate legal and regulatory constraints on a product or service.

Value of Customer-Driven, New Product Risk Management

Customer-driven risk management captures the critical importance of seeing risk in the eyes of the customer and client. A focus on customer risk can uncover uncertainties and risks in a project that are not apparent in an internally focused risk management process. For instance, in developing a software product, the customer focus may be grounded in compatibility or interface of systems, while the internal, project-oriented focus for software development might well be conformance with performance requirements without much attention to interface issues.

Risks in Customer Expectation, Need, and Requirements

Program and projects face customer risk is three areas, and each challenges the risk management process:

- Customer expectations
- Customer needs
- Customer requirements

Customer expectations

Sometimes customers expect more than they specify in written specification documents. And expectations change from new information uncovered in the project itself. The risk associated with customer expectations reflects the inherent value of projects themselves; as products and service outcomes are produced and information is made available to the customer that uncovers new opportunity, customers sometimes change their minds.

Customer needs

What the customer needs is not necessarily what the customer expects or requires. Needs suggest analytic data on customer needs; an objective view of needs underlies the project process, but needs are seldom differentiated from expectations and requirements.

Customer requirements

Requirements are the actual specifications for the product or service outcome. Sometimes requirements are drawn up by project teams based on what is feasible rather that what is required. The risk here is that the requirements document does not adequately capture customer need.

Risk Lessons Learned and Project Risk Audit

There are two kinds of "post mortem" on a project, and both open up opportunities to look back at the risk management process to see what worked and what did not. This chapter addresses how to do a "risk lessons learned" review and a project risk audit, gives examples, and provides insights on corrective actions based on actual risk feedback from post-mortem sessions. Figure 7-1 shows the transitions from lessons learned to audit.

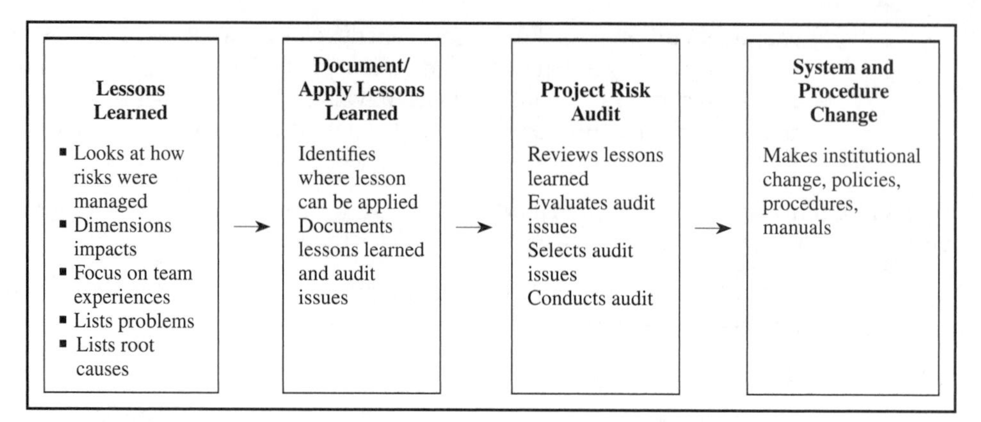

Figure 7-1 Transition from lessons learned to audit.

In Figure 7-1, lessons learned is the first step. This is the process of bringing the new product team together for an informal discussion of what went wrong and what went right in the project, and what the impacts were. Then the team identifies where such lessons can be applied in the risk management process, and identifies possible institutional or organizational problems associated with the lessons learned. Potential audit issues are identified; then an audit is conducted on those issues top management would like to earmark for policy, system, or organizational change.

The value of a timely post-mortem "lessons learned" review lies in the capturing of insights of the project team members and stakeholders and documented information on the project cycle, fresh from "combat." The process is like a military debriefing in the sense that it is well focused and tries to capture the intensity of feeling and information all at once. The implications for successful risk management in future projects are major because it is in those insights about what did and did not work that the organization and management team gain valuable information on future project risks and opportunities.

Project audits

New product project audits are not the same as lessons learned. Project audits are performed by external teams (objective outsiders) who have not been part of the project process. The audit has been described as "the process of coming down off the mountain after the battle to kill the wounded." Such audits can be useful if they address issues that have already been identified by the project team. The purpose of an audit is to connect factors that contributed to project successes and failures based on documents and recorded information, and to match the project outcomes with project goals and objectives.

Performing a risk lessons learned review. A lesson learned project review should be conducted soon after project closeout, and should include all project team members, a member of top management, and stakeholder representatives. The focus should be on identifying what went right and what went wrong—and dimensioning what went wrong in terms of unanticipated or unmitigated risks and uncertainty. This can be accomplished by scheduling and facilitating the meeting around key risk issues or topics that help to capture the risks inherent in the project. The following is an example of an outcome report from a risk lessons learned session in a modern avionics instrument product development process.

Example of a lessons learned report. What went right (things that we want to recreate for future projects)

- This was a focused team with largely full time assignments
- Our team was autonomous: we drew the line with the customer for appropriate "windows" for changes, disruptions, etc.
- Little or no scope creep
- Resources (e.g., test equipment) allocated to resolve problems quickly
- Contingency plans in place when necessary
- Team defined its approach to documentation, etc., together, "as we went" (also noted as a weakness below)
- Communication within team was good
- Consistent high priority on this project from corporate: this was a high visibility, high priority project from the beginning, and we knew it
- Responsiveness of team members to each other very effective
- Project itself was not technically insurmountable; success was feasible
- Team able to separate what was controllable from what was not, e.g., flight test, and managed accordingly
- Team was highly proficient; good professional and technical skills
- Scheduling process handled resource issues somehow
- Program manager was open to change; flexible in responding to team issues
- Program manager used schedules as guides, but was very task and action oriented in meetings

What went wrong (things that we want to correct for future new product projects)

- Company process and procedure requirements, and differing interpretations of document requirements, sometimes acted as barriers to necessary work, and did not always facilitate successful completion of the project, e.g., software documents, reference documents, tables

- We sometimes described procedures *after* we completed them instead of before; documentation often followed verification rather than guiding it, e.g., STD checklist
- Confusion and uncertainty in the actual application of ISO, FAA certification procedures on the one hand, versus "actual" procedures that the team decided were necessary to get the product out on time
- Company changes (split) created resource issues; we had to "ad hoc" it in handling engineering change notices, assembly drawings, etc., because of resource problems created by the split and loss of staff
- Document numbering system created a lot of tension and uncertainty
- Training needs: staff involved in the project were not always trained to carry out procedures, e.g., staff loading the boxes did not have good guidance and training
- Ineffective version control on some documents and configurations
- Stress created by long hours was a problem; can't stretch people and expect them to stay on
- Schedules were not accurate in many cases, compared to the real work, e.g., the sequence of tests and dry runs
- Sometimes the team did not have the "big picture" on the project; sometimes the big picture helps to facilitate doing your job
- Wasted time in some meetings; meetings had agendas, but there were times when the team "blew off steam" and wasted a lot of time
- Some residual issues rooted in "military" versus "commercial" approach had an impact on the project, which came in the middle of the transition from military to commercial methods

Contingency actions

In every new product lessons learned exercise, solutions and advances in process management are identified. For instance, here is a list of lessons learned follow-up actions from an electronic instrumentation company:

Issue #1: Document numbering system

Recommendation: Set aside time and develop a new document numbering scheme that reflects the way we want to do business

Issue #2: Big gap in documenting actual procedures and processes; problems in sequence and review of documents

Recommendation: Develop and document how we followed or created procedures for ISO audit (they will want to see how we followed our own procedures); review current processes for sequence and review of documents

Issue #3: No accountability for master charting function

Recommendation: Decide clear accountability for master chart function

Issue #4: Don't have the right tools to accomplish new product versions, process definition, and configuration control and documentation

Recommendation: Decide what the right tools are to facilitate documentation, and acquire them, e.g., *Framework* software replaces *Word, Agile software* used to conduct configuration management and control versions, and *Requisite Pro* software will be used to document requirements

Issue #5: Ineffective management of document and procedure revisions

Recommendation: Develop a referenced master list of documents and acronyms to assure effective revision management

Issue # 6: Some staff members don't follow established procedures and processes

Recommendation: Develop staff accountability for following established procedures and documentation requirements; follow up with consequences, if necessary

Issue #7: Inefficient acquisition of needed equipment

Recommendation: Plug required test equipment early into the scheduling process; anticipate asset issues before they happen

Issue #8: We don't capture observations, problems, and corrective actions along the way—lessons learned are lost

Recommendation: Establish a way to identify and capture project issues, processes, and corrective actions along the way

Issue #9: Conflicts in doing product development in a manufacturing environment; can't build the same thing twice

Recommendation: Establish a small prototype manufacturing unit in engineering to produce prototypes without leaning on product facilities already over-taxed with volume production requirements

Issue #10: There is a widely held view—that we need to turn around—that the company is good at identifying weaknesses, but does not follow up to fix them: there is a concern that issues such as those coming out of this session will not be addressed because of the press of time and work

Recommendation: Make presentation to top management to get personal commitments from key executives on corrective action to mitigate risks on future new product projects

A postscript to lessons learned

Risk management is, in the end, a people-centered process, and it is in the key decisions that people make daily that the conditions for effective risk reduction and response are created. Because the lessons learned process focuses on the people who actually did the project work and gleans from them a realistic and practical perspective of risks as they played out in the project, the lessons learned process can be very valuable.

Project Audit

The project audit starts with the appointment of an auditor and audit team. The auditor is typically another project manager who assumes the role of auditor with some experience in project planning and control.

The audit, unlike the lessons learned session, is focused on an independent gathering of information and documents from the project, and reviewing them against the goals and objectives of the project and best practice criteria. This involves the question, "did the project produce what it intended to produce, and how effectively and efficiently?"

The project audit focuses on key aspects of the project, as follows:

- Business planning: Were the risks that actually occurred, and that impacted the project, identified adequately in the early business planning process?
- Follow-up response: Were those same risks monitored and controlled?
- Organizationwide culture: Did top management support effective risk management as part of the project cycle?
- Project team: Did the project team members perform their roles as "risk managers" during the process and integrate risk into their daily work?
- Risk identification, assessment, and response: Was a systematic process used to identify, assess, and respond to risks?
- Key processes, decisions, and milestones: Were key risk-related processes, e.g., testing and reliability, quality assurance and control, decision trees, and product development integration milestones, actually followed?
- Resources: Did the project experience resource constraints, and were the constraints managed with buffers?
- Safety and reliability: Were the correct tests and reliability processes in place?
- Risk-based scheduling: Were project schedules adjusted using risk inputs and using the MS Project PERT analysis tool?
- Monitoring: Were risks followed in the project process and decisions made on risk mitigation and contingency that reduced risk?

The question of efficiency is reviewed through earned value and cost variance calculations. The issues would be, "did the project stay consistent with the schedule and budget; did the project manager make adequate adjustments based on variations from risk events; and did the project make its quality, schedule, and budget goals?"

Scheduling Contingencies and Improvements

Now we turn our attention to the importance of scheduling improvements, such as those uncovered in lessons learned exercises in new product development. With all of the emphasis on quality planning, quality assurance, and quality control in new product development, there is a tendency to forget that in project management it is the scheduled work—the tasks that are built into the work breakdown structure and made part of the baseline schedule—that actually gets done. If the quality functions and activities covered in this book are not actually scheduled, or if they are treated as "boilerplate" functions to be performed by the quality office, they will not be effective. Thus the reader is introduced early in the book to the scheduling process to emphasize the importance of scheduling in project quality management.

With the increasing availability of effective project management scheduling and resource management software, the scheduling process has become more useful as a project quality management tool. Software has enabled the integration of quality and resource planning as a single process because all tasks, including those specifically focused on quality such as failure assessment, can be scheduled into the baseline schedule. The purpose of this chapter is to orient the reader to the critical importance of *scheduling* as a practical quality tool and to point toward the useful and effective application of the quality tools covered in later chapters of the book.

Quality Tools and Techniques

It turns out that total quality management was a good idea, and one that has had profound impact on American industry. And the integration of project management and total quality made sense, evidenced by the Project Management Institute's development of a separate section on project quality management. But there has not been much headway in putting these concepts into action in the project process. One reason is because project management is not seen as a planning tool; rather, it is seen as a scheduling and action-oriented tool. Quality plans do not get translated into project schedules as easily as product specifications.

Project managers typically see quality as an external aspect of the process because quality has been incorrectly "sold" in the workplace as something different from and external to the core product design and development process. Of course, we have found in experience that quality must be an integral part of the process, not an external force to be dealt with later when the quality office appears to inspect the products and audit procedures against ISO standards, and second-guess the process. Quality is what work is done, and the way that work is done and when, not how, the work is inspected. We have mistaken the measure for the process itself.

We also believe that quality is the key objective in new product project management, while cost and schedule are indicators. The theory is that the way to ensure successful integration of cost, schedule, and quality is to first satisfy quality requirements; cost and schedule will take care of themselves because achieving quality is less expensive than not achieving it. This means in practice that the project manager must first assume a mindset that places the qualitative aspects of the product out front. If quality tasks are scheduled first, not last in the process, they take on more significance. The project manager integrates quality by scheduling quality tasks and milestones to ensure that quality is built into the project process. In effect, we are saying that the classic triumvirate of cost, schedule, and quality is not a triumvirate at all, but three sides of the same quality concept. Quality means customer satisfaction, and conformance to specifications with budget and time constraints. In a sense we can conclude that quality is the key effectiveness goal, with cost and schedule mere measures of efficiency.

Bridging the gap between customer requirements and scheduled work is no easy job. In writing the specification, the project manager points forward to the scheduling and monitoring phases, but is largely responding "out" to the customer's requirements. After specifications are clear, the project manager turns to the scheduling tool. But how can work definition and scheduling ensure that the work itself will achieve conformance to specifications? The answer lies in the scheduling process, which can now be integrated with work definition in one seamless process.

But, as indicated earlier, there are two quality objectives, quality as conformance and quality as customer satisfaction. Both must be achieved before the project can be considered successful. If quality as conformance can be measured and traced to specification, and then scheduled into actions, then conformance is ensured. But how does quality as customer satisfaction get translated into action? The answer requires some discussion of customer satisfaction.

"Quality as customer satisfaction" is relational rather than absolute and tends to get lost in the shuffle of *real* work. Therefore, the achievement of this aspect of quality must be planned, and scheduled, in a different way. Customer satisfaction is a function of four key forces:

1. *Expectation*, Documented project developments on user expectations and insights gained prior to or during the project
2. *Feelings* about the project manager and team, e.g., trust, reliability, loyalty, and other emotional responses to the project organization and team
3. *Feedback from stakeholders*, e.g., feedback the customer is getting from key stakeholders such as a standard setting association or government regulating agencies that would change their views of project progress
4. *Project performance*, e.g., schedule and cost variance, early indications of product and service quality that will give the customer a sense of confidence about progress or alternatively a sense of discomfort that will affect their satisfaction with the work

The issue becomes how to ensure that the soft objectives of customer relationship management are achieved while satisfying the customer's hard, product requirements. This is accomplished again through defining customer satisfaction into the work breakdown structure and schedule for the project. An illustration of this approach is the scheduling of a customer quality survey to be administered monthly to build a running account of customer reactions to the progress of the project work. Such a survey would be scheduled along with meetings to discuss the results as part of the original scheduled work.

The following are key quality tools in new product development. These tools are positioned in the product development process to ensure a quality product or service.

Quality function deployment (QFD)

QFD is applied in the design phase to translate customer requirements into project methodologies, e.g., design and specification. It is a translation of *what* to *how*. QFD is useful both as an agenda and a framework for discussion as well as a methodology. QFD suggests a way of thinking, an approach to asking questions of the customer, brokering answers to those questions to figure out how the customer's needs are going to be reflected in internal processes, designs, and production capabilities. QFD is a process, not simply a quantitative template for analysis, which assures that the project team is in dialogue with the customer and can assure that it can employ the appropriate systems and processes to produce what the customer wants. The QFD exercise should create standards or criteria for evaluating later the extent to which a project process has produced a product in the range of acceptability to the customer, e.g., that it conforms.

QFD is scheduled as a series of customer requirements activities in the initial concept phase and includes three summary tasks:

1. Obtain customer requirements in writing
2. Complete the QFD "house of quality" with the customer
3. Translate QFD results to product specifications.

Throught these tasks, the project manager is ensured that the team can verify or trace the scheduled work directly back to the customer requirements.

Statistical process control (SPC)

SPC is used to identify processes that are out of control, or are varying unacceptably from the standard. Process stability is important in project quality management because it indicates that a product is performing within customer expectations and requirements, and within specifications. SPC is also a way to educate the customer on how it can be used to help them articulate requirements right up to "Six Sigma." It is a planning tool as well as a quality control tool.

In addition to including SPC in quality control downstream in the project, SPC can also be scheduled early in the concept phase as an activity performed with the customer. The scheduled task would include defining the customer's perception of key product tolerances and performance standards in terms of SPC.

Pareto analysis

Pareto analysis is the analysis of frequency, and is another tool to confirm that the product or service is meeting customer expectations. Frequency measurements are made to confirm the statistical significance of occurrences in the project process that might affect quality. For instance, the number of schedule variations in a project schedule due to various root causes would be a good target for frequency analysis.

Pareto analysis can be scheduled into the project as a way to follow rework and errors in a product from design to assembly and testing. Frequencies of errors at each stage in the product development give the project manager a good way to identify key points in the design and development where errors occur, then display the results on a graph to show the relative frequency of errors at each point. Thus this kind of analysis should be scheduled at several intervals in the process.

Cost of quality

The cost of quality accounts for quality control and audit costs, such as the *unnecessary* costs of correcting for bad decisions. Costs of appraisal, inspection, and rework involved in the design and production process are covered in a cost of quality analysis. As such, cost of quality is an important tool for the project manager to identify failures in quality assurance in the early processes of design that evolve into costs downstream in inspection and rework.

Cost of quality analysis is scheduled late in the project process to identify the costs of quality control and inspection. The task involves collecting the actual labor and equipment costs involved with such activities.

Quality assurance (QA)

Quality assurance is the process of building quality into the project process—doing it right the first time—and is an important method of cost avoidance. But further, quality assurance is the integral aspect of project processes that create a safe and reliable product. For instance, the imposition of reliability testing in a product is a good indicator of the internalization of quality assurance into the design and production process. QA is not an external concept; it is the total integration of procedures and processes that test, validate, and verify the functionality of the product according to customer needs.

Quality assurance is scheduled early in the project in the form of a set of quality assurance tasks, e.g., design reviews that examine how initial design tasks were performed and whether those processes were consistent with professional design standards. New design software in the engineering field now provide for these tests and ensure that computer-aided design work is performed right the first time.

Earned value

Earned value is a key project quality management tool when used effectively. Earned value measures whether the work performed on a project has been performed consistently with customer requirements as well as product specifications. Effective use of earned value for quality purposes requires that the project manager explicitly define earned value and percent complete reporting in terms of customer or user acceptance. In other words, when a task manager reports that a task is 50 percent complete, that must mean the work completed meets the quality specifications set out for the product or service as well as satisfy the customer. That inherent definition of earned value, backed up by a corporate culture that places primary priority on quality, then becomes the insurance policy against a high cost of quality.

Earned value is both a front-end function and a monitoring function. Earned value is scheduled into the front-end definition phase by describing key milestones in the project schedule where earned value assessment can be made. The scheduling process then accounts for these scheduled milestones in the baseline schedule.

Project review

Project review is another quality tool because its purpose is to assess progress in terms of customer value, and provide data and information for a new product go or no-go decision. This requires a mindset that defines the program and project review process as a series of questions and answers that periodically pins down the quality of the work in progress. Quality questions must be asked before scheduling and cost questions. For instance, in earned value assessment during project review, the project team must struggle with the issue of whether the percent complete is an accurate representation of what the customer would say. Better still, the customer should be integrated into this decision so that they can probe earned value in terms of customer satisfaction. This process can be implemented easily by involving the customer in project review meetings, either in person or through virtual conferences that give the customer total access to the decisions process.

Project review is scheduled weekly or biweekly into the work, as well as at the end of each phase and at any other time when the customer or the project manager anticipates the need to assess progress and review key indicators of success. Again, if project review is not specifically scheduled as an activity in the baseline schedule, it is liable to be missed.

Documentation

Documentation is an essential project quality management tool because it disciplines the process and ensures that quality methods have been built into the design and production process. The project manager should see documentation in terms of a front-end process of driving work rather than as a back-end process that records what work has been done and how. In other words, documentation is a quality tool if it is used to define an early process (one that occurs in the concept and definition phases of a project) that has been proven to yield a quality product or service. If used to simply record what happened in a project or as boilerplate by an intervening and offline quality office, documentation has marginal quality benefits.

Documentation reviews are scheduled at each point in the project process where a document is to be internally approved or reviewed by the customer. Scheduling documentation reviews ensures that documentation is not delayed until the later phases of the project and treated as an afterthought.

Scheduling as Team Motivator

Team complacency is the enemy of quality; therefore, any strategy that addresses complacency will yield quality benefits in the project management process. The way project managers deal with complacency and maintain focus is to understand the tendency of the system to go to disorder and complacency unless regularly charged with new energy and purpose. If complacency is the enemy of quality, purpose and scheduling are the enemies of complacency. New purpose and energy are introduced to the system by focusing on the scheduled work and immersing the project team in the task interdependencies and interrelationships designed into the project process.

Quality has a unique structure and texture in each project—a project quality persona—that is built into the project. Quality must be structured into the work, and each project presents a different challenge in finding ways to do so. Tasks and subtasks that ensure quality should be made integral to the work breakdown structure and the schedule for the work. Scheduling quality into the work means addressing quality in the definition of summary tasks, subtasks, linkages, and work packages. For instance, for an electronic instrument project, the actual task of drawing an electronic version of the product is linked back to the baseline specification of the deliverable and to customer expectations. The drawing must meet specifications—and, conversely, specifications must be structured to allow alignment with drawings—before quality is actually structured into the task. If the potential gap between specifications and the drawing cannot be closed by a scheduled task, e.g., "Check Drawing Against Spec," there is no linkage from the drawing to a functional requirement. This is the way scheduling is used to ensure project quality.

The advantage to using scheduling as a team motivating activity is that the schedule is a way to refocus individual work into the context of the team effort, to continually remind team members of the interdependencies in their work. Schedules projected on a screen for team review and discussion create the sense of a common purpose for the team. This motivates team members to see how their work fits into the big picture. Customers respond positively to project schedules because they can fit key milestones and deliverables into their own calendars.

As a special note, there is wide variation in the extent to which project teams take the time to plan and schedule projects, especially across international lines. These differences are attributable to culture differences as well as style. European project firms, for instance, appear to place much more emphasis on front-end project planning, perhaps up to 40 percent of the project timeline, than American or Canadian project managers. American project managers minimize up-front planning and scheduling, and rely more on down-the-road change management strategies to deal with issues not addressed in the planning and scheduling process. We believe that this tendency to downplay project scheduling is being reversed by the competitive forces of the global economy. American project firms are giving more emphasis to planning and scheduling as a result.

Quality Must be Translated to Scheduled Tasks

While we emphasize the overwhelming importance of quality in the quality, cost, schedule triumvirate, quality is the most difficult of the three to pin down. The key integrating principle is that *customer involvement and quality assurance* are mutually supporting activities that increase the probability of producing customer satisfaction. There are fundamentally two basic ingredients to quality—conformance to specification or requirements and customer satisfaction—and one does not necessarily produce the other. Conformance to specification involves controlling the development of the deliverable so that it can be validated and verified, e.g., that it meets the specifications as stated and "works" in its systems environment, whatever that is. This is where the application of quality tools and techniques is important within the project process.

Customer satisfaction, on the other hand, is tied to customer expectations. While one would assume that customer satisfaction is related to conformance to customer specifications, that is not always the case. Typically customer satisfaction is related to conformance only to the degree that the customer's expectations, needs, and wants are reflected in the specification and that any fundamental changes in the customer's view are somehow reflected in the specifications through change orders and modifications to the project scope. In other words, while conformance is somewhat controllable in the sense that it can be quality assured and measured, customer satisfaction is not. Customer satisfaction is a feeling, a perception, and a disposition that is based on the continuing relationship or project firm and customer/sponsor. That is why it is important to keep an eye on customer expectations during the project tenure just as closely as the other eye is on conformance to specification.

Because the most useful measure of quality is earned value (whether a project is on planned cost and schedule), it is important for project managers to educate customers on earned value as an indicator of quality. We believe that customer satisfaction derives from continuous project involvement and education. Therefore, to the extent that earned value reports can be presented regularly to customers—and interpreted by the project firm—customer satisfaction can be regularly gauged. In the end, it is the quality of the product or service—its very features, functionality, performance, and capability to create value—that drives the customer's expectations.

There are several key points or windows in the project management and scheduling process when *quality as customer satisfaction* and *quality as conformance* can be expressed and integrated. It is in the project manager's interest to use these key windows explicitly for the purpose of assuring quality, and to share each schedule version with the customer to ensure that quality is translated into the real work breakdown structure and timeline. These key windows are:

1. Front-end customer process analysis
2. Concept development
3. Generation of alternative candidate projects
4. Scope of work
5. Scheduling
6. Budgeting and earned value planning
7. Quality assurance

8. Project metrics
9. Prototyping
10. Quality audit

We have learned that the project manager must find ways to practice quality at every turn and achieve *continuing closure* with the customer on all the aspects of scheduling quality work as the project takes shape and progresses to the final deliverable. Quality is not the preserve of the quality office; it is integral to the project planning, scheduling, and controlling processes.

Project dynamics often work against achievement of quality objectives. *Without constant vigilance, the project system moves naturally away from the customer; the dynamics are in favor of isolated, narrow project tasking and insulation of the project team—and therefore of quality issues—from the customer.* That fragmentation occurs *naturally* unless managed, almost as a function of the system itself, because the project is a system seeking entropy or increasing disorder. Many a project has failed because there was a major gap in the dialogue between the customer and the project team on quality issues—attributable to the lack of scheduled tasks to close those gaps. So these windows become strategically important, if they are scheduled and monitored, as an insurance policy against the forces that isolate the customer from key quality issues and decisions.

Let's consider these windows and the tools and techniques associated with them, and see how they are scheduled so that they can contribute to project quality management.

Front-end customer process analysis

Although there are always variations on the theme, projects typically go through five phases: concept, definition, production, operation and testing, and closeout. Prior to the concept phase, however, there is a key step that ensures an understanding of the customer's business processes, work setting, and market forces. That step is front-end customer process analysis, and it should be a scheduled summary task in every project schedule.

It is important to know the customer's business and organizational setting and key processes before a project relationship is created. This is because a project manager and team must understand the customer's markets and customer base *better than the customer does.* Customer expectations for a project deliverable come from the customer's own markets and opportunities and must be seen in the context of the customer's milieu. Value is created out of understanding and thinking ahead of the customer to anticipate wants, needs, expectations, and requirements and translating those needs into the project schedule. For instance, if during this initial window, a project sponsor is clearly expressing the need for a given product to create value in a given working environment or setting, the scheduled work includes a set of tasks that specifically verifies and validates the product prototype *in that specific environment*, early in the project process. The actual setting of this analysis with the customer includes continuous reviews of a draft project schedule projected on a screen in the customer's office. We favor direct translation of quality issues into a project schedule *as they come up* in this initial step. This process assumes that the project manager has acquired proficiency with project scheduling and presentation software to integrate analysis of customer needs with initial planning and scheduling of the project.

Even if a project has not yet been selected, it is good practice to schedule candidate projects out through a high-level scheduling process in order to give the customer a clear idea of expectant deliverables, milestones, quality issues, and cash flows. Cash flow analysis of alternative projects allows the customer to assess net present value of alternatives as one indicator of potential project value.

This is the point at which quality as customer satisfaction and quality as conformance begin to take shape. Customer satisfaction is grounded in a high confidence level that is derived from the customer's feeling that a project team can literally *walk in their shoes, that they know the customer's business milieu, are scheduling appropriate quality steps in response, and can help evaluate and select appropriate projects that promise value to the customer.*

This process analysis phase involves the use of process assessment, market analysis, discounting and net present value analysis, and scheduling tools. Strategic planning exposes the customer's broad business vision and strategies for long-term success. The customer's values and principles of practice are surfaced here. Process assessment is accomplished through strategic planning, process improvement techniques, stressing identification of key processes, and beginning to look for opportunities for improvement in the customer's work processes. Discounting is a quantitative approach to evaluating the financial viability of alternative projects, using net present value.

Scheduling is accomplished through project management software, using GANTT charts, resource assignments, and "what-if" analysis. Scheduling is both an art and a science, and proficiency in the use of scheduling software enables a project manager to continuously *center* the project on the key issues and activities and to stimulate discussion. Because a schedule embodies quality, cost (resources), and time, the schedule is an integrating vehicle for the project manager. The process of customer relationship management involves the use of schedules to highlight current or future problems or opportunities and to place time and cost dimensions on them as well.

Concept development

Following initial process analysis, the initial concept phase involves the flushing out of alternative project ideas and opportunities after full immersion in the customer's key processes and product/service mix. Here is the point where issues and needs are translated into conceptual solutions and potential project ideas. A conceptual solution is an idea that shapes a need into a working vision. For instance, a customer may see a particular product serving a particular need; the concept is a visual picture of that product actually serving a customer need.

The ideal setting for concept development is a collaborative process involving the customer and the project team, sharing information and collectively exploring potential projects and benefits. This is where new technologies are born, through the stimulation of project teams interacting with customer representatives to encourage out-of-the-box thinking.

Concept development draws on QFD tools, addressed earlier in this chapter. QFD translates needs and expectations to design concepts, which can be detailed later into product and service specifications. The key issue here is that concept development is a group activity conducted through brainstorming and other group techniques. Brainstorming provides a forum for identifying alternative ideas and concepts and prioritizing them so that the customer can make decisions to proceed with the best possible project concepts. Brainstorming is facilitated by the project team and starts with the information gained in process improvement and immersion in the customer's business.

Generation of alternative candidate projects

The generation of alternatives is a delicate process of creating ideas and options from free flowing discussions in brainstorming sessions *and from current projects.* Since the majority of projects come from current ones, or are related to current project work, the customer-driven project management firm ensures that creative ideas are captured from current work to feed the pipeline. This is accomplished in program reviews, where new technology and marketing ideas are made a regular part of the agenda, and through closeout project team meetings, which glean lessons learned and new marketing concepts from team members.

Alternative projects are compared through a number of tools and techniques, including net present value and weighted scoring models. Net present value is the process of estimating cash flows over the life cycle of the project and calculating the net cash flow and net present value of candidate projects. The weighted scoring model scores candidate projects against a series of goals, gives weights to those goals, and then multiplies the weights by the scores to get a weighted score.

Scope of work

The scope of work provides an effective window for emphasizing quality, but project scopes rarely include reference to quality management. In project quality management, the scope is written to include all quality and performance testing for the product or service, thus ensuring that the key quality aspects of the deliverable are visibly addressed in the scope.

The scope of work is used as a tool to confirm not only the work to be accomplished but also the approach to quality assurance and quality control. The scope template is typically included in the project plan and is associated with the project schedule. Key milestones from the schedule are referenced in the scope document.

Schedule

Project managers must *schedule the scheduling task.* The scheduling process may be the most useful way to promote project quality management because the task structure and scheduling of the project are the usually the most consistently used management tools. Scheduling is accomplished by first developing a work breakdown structure and then scheduling the tasks built into that structure. In project quality management, the top-level summary tasks not only include a quality assurance task that is decomposed to the same level as all other tasks, but it includes quality steps as an integral part of each task breakdown to the lowest level. This requires that the project manager specify in the company quality policy that all tasks will be defined in terms of quality to the extent feasible and appropriate.

The basic tool of scheduling is a project management software package. The scheduling tool is used to take the work breakdown structure and plan it out over time, identifying predecessors, durations, and resource assignments. Scheduling the scheduling task involves two key subtasks: developing a preliminary schedule with resource assignments and key linkages, and getting approval of the schedule from the customer, stakeholders, project firm leadership and the project team, and saving as a baseline schedule. While this may seem a mechanical process to some, the schedule becomes the key *contract* between the project team and the customer and thus the scheduling process is considered a major customer relationship management tool.

Budgeting and earned value

The concept of earned value is an important tool of quality if it is planned into the project correctly from the beginning. Earned value is an indicator of how much work that has been accomplished at any given time in the project has earned its value, that is, meets the dual test of quality, customer satisfaction, and conformance to specification. When a percent complete for a task is estimated by a task leader, the percent complete represents an assessment not of all the work accomplished but all the work accomplished that will be acceptable to the customer, e.g., quality work. The earned value is the dollar amount from the original budget that *should have* been spent for the work, which earned its value.

It is important that the project plan, scope of work, and schedule all reflect the earned value concept because if the project is not set up to relate percent complete to the original budget, then the dollar indicator of earned value will be difficult to estimate.

Earned value is a monitoring tool calculated automatically by any professional project management software if budget costs have been entered into the baseline schedule. If budget has not been entered into the schedule, earned value cannot be calculated. Earned value analysis is scheduled into the project review and monitoring process at key points in the project.

Quality assurance

Quality assurance is the process of building quality into the definition, design, production, and testing of the product deliverable. Quality assurance procedures provide for testing, verifying, and validating work as its progresses, not later in project monitoring and appraisal. In effect, quality assurance builds quality into the deliverable so that quality practices are indistinguishable from all other design and production practices.

Quality assurance is implemented through a variety of statistical, testing, verifying, and validating procedures to ensure that work is done right the first time, that it meets necessary tolerances and standards, that it performs consistently against both the specification and the expectation of the customer, and that it works in the customer's work setting.

Project metrics

Project plans and schedules typically include the application of a set of generic and tailored metrics or measures that are to be used to monitor progress and the final deliverable. Generic metrics include earned value, budget variance, and a wide variety of verification and validation measures. Tailored measures, however, include indicators that are unique to the deliverable.

Metrics are tailored to the unique performance requirements of the product. For instance, in the design of new electronic products, which require embedded software, a useful metric is a measure of software design life cycle, an accurate indicator for the whole product life cycle. This kind of metric is often regulated or standardized by an industry or professional trade association for that product or family of products. A good project quality manager will use these kinds of standards to discipline design and production work, which then can be marketed as regulatory compliant. Review of project metrics is scheduled into every project review meeting.

Prototyping

Prototyping is a useful approach to project quality management because prototyping often addresses the expectations—and the vision—that customers often have of the deliverable, but may not have expressed in requirements and specifications. Prototyping is the process of demonstrating an early model of the deliverable and how it will work without having made major investments in its design and development.

Prototyping can be accomplished through electronic and visual representations, computer screens, models of products, and graphics. Prototyping is scheduled into a typical project as an early summary task with subtasks addressing prototype development, presentation, testing, and approval. Eventually, a prototype must be approved by the customer before proceeding, thus it is a key milestone that should be scheduled into the project.

Quality audit

The quality audit is a post-mortem review of the project process to ensure that the experiences and documentation are captured and assessed for the purpose of improving future projects. While the quality audit is typically performed by the quality office, the quality engineer, or finance/administration office, the principles of project quality management would include the project team itself in the audit.

Auditing tools are document reviews, interviews, and internal control analyses that ensure that planned procedures and practices were followed and that the project accomplished what it set out to accomplish efficiently and effectively. Project audits must be scheduled during project closeout.

Transform customer expectations to requirements

Customer expectations must be "teased" out and transformed into customer requirements for new products because projects cannot succeed in meeting vague and unexpressed expectations and undefined specifications that typically follow. The quality function deployment tool is useful in transforming expectations to specifications. But the essence of the process is the will and determination of the project manager to educate and be educated by the customer in the initial concept phase of the project.

After they are developed, customer requirements are typically embodied in a scope of work statement, or other documentation of the customer's performance and functional standards. Detailed system specifications—derived from customer standards—then serve as the basis for development of the prototype with the customer.

Follow a defined development process and work breakdown structure

Project management is a generic management system laid on a technical or product development process. If that development process is not defined in terms of the technology and language of the product or service, good project management practices will be wasted. This is both the strength and weakness of project management systems. Such systems focus on quality, resources, and time management, but sometimes these systems cannot alone carry a new product development project with a complex technical process. Projects must have strong subject matter experts doing the technical work. For instance, good, scrubbed project schedules are important, but they must be built around generic work breakdown structures tailored to the product or service being developed. A work breakdown structure for electronic instrument development is quite different from a work breakdown structure for a new telecommunications product. But the project manager is the facilitator of the process, pressing for a useful and workable WBS and keeping the customer involved through email and Internet pathways on progress in defining the product.

To the extent possible, scheduled work will be broken down into tasks, which are defined and specified. For instance, for a product development process, the basic structure of it is specified to four levels:

- Stage: These are generic development activities in the project management system, e.g., summary tasks in the schedule structure, including customer requirements, concept development, detailed design, prototype development, design validation, product transition, and manufacturing
- Project/Phase level: This is the next level of development, tailored to specific project phase features
- Systems: This is the product systems or functional systems level, defining how parts integrate with each other to produce product functionality
- Task: This is the operating component level, where work packages are put together and achieved through individual or small team activity serving as the basis for design reviews

Schedule customer and quality early

This concept stresses the importance of building the customer and quality into the work breakdown structure and schedule. For instance, customer reviews are identified in the project schedule as separable milestones. Customers are brought into the development of the work breakdown structure through Internet exchanges and meetings in which the customer is educated on all the details of the WBS and the preliminary schedule.

Quality is integrated into the early development process. For instance, product quality and reliability testing are specified as product features, not as a separate testing process. Quality data on variances and test outcomes are shared with the customer as with all project team members and stakeholders. ISO standards are written and scheduled not by separate quality staffs, but by product engineers who are actually doing the work. Documentation of product quality and safety is verified in a scheduled task to ensure that the product meets the specifications, and validated to ensure that the product works in the customer's operating environment.

Customer-driven teamwork

Project teams are established with the objective of including the customer in all team deliberations and building an environment of high performance teamwork with the customer. The customer is seen as a team member, not simply a sponsor. To the extent possible, staff will be assigned tasks that are consistent with their backgrounds and expressed professional interests. In product development firms, project teams are composed of engineering, software, and technical professionals who are suited to the work they are expected to perform. Team staff members will be trained as necessary to enable them to perform assigned tasks. Team support functions are scheduled.

Define and communicate the scope of work and assignments clearly

Product requirements and job assignments, including assignments *to the customer* for design clarification, customer data, and other necessary inputs to the project, are defined and communicated to the program team clearly. This will allow staff to understand how their work contributes to product development, and prepare for work assignments with necessary training and development.

Project management software resource reports are shared weekly with project team staff *including the customer* to indicate how tasks are assigned, to whom, and to identify key milestones. While this practice of sharing resource and schedule information with the customer as a matter of doing business creates a bond with the customer that ensures trust and reliability in the relationship, some customers may have difficulty with this level of involvement. For this reason, the project manager must make a strong pitch in the initial phases of the project that the only way to truly integrate the work with the voice of the customer is to place the voice of the customer in every discussion involving the project. This will require some education and advocacy, but customers quickly see the value of such an arrangement and learn when to play a passive role and when to interject themselves in key decisions in which they have a stake.

Collaboration across the organization

Collaboration among project managers, department managers, and the customer- driven team is the essential ingredient to the success of customer-driven project management. This arrangement encourages constant, professional communication and information exchange among project, departments and system managers in the process so that there is both individual team member accountability and collective ownership of schedules and requirements. In addition, an interesting dynamic occurs when the customer is involved; project team and department managers pay particular attention to customer reactions to project issues and gain valuable insights, which help them with other customers and other products.

Work will be quality and schedule driven

Since the heart of project management is the scheduling process, maximum emphasis will be placed on preparing tight program schedules that incorporate all the work necessary for program success. Program schedules will be planned and "scrubbed" in a collaborative process that ensures that all necessary work is included and all durations and resources are tightly planned and estimated. Schedule baselines will be established and work initiated only after schedules have been tightened through this process.

Ensure timely procurement of product components

Project managers pay special attention in early scheduling of activities that involve customers, suppliers, and stakeholders. These activities include procurement actions, testing actions, and technology challenges. Hardware specifications, parts, test equipment, and supply items will be included in schedules, and appropriate lead times established. Procurement actions are taken in a timely way to avoid schedule delays attributable to lack of inventory, but also to bring suppliers and vendors into the scheduling and planning process so they can share ownership on the timing and quality of the deliverables.

Change is managed

Customer-driven firms administer a formal engineering or product change notice process and a documented configuration management process that ensures that requirements and product component changes are managed and controlled. A systematic change management process will be maintained through sound documentation and configuration management in these organizations. For the customer, this ensures that orders and reorders can be made online with a minimum of assistance from the project firm. The more the customer knows the parts list and numbering system of a product, the more the customer feels a part of the process and interfaces with it seamlessly. Thus configuration management is an important customer strategy as well as a key internal control on inventory and procurement.

Program progress will be tracked and periodically reviewed

Project reviews involve the customer as well as other key stakeholders. Project managers track program progress using software schedules showing "baselines and actuals," prepare weekly presentations and earned value reports, and prepare for weekly program reviews *conducted by the customer* with the assistance of the project manager. This kind of customer role is essential if the key issues are to be addressed in project reviews. Typically, proprietary information is not surfaced in these meetings to ensure no sensitivity in the agenda or discussions.

Involve the customer in designing the management support system

A project management support system is an institutional set of supports, processes, and resources to ensure successful projects. Such a system involves a consistent approach and set of tools and techniques for planning and implementing programs in close association with the customer. When implemented across the division, such a project management system provides an organizationwide program management culture that places emphasis first on product quality and timeliness. Such priority is confirmed with customer involvement. Formats for planning, information exchange, and project review are governed by distinct work breakdown structures and project phases, reporting requirements, and schedule, cost, and quality controls. Project managers are trained and developed in a consistent way—both in technical project techniques and team leadership, as well as customer interface—and manage their programs using standard project management software and Internet tools. Program teams are established as formal groups with charters in writing, and with performance guidelines and criteria for evaluating team members. A project management guide is used to communicate policies, procedures, and support systems, and the system is continuously reviewed. A common project management language governs communication on programs.

Quality as Driver

Quality must be built into the deliverable, not stamped on during inspection. That means that the deliverable specifications are defined in terms of quality and subject to classic quality testing and measurement during the development stage of the project. As is evidenced in the following chapters, customer involvement is the best quality assurance mechanism because it combines two critical forces in the determination of quality: the quality tools and metrics deployed in the project are appropriate to the customer's needs for quality as stated by the customer; and the development of the deliverable reflects the customer's changing views of a quality product and/or service as the customer sees it *now*. This is a difficult lesson for project managers to learn—quality is not a static concept, even though it might be served by a series of static processes and procedures. Quality is the combination of quality processes and customer learning, challenging the project manager to balance the application of quality standards to the appropriate user view of what is important. This requires that the customer be educated by the project team on the appropriate quality testing and standards *for the customer's planned application*. This suggests the relative nature of quality in the context of the wide variety of quality objectives or features for a given deliverable.

This relativistic view of quality does not diminish the importance of reducing and eliminating waste and keeping the cost of quality under control. Whatever the quality standard, the project must continually strive to improve processes and reduce rework as a basic proposition. But the project manager must exercise discretion in the actual application of quality measures and standards simply because there will be a limit to concepts, say of six sigma, to project deliverables that do not require it.

Reviewing Program Progress and Resolving Conflicts

The project manager conducts weekly review meetings with the program team to ensure a unified and informed effort. Sometimes these meetings generate substantial changes. Such change inevitably brings tension and conflict, particularly if its source for some reason is the customer. To dispel this tension, the project manager provides for a weekly agenda of discussions with the customer on anticipated changes and strategies to deal with them. This includes discussions of risks and risk mitigation strategies focused on the customer's perspective of risk. As project requirements change, the project manager ensures that the plan, schedule, and configuration management documents are kept current and distributed.

The project manager resolves resource and program schedule conflicts if possible within the team. In the event of issues and resource conflicts that cannot be resolved by the project manager, the customer, or the departments, the project manager is responsible for raising issues to a joint team of executives involving the customer for resolution.

Project planning

The project manager has the primary responsibility for creating a project plan and a schedule using project management software and composed of tasks and milestones down to the fourth level in the WBS hierarchy, tailoring the actual work breakdown structure to the unique needs of the project. The plan is created with support from the customer, team members, and department managers. The project manager is required to keep the program schedule current, track progress, and incorporate changes as required, and for sharing updated schedules and earned value reports with the customer through the Internet.

The project manager utilizes project management software to produce and update schedules and resource reports, and is expected to be proficient in the use of such software for control and presentation purposes.

The project plan must include, as a minimum:

- An overview of customer requirements, program scope of work, and program objectives
- Specifications derived from customer requirements
- Schedules, including major schedule tasks and milestones
- Resource assignments, linked to the central resource pool file to allow analysis of overall resource impacts
- Identification of test equipment and components needed for the program (special test software, special test sets, fixtures, jigs, cables, and so on)
- Identification of any special tests required for the program
- Procurement requirements for development efforts
- Manufacturing requirements for development efforts
- Outside integration (A/C, customer lab, and so on)
- An estimate of other direct materials required for the project, including materials for test assets, test equipment, outside facilities, travel, and any other pertinent costs
- Risk assessment and risk mitigation plans

The project manager has the primary responsibility to create and maintain a detailed schedule that meets the needs of the customer, team members, and department managers. The schedule must contain, as a minimum:

- Summary tasks, task structure, and key milestones that correspond to all major program objectives contained in the plan
- All product or service development activities and tasks required to execute a given project, including systems design, detailed design, certification, test equipment, reliability, safety, design reviews, manufacturing, procurement, test assets, and so on
- Tasks detailed to the lowest practical level; activities and tasks should generally be built four levels down
- Resources assigned to activities and tasks and leveled to reflect a realistic workload
- A central resource pool which indicates the combined effect on the workforce of all active project schedules

Departmental manager roles

In a customer-driven organization, department managers for functions such as systems engineering, mechanical design, electrical design, and software certification are responsible for building and maintaining the resource and technical capacity of their departments to support of the product development process. Here are some key functions of these managers.

- The department's organizational structure and reporting relationships, roles, and performance expectations are communicated
- Performance evaluation; all staff receive annual performance evaluations reflecting department and program manager feedback
- Hiring, training, and career development plans; the department's technical capacity to perform is a function of its people and their ability to keep up in their fields; technical training is an integral part of the department manager's job
- Development of a department budget, which is used to guide and schedule implementation of the department's program; the budget includes labor and equipment acquisition
- Preparation of a staffing plan that projects future resource requirements
- In a matrix organization, department managers assist project managers in bringing program schedules to baseline, determining resource requirements, and supporting delivery of program products

Project team roles

Each project team member is responsible for understanding their assigned tasks, their interdependencies with other tasks, and for general support to overall team performance. The primary given is the team member's technical competence; the primary value-added is the team member's dedication to the project team's objectives and the project deliverable. Team members are accountable for keeping technically proficient and performing their assigned tasks in a timely way, consistent with the schedule. Team members collaborate with each other and the project manager, promptly attend team meetings, and report to their project managers and department managers on technical and schedule issues.

Role of a project management office (PMO)

The role of the project management office is to promote best practices and consistency in project management. The office provides administrative support to project managers and departments with scheduling, resource planning, and reporting services and activities. A key role is the analysis of all resource impacts to identify and resolve conflicts. In addition, the support office prepares management guidelines, provides training, and develops project evaluation metrics. In addition, the office implements standard methodology through project management guidelines, implements tracking software, supports project review, and distributes all project reports and meeting minutes to the customer and the team. The office assists with estimating costs, manages the project documentation process, and produces resource usage reports to all effected staff. The project management office helps management review alignment of the project with customer strategies as well as project firm strategies, and helps plan for project process improvement initiatives and future resource needs.

Scheduling

Good scheduling is at the heart of the customer-driven project management process. Schedules created in appropriate software are scaled to the project effort, linked to a central resource pool file, and posted on a network including the customer. Schedules constitute the basis for program development, tracking, and review, and project review meetings involve projections of schedules on standup screens to encourage group discussion and understanding. Customers are involved directly in schedule reviews or through conference calls participate actively. A good scheduling process provides adequate time to ensure that the work breakdown is comprehensive, that scheduled task durations and predecessors are as accurate as possible, that key linkages are made, and that resources, once assigned, are actually available and committed to the program when they are needed.

Scheduling and associated resource planning are accomplished collaboratively by the project manager, working closely with department managers and the project team. Communication and sharing of schedule and resource information, check points, approvals, and feedback are managed on the network to the extent feasible, with meetings and review of hardcopy schedules or action lists as necessary.

The project manager carries out the following basic scheduling procedures with support from the program planner:

- Ensures that customer requirements, expectations, and reviews for product functionality are clearly represented at four levels
- Establishes project team "sign off" of the schedule before baselining
- Develops the top-level work breakdown structure to serve as the basis for scheduling, using current WBS templates or creating them in consultation with the customer and key departments
- Prepares a top-level schedule showing basic task structure, durations, start and finish dates, linkages and predecessors, and links the schedule to the central resource pool
- Integrates top-level tasks into a more detailed schedule
- "Scrubs" the schedule, involving four steps:
  - Drafts an initial schedule, with a work breakdown structure to the fourth level, and makes resource assignments for all tasks
  - Works with department managers to assign resources and resolve conflicts
  - Links the schedule into the central resource pool to identify resource issues
  - Manages meetings with department managers and customers to work out final schedule structure, task definition, linkages, and resource assignments
- Enters hourly, fixed, and equipment costs and produces a project budget that is shared with the customer
- Gets sign-offs from all department managers before proceeding to schedule baseline
- Saves the project baseline as the key point of departure for the work
- Kicks off project with team meeting; hands out hardcopy schedules and individual task assignments
- On a weekly basis, collects and enters actual performance data from team members on: (a) percent completion, (b) actual hours spent on tasks (from time sheets), (c) actual start and finish dates, (d) actual durations, (e) remaining durations, and (f) actual ODC costs, and updates schedule on the network
- Prepares summary presentation reports, tables, and narratives on earned value, estimate to complete, and estimate at completion and submits them to the customer and management, along with schedule updates with all tasks updated and conflicts resolved
- With the assistance of the project management office, analyzes resource usage from central resource pool file and identifies conflicts, issues, and problems for current and future projects

Baselining the schedule: A quality management action

Establishing the baseline schedule is a significant quality action in the project management process, signifying the official kick-off of the work and indicating a strong commitment to the quality objectives, quality procedures built into the schedule, and to the schedule and resource plans themselves. The baseline is the point of departure for monitoring and tracking the process. When a project is baselined, the project schedule is complete. The following are some rules of thumb for baselining:

1. The purpose is to get to a baseline schedule that captures all the work to be done, particularly quality procedures as summary tasks. This includes key documentation and procurement tasks. The baseline schedule does not change unless the basic scope changes. After agreement is reached, the project manager confirms the baseline by saving it and making it available on the network *as the baseline.* There is no uncertainty where the baseline is and how to access it.
2. The baseline schedule is the agreed upon, scrubbed schedule for the project, linked to the resource pool. The baseline shows all interdependencies, linkages; and resource requirements; includes all tasks necessary to get the work done; and shows impacts on parallel programs and resources. All procurements and test equipment are covered in the schedule.
3. The baseline schedule is resource-leveled—the schedule can be implemented with current, available resources. Assigned staff and the customer are aware of the commitments and have "signed-on" to complete their tasks to meet the schedule milestones.
4. Getting to the schedule baseline involves collaboration between the project manager and all departments and staff involved in planning and implementing the schedule, as well as the customer. A baseline meeting is held to arrive at final agreement on schedule and resources committed before the baseline is saved to the network. The project manager facilitates the meeting, and all department managers attend and come prepared to commit their resources to the final, agreed upon baseline schedule.

Schedules on a network

Managing schedules in a network environment involves ground rules for access and change, and for preserving the project manager's control of the baseline schedule. The following are the steps involved:

1. All baselined and planned schedules are housed on a server that can be accessed by the customer and the team. Separate folders will be located within this directory for each schedule. There are no dates in the file name. The central resource pool file will continue to be named "Central Resources," and will be housed in the same directory. Archive versions of schedules will be housed in a separate folder.

2. The project management department controls access to schedule files. Customers have "write" access to the schedules. Department managers, systems engineers, and other engineering and manufacturing staff have "read" access to program schedules.
3. The project management department is responsible for maintaining and updating program schedules on the network. The project manager will then save the schedule as a baseline schedule. Once baselined, the schedule will be placed in an administrative directory. The baseline schedule will be the only version of that schedule housed on the network and will serve as the source of "planned versus actual" tracking information.
4. In preparation for the weekly project review and weekly report, the project manager will update the schedule in consultation with department managers and the program team. To update the schedule, the manager will update percent complete for all tasks, and update all start and finish dates as appropriate. All updates will be made to the network schedule version.
5. Department managers will be responsible for maintaining their own department schedules and resource pool files, and assisting project managers in updating program schedules. Project managers and department managers will share schedules through email or hardcopy until updates are agreed upon.

Resource Planning

Customers are interested in resources, resource planning, and scheduling for projects they sponsor. They want to know who is working on their deliverable because *staffing is a major quality issue.* Because scheduling is essentially the process of planning for use of personnel and equipment resources, schedules and their resource files are of direct interest to customers. Good project management requires that there be a process to plan for future resources, to allocate current and projected resources to schedules, and to make shifts in resource management as required. The process must provide for a central resource pool to identify impacts of project schedules and ensure the efficient utilization of the workforce. The resource pool information on the network is shared with management staff and all team members to allow each team member to evaluate the scheduled work assigned and to provide guidance on task definition, durations, start and finish dates, and interdependencies.

Resource conflicts are identified through the resource pool. The project manager and program planner are responsible for identifying conflicts and root causes and involving the team in resolving issues. Corrective actions are taken by the program manager in consultation with the customer.

Long-Term Staff Planning

Involvement of the customer in long-term planning can help to build longer-term relationships with the customer. Long-term staff planning begins with some understanding of what the current workforce is doing, e.g., its current capacity. Standards are then developed that relate current capacity to performance standards. The process involves relating staffing levels and/or staffing mixes to standards for work and output. For instance, such a standard might define what a given level and mix of software certification engineers can produce, based on past history, e.g., five engineers are supporting three concurrent product development projects with an average turnaround on software certification documents of three weeks.

These standards are used to estimate what various alternative staffing levels might produce in the way of more capacity to produce certification documents sooner or produce more such documents in a given period of time. The issue in this case would be to see what new staffing levels would be required to put out twice as many software documents in the same three weeks.

The more immediate workforce decisions are made on the basis of schedule conflicts created by current scheduling impacts on the current workforce. Whenever a current conflict is created by current and planned schedules, program managers are expected to collaborate to resolve such issues or bring them to management.

Good workforce planning involves the staff themselves in the planning process. Providing team members with information on assignments helps to give them ownership on key project issues such as task interdependence and resolving resource conflicts. It also allows them to assist in estimating levels of hiring necessary to raise the capacity to improve performance to various levels to meet future demand. To provide access, program managers provide information from the central resource pool to all staff, reflecting assignments for all scheduled programs to team members. If staff confirms that these resource plans accurately reflect the work actually going on and anticipated, that would tell the manager that he or she is on the right path in the scheduling process. Positive staff reaction to getting this information would tell the manager that staff needs more information on expectations and assignments than they are getting.

Preparing Staffing Policy and Plans

This section defines the process for meeting staffing requirements through the development of staffing plans. Staffing plans are part of the project planning process to prepare for future customer and program requirements.

The objective is to initiate a planning process to prepare and implement staffing plans based on forecasted programs and workload requirements. Staffing plans are produced through a process of accounting for current staff activity, then forecasting program and staffing needs.

It is the responsibility of department managers to periodically review staffing and resource requirements for their departments, and to develop staffing plans to meet future program requirements. Based on these plans, department managers are jointly responsible for acquiring the necessary staff resources and building their capacity to meet projected needs.

The planning process involves six steps:

1. Determine department staffing levels and assignments
2. Develop staffing/workload standards
3. Forecast future requirements
4. Develop department staffing requirements
5. Develop department staffing pattern
6. Prepare staffing plan

Step 1—Determine staffing levels and assignments

This step describes and documents current department staffing levels and staffing mixes, and relates current staffing levels to current assignments. The purpose of this step is to describe the current department workforce, functions, and assignments. One source of information for this review is the ID resource pool that reflects all program assignments from active program schedules.

The report from this step is shown in Table 7-1.

Step 2—Develop staffing standards

Staffing standards help compare staffing levels with performance goals. This step involves developing staffing standards that relate staffing levels and mixes of technical expertise to specific workloads or performance goals. Standards compare staffing with particular product development workloads, schedules, activities, documents, and other outputs. This allows the department manager to forecast what kind of staffing increases are necessary to meet future needs for timely product development. Functional areas, workload factors, and standards should be tailored to the department's functions, but should relate performance, e.g., meeting a functional requirement within a given schedule requirement, to staffing levels.

Table 7-2 shows a recommended sample format.

TABLE 7-1 Sample Staffing Report

Department role/ function				
Software department	Position	Current staffing level	Task summary	Job assignment
Software certification	Software certification engineer	3	Support program management	Requirements definition Software code development Software documentation

TABLE 7-2 Recommended Staffing Standards Format

Staffing level	Functional area	Workload factor	Task duration	Standard
3 Software certification engineers	Software certification documentation	Concurrently prepare all software certification documentation requirements for three program phases	45 days	Meet certification documentation requirements in 45 days for three programs concurrently, with three engineers

Step 3—Forecast future requirements

This step identifies future program and project requirements for the department, e.g., what workload the department is likely to be required to do in the future, and forecasts the appropriate staffing level and mix to meet those requirements. Future requirement information will be obtained from program planning forecasts and other information on future project demands, as well as analysis of current assignments and durations.

Step 4—Develop department staffing requirements

This step involves developing a department staffing pattern, which defines the level and mix of staff required to meet current and future needs, and placing them into a department organization chart. The staffing pattern identifies core positions, e.g., positions that are essential to carrying out the department's role and function, as well as support and junior/training positions as required.

Step 5—Develop department staffing patterns

This step produces a plan to acquire the levels and mix of staff necessary to meet future department staffing needs. This step determines various sources of the expertise needed, various employment arrangements, e.g., employee, contractor, etc., and outlines a hiring and/or internal career development approach.

Step 6—Prepare staffing plan

A department staffing plan includes the results of previous steps. It includes current staffing levels and mix, and assignments; staffing/workload standards; a forecast of future program requirements; future department staffing requirements; a department staffing pattern; and a staffing implementation plan.

Program Review

The manager of projects or program manager holds weekly program review meetings to discuss broad program issues; detailed technical, resource, and schedule problems; project team performance; and risks. Customers are invited to attend. Project managers are responsible for preparing for these reviews and anticipating key agenda items. Department managers attend program review meetings as appropriate.

The project manager is required to track the progress of the program on an on-going basis and update the program schedule on a weekly basis. The program plan is updated as changes in plans warrant. Project managers are required to hold periodic reviews with the program team in preparation for reporting and to support task assignments and feedback, either as a single meeting with all functions represented, or as a series of meetings with major functional areas represented at each meeting. The project managers are required to report progress to customers and management at a minimum of biweekly intervals.

Problems that significantly impact cost, schedule, or product performance must be identified, investigated, and reported to management so that decisions can be made on a fully informed and timely basis. As part of the program tracking and reporting, it is the program manager's responsibility to identify those problem areas, and to assume responsibility for resolving them.

A good tracking system and weekly program review meetings are keys to the success of the project management system. The manager follows up on program reviews, and the program planner is responsible for maintaining program review agendas and documenting action follow-up. In support of program review, the project management office:

- Flags current and new issues for the week
- Distributes assignments to staff and gathers feedback
- Identifies conflicts and facilitates resolution
- Provides weekly hardcopy updates of all schedules before program review meetings with summaries of variance between planned and actual, by schedule (text and graphics); problem areas with a plan for correcting the problems; risk management issues report; 12-month lead time resource assignment reports; labor utilization and earned value; ETC (estimated time to complete); and EAC (estimated budget at completion)

Development of Customer-Driven Program Manager Competencies

Project managers must be more than administrators of their schedules. To build and improve the program management capacity of the organization, project managers are trained and developed, emphasizing key competencies. Training in project management approaches, tools, and techniques is provided to all such managers, as well as team facilitator training, and project management software training. A typical training and development program consists of workshops on:

- Customer interface, which covers an understanding of the business value of customer relations, the particular customer processes and key markets, key contact points, negotiation and presentation skills, and communication
- The project management organization, which covers the matrix relationships in the organization, roles, functions, and responsibilities
- Program planning, which covers the planning process, planning outputs, and the structure of the project plan, e.g., objectives, scope of work, schedule, and budgeting
- Scheduling and resource management, which covers proficiency in project management software, task linkages, resource planning, and tracking practices
- Budgeting, which covers budgeting techniques, standard labor and overhead costs, fixed costs, and accounting codes for capturing project costs
- Staffing skills, which covers workforce and staff planning approaches, staffing patterns, job definitions, and hiring and interviewing
- Team facilitation, which covers team and group facilitation skills, including how to enable the customer and the team to interact productively

Agile Project Management

Gary Chin, in *Agile Project Management: How to Succeed in the Face of Changing Project Requirements* (AMACOM 2004), challenges traditional project management in his discussion of keeping the process flexible. His admonitions are especially applicable to quality issues in new product development.

Here are the salient points in Chin's book:

1. Estimates of resources and work versus commitment. While classic project management focuses on resources and time, agile project management focuses more on team commitment and key milestones. Less attention is paid to tight resource allocations, and more to ensuring that team members do whatever they need to do to get the whole project completed.
2. The project manager takes an external perspective, not internal, and sees business risks. Program and project managers are seen not simply as "task masters" over schedules and budgets, but rather representatives of the business, extensions of the business in the project environment. The external perspective places the project and program manager both in the world of translating, communicating, and negotiating with project sponsors, sensing market and global changes affecting the project deliverable, and adjusting to change.

3. Achievements versus activities. The emphasis is on achievement of the project objectives, with less attention on strict enforcement of work activities and tasks. In other words, team members are empowered to change the way work is conceived and done in the agile environment. This gives team members the challenge of owning the key deliverables and outcomes of the project and shifting job tasks and redefining and reinventing work processes to get the job done faster and better.
4. Shorten the time horizons. Because of the changes in project variables and competitive forces, the time horizon for planning becomes shorter and more focused. The 80-hour rule becomes the approach, while longer-term planning and milestones are left more vague and undetermined. The assumption here is that project risk is increased if a project locks into a structured approach over the long term and cannot adjust when necessary. The focus is on "cost to complete" and remaining work, looking ahead to redefine and redirect the project work given the situation at any given project review point.
5. Technical skills versus adaptability. Agile project management requires a stronger concentration on adaptability than technical skills. Team members are recruited and developed not simply to perform technical work, but to be able to adjust and play a variety of roles across many technical interfaces, e.g., electrical engineers are expected to understand software engineering; mechanical engineers work with electrical and software engineers in "seamless" teams to integrate products and get them to market.
6. Variances in external forces versus variance in plans. More attention to external influences, e.g., the price and quality of contract supplies, changes in market demand and customer requirements, and less attention to variance in plans, schedule, and cost. This does not mean that earned value and variance are no longer useful; rather, that short-term shifts and variances are seen as indicators of the future rather than strict guidelines.
7. Achieving business results versus managing schedule, scope, and resources. Program and project managers are encouraged to look at business results, e.g., cash flow projections for marketing the project deliverable, profitability and cost reduction, and sponsor satisfaction. Team members are delegated more responsibility to manage day-to-day work schedules and variances with their eyes on the project outcome.
8. Achievement-based networks rather than Gantt charts. Track network diagrams and "project paths and interdependencies," rather than standard Gantt chart schedules. Achievements are seen as key "gateways" or milestones that must be reached.
9. Look at decisions rather than activities. More emphasis on decisions made in a project and less on activities per se. This trend has more project managers looking at decision *trees* and key options at various project crossroads, rather than assuming project tasks are settled and pose no decision challenges. In other words, if a project manager must make a decision to deal

with a future risk and there are important implications for taking one or another decision path, those decision paths are mapped into the schedule and team members contribute to the decision-making process. Chin's "Project Data Sheet" helps to integrated network diagram and decision milestones into project planning.

10. Successful product versus successful project; more emphasis on the outcomes and products of the project than the project itself. Projects that meet schedule and cost goals but do not produce marketable and profitable deliverables are not considered successful projects.
11. Integrating the project and the business. Rather than seeing projects as separate from the business, agile project managers *consider themselves the business*. In practice this means that project managers are aware of business considerations, e.g., market share, product cost and pricing, competition, quality, and customer satisfaction, and reflect them in project decisions.
12. Build contingencies into schedules and plans. This approach gives more priority to looking at key risk decisions and options and building contingencies into the baseline schedule rather than keeping them outside the project circle until needed.
13. Alternative pathways to deliverable versus shortest path. Program and project managers encourage team members and support staff, including the project management office, to identify shorter pathways to task and project outcomes, loosening up the work breakdown structure and schedule and "authorizing" different ways of accomplishing the work.
14. Expect versus discourage change. Project changes are encouraged as the project team and the customer learn from the project, and change management practices and procedures take on more significance. The built-in bias against change is transformed into a "learning environment" that adjusts to change and new insights especially in new product development.
15. Reinventing project boundaries versus enforcing them. Boundaries between functional managers and project managers are blurred as team members cross technical and project boundaries to get the job done, e.g., professional engineers help technicians understand and complete testing requirements; purchasing agents spend time with team members to understand supply and equipment needs and issues to allow them to serve the project team more effectively.
16. Risk-based infrastructure for the agile company. The agile project management company builds a supporting culture for risk-based scheduling and decision making. Support systems encourage tight short-term management but provide for analysis of future options in project reviews.
17. Balance innovation and process. Barriers to innovation and creativity are removed in the project planning and execution process; premium is placed on finding better ways to get the work done and to redefining the work with "work arounds" and "out of the box" thinking.

18. Emphasis on the execution stage versus emphasis on the planning stage. Risk and uncertainty encourage more focus on execution, corrective action, and agility, less on long-term planning. Rather than spending 20 percent of the project life cycle on planning and getting to a baseline, work is initiated with skeletal scheduling (5 percent) and cost information. More emphasis is placed on midstream adjustments and realignments of tasks to new forces that surface during project execution.

Chapter

8

Measuring New Product Development

Tools and Techniques

This chapter presents some of the advanced measurement and quality tools and techniques for new product development and testing. As we stressed in Chapter 7, because new products often require new design and testing processes, there is more to lose if these processes are inaccurate or unsuited to the product and customer specification.

Again, this is where quality and Six Sigma become important in new product development. The process must ensure that measurements can be made of three aspects of new product development:

1. Design to quality
2. Design to Six Sigma
3. Design to cost

Design to Quality

Design to quality in new product development means crafting the product solution around customer requirements and then making sure the prototype tested meets those requirements. What makes this especially difficult in new product development is that customer requirements are often uncertain simply because the product and associated processes have not been implemented or tested in the field in a customer setting. The risk here is that the product will test to requirements, but that the requirements do not link to the customer when market tested in the field. Thus design to quality requires a linkage from early design through to market field test and launch, a seamless interface that allows design to respond to real-time market testing.

Design to Six Sigma

Six Sigma is a concept grounded in total quality and continuous improvement. The six sigma approach focuses on quality assurance in the design and development process, measuring for continuous improvement of product performance, and empowering the workforce to understand their *systems* and *processes* and fix them when appropriate.

Six Sigma is a measure of quality that drives an organization to achieve near perfection through a *management-by-fact and data-dr*iven process that defines a *defect* as any product performance measurement that is outside customer specifications. A defect in Six Sigma terms is six standard deviations between the mean and nearest specification limit.

Designing to Six Sigma means designing the product so that components, subsystems, total system, and interfaces all perform exactly to specifications directly linked to customer requirements. Designers and developers—and partners—work from the product breakdown structure up to total system performance to ensure that each component is tested to specifications.

Six Sigma is also a way of thinking and approaching the development of a new product. The effort is driven by a sense of integrity and professionalism. Work is aimed at doing it right the first time and striving for perfection in product development. The tone of an organization that walks the talk of Six Sigma is professional, collaborative, and disciplined.

Design to Cost

The new product is designed to cost when each component and the product itself is configured to reduce costs of production to the lowest possible point while still meeting requirements. This is a special burden on early engineering and architecture of the product, stressing the need to find low-cost, alternative components that will reduce the cost of production and thus allow more competitive pricing.

Design to Process

Design to process means that the product is not only designed to meet product specifications, but also to meet process specifications. Process specifications address the process or context of product use, and identify all the necessary parties in the whole process of successfully implementing market launch. In other words, if a company is designing a new, automated self-service checkout system for a grocery chain, the design also addresses how the system is embedded in a grocery store, how the system will be introduced to customers, how the system will be supported by employees, and how the system could change current employer-employee relations in those stores. All of these factors are process issues that can impact successful launch of the product.

More Tools

Each of the following tools can help in establishing and implementing a measurement system for new product development:

- Concurrent engineering
- Quality function deployment
- Robust design
- Statistical process control
- Cost of poor quality
- Miscellaneous other methodologies

Many times a system or new product process must be developed or completely redesigned to adapt to a new product or service concept. System improvement focuses on the development or redesign of systems. The system can be as complicated as a new way to design a soft drink bottle or dispenser or as simple as a car door. The system improvement tools and techniques can be used for any system—a system, subsystem, or part. In fact, some of the tools, such as statistical process control (SPC) and quality function deployment (QFD), have been used successfully for the continuous improvement of entire organizational systems.

Because the performance of a product is critical to customer satisfaction, this chapter focuses on the system improvement of a product. The product is any output to a customer including a system, subsystem, or part. A major impact on product performance is the product design, process design, and production processes.

The tools and techniques described in this chapter have specific application in the product design, process design, and production processes. However, these tools and techniques can be used to improve any system in the organization. They are applicable for improving whole systems, subsystems, or parts.

System Development/Improvement

As stated earlier in this chapter, system development/improvement focuses on improving the actual performance of a product through product design, process design, and planning of the production processes.

System improvement starts with the customer. Next, the new product and processes are designed. The voice of the customer carries through the product design and process design to the actual production of the product. Within the product design, process design, and production processes, specific tools are useful for ensuring customer satisfaction. The specific tools are concurrent engineering (CE), robust design (RD), quality functional deployment (QFD), statistical process control (SPC), and cost of poor quality (COPQ).

Concurrent engineering is useful during the product and process planning and design phases for reducing the time and cost of product development.

Quality functional deployment is beneficial for carrying the voice of the customer throughout the entire process. Robust design focuses on designing in quality by eliminating loss. Statistical process control is a technique for measuring process behavior during production. Cost of poor quality emphasizes eliminating waste in all the processes.

Concurrent Engineering

Concurrent engineering (CE) is a philosophy and set of guiding principles where product design and process design are developed concurrently, i.e., with some product design and process development overlapping. With sequential engineering, the engineering phases are accomplished one after the other. Concurrent engineering overlaps the engineering phases.

Concurrent engineering is a system of new product development in which design and development tasks occur in parallel, with a high level of interaction between functions and tasks. This process requires a management and cultural environment, teams, and an improvement system focusing on collaboration and a commitment to customer satisfaction.

The concurrent engineering philosophy emphasizes a customer focus. It advocates an organizationwide, systematic approach using a disciplined methodology. It stresses the never-ending improvements of product, processes, production, and support. It involves the concurrent, simultaneous, or overlapping accomplishment of the phases of the project. For instance, the concept and design phases are accomplished concurrently. The design and development phases are performed simultaneously. The development and production phases are done with some overlapping activities. In most cases of concurrent engineering, all the phases contain some overlapping activities. Concurrent engineering requires upper management's active leadership and support to be successful. It accents robust design that decreases loss. It aims at reducing cost and time while improving quality and productivity. It uses the latest engineering planning initiatives including automation. Concurrent engineering forges a new reliance on multifunctional teams using tools and techniques such as six sigma, quality function deployment, design of experiments, the Kaizen approach to continuous improvement, statistical process control, and so on.

A more formal definition from the Institute for Defense Analysis Report R-338 states, "Concurrent Engineering is a systematic approach to the integrated, concurrent design of products and their related processes, including manufacture and support. This approach is intended to cause the developers, from the outset, to consider all elements of the product life cycle from conception, through disposal, including quality, cost, schedule, and user requirements."

The concurrent engineering steps are:

1. Establish a multifunctional team. Ensure representation from all required disciplines. The team should include representatives from such functions as

Systems/Design Engineering, Reliability and Maintainability Engineering, Test Engineering, Manufacturing Engineering, Production Engineering, Purchasing, Manufacturing Test and Assembly, Logistics Engineering, Supportability Engineering, Marketing, Finance, and Accounting.

2. Use a systematic disciplined approach. Select a specific approach using appropriate tools and techniques.
3. Determine customer requirements. Be sure to communicate with customers.
4. Develop product design, process design, and the planning of production and support processes together.

Robust Design

Robust design means designing a product having minimal quality losses. There are several methodologies associated with robust design. The major ones are traditional design of experiments (DOE) and the Taguchi approach. Traditional design of experiments is an experimental tool used to establish both parametric relationships and a product/process model in the early (applied research) stages of the design process. However, traditional design of experiments can be very costly, particularly when it is desired to examine many parameters and their interaction. Traditional DOE examines various causes of performance for their contribution to variation, with a focus on arriving at the most influential causes of variation. Traditional design of experiments may be a useful tool in the preliminary design stage for modeling, parameter determination, research, and establishing a general understanding of product phenomena.

A major approach to robust design is the Taguchi approach. Born January 1, 1924, in Tokamachi, Japan, Gen'ichi Taguchi is an engineer and statistician. From the 1950s onwards, Taguchi developed a methodology for applying statistics to improve the quality of manufactured goods by focusing on robust design. The Taguchi approach focuses on quality optimization. "Quality optimization" is based on Dr. Taguchi's definition of quality. Taguchi, in his book *Introduction to Quality Engineering*, states, "Quality is the (measure of degree of) loss a product causes after being shipped, other than any losses caused by its intrinsic functions." Simply put, any failure to satisfy the customer is a loss. Loss is determined by variation of performance from optimum target values. Loss, therefore, in the form of variability from best target values, is the enemy of quality. The goal is to minimize variation by designing a system (product, process, or part) having the best combination of factors, i.e., centering on the optimum target values with minimal variability. By focusing on the bull's eye, the product, process, or part is insensitive to those normally uncontrollable "noise" factors that contribute to poor product performance and business failures. The Taguchi approach is not simply "just another form of design-of-experiments." It is a major part of a successful quality philosophy.

Loss Function

The loss function is a key element of the Taguchi approach. The loss function examines the costs associated with any variation from the target value of a quality characteristic. Any variation from the target is a loss. At the target value, there is little or no loss contribution to cost. The further away from the target, the higher the costs. Costs get higher as values of the quality characteristic move from "best" to "better" to "poor" levels.

Robust Design Phases

In the so-called Taguchi approach, variation is the enemy of quality, and the design of a product or a process is depicted in three phases to address variation:

- Systems (part) or concept design—This phase arrives at the design architecture (size, shape, materials, number of parts, and so on) by looking at the best available technology and picturing the whole, target customer system in operation, not just the new product.
- Parameter (or robust) design—This stage focuses on establishing parameters or measures of robust design that can serve as surrogates for perfection, e.g., mean-time-between-failures.
- Tolerance design—This stage focuses on setting tight tolerances to reduce variation in performance. Because it is the phase most responsible for adding costs, it is essential to reduce the need for setting tight tolerances by successfully producing robust products and processes in the parameter design phase.

Statistical Process Control

Statistical process control (SPC) is a statistical tool for monitoring and controlling a process. SPC monitors the variation in a process with the aim to produce the product at its best target values. SPC uses process charts consisting of data plots, upper control limit (UCL), lower control limit (LCL), and the mean for the process. The variation is the result of both common and special/assignable causes. Common causes produce normal variation in an established process. Special/assignable causes are abnormal causes of variation in the process.

There are four steps in SPC:

1. Measure the process. Ensure data collection is thorough, complete, and accurate.
2. Bring the process under statistical control. Eliminate special/assignable causes.
3. Monitor the process. Keep the process under statistical control.
4. Improve the process toward best target value.

Cost of Poor Quality

Cost of quality is a system providing managers with cost details often hidden from them. Cost of quality includes both the cost of conformance and the cost of nonconformance to quality requirements. Cost of conformance consists of all the costs associated with maintaining acceptable quality. The cost of nonconformance or the "cost of poor quality" is the total cost incurred as a result of failure to achieve quality. Historically, organizations looked at all costs of quality. Today, many excellent organizations are concentrating strictly on the nonconformance costs. This highlights the waste, or losses, due to deviation from best target values. After these costs are determined, they can be reduced or eliminated through application of the continuous improvement philosophy.

Typically, the cost of nonconformance includes items such as inspection, warranty, litigation, scrap and rejects, rework, testing, re-testing, change orders, errors, lengthy cycle times, inventory, and customer complaints.

Other Measurement Tools

There are many tools and methodologies that can be used for system development/improvement in new product development. Some of the more common systems include: Just-in-time (JIT); Total Production Maintenance (TPM); Mistake-proofing; Enterprise/Manufacturing Resource Planning (MRPII); Computer-Aided Design, Computer-Aided Engineering and Computer-Aided Manufacturing (CAD/CAE/CAM); Computer Integrated Manufacturing (CIM); Computer Systems; Information Systems (IS); and Total Integrated Logistics (TIL).

Just-in-time

Just-in-time (JIT) is a method of assuring that the right material is available just in time to be used in production, thus eliminating the need for inventory and associated costs. JIT reduces inventory and allows immediate correction of defects. This methodology is used for reducing waste, decreasing costs, and preventing errors.

Total production maintenance

Total Production Maintenance (TPM) is a system for involving the total organization in new product support and maintenance activities. TPM involves focusing specifically on equipment maintenance. TPM emphasizes involvement of everyone and everything, continuous improvement, training, optimum life cycle cost, prevention of defects, and quality design. This methodology is effective for improving all production maintenance activities.

Mistake-proofing

Mistake-proofing is a method for avoiding simple human error at work. The application of mistake-proofing frees the workers from concentrating on simple tasks and allows them more time for process improvement activities. This is a major measure in the prevention of defects.

Enterprise and manufacturing resource planning

Enterprise and/or Manufacturing Resource Planning (MRPII) is an overall system for planning and controlling a manufacturing company's operations. MRPII is used as a management tool to monitor and control manufacturing operations.

Computer-aided design, computer-aided engineering, and computer- aided manufacturing

Computer-Aided Design, Computer-Aided Engineering, and Computer-Aided Manufacturing (CAD/CAE/CAM) are automated systems for assisting in the design, engineering, and manufacturing processes. CAD/CAE/CAM are used to improve systems and processes, enhance product and process design, reduce time, and eliminate losses. In new product development processes, new design software now allows an almost unlimited variety of graphical representations of a new product, and capacity to share widely through the Internet with project stakeholders. Now mobile phones with Internet capacity can be accessed for product designs and comments.

Total integrated logistics

The concept of logistics covers all the resources and materials used in producing a new product, as well as solutions to supply, distribution, and inventory issues. Total Integrated Logistics (TIL) is the integration of all the logistics elements involved in the inputs to the organization, all the processes within the organization, and the outputs of the organization to ensure total customer supportability at an optimum life cycle cost. This method aims at total customer satisfaction by supporting the operations of the organization and the customer. Total Integrated Logistics can be a major differentiator.

System Development/Improvement Methodologies within the DoD

There are many methodologies that are used specifically by the U.S. Department of Defense (DoD) that have application in the commercial world. Many of the tools and techniques described in this book can be attributed to the DoD or defense-related federal government agencies. Some examples of the more specific DoD TQM methodologies include: Six Sigma, Computer-Aided Acquisition and Logistics Support (CALS); In-Plant Quality Evaluation Program, R&M 2000, and Value Engineering. This is not an all-inclusive list. The DoD has many TQM methodologies being used in all its agencies to continuously improve its processes focusing on customer satisfaction.

Computer-aided acquisition and logistics support

The Computer-Aided Acquisition and Logistics Support (CALS) program is a strategy to institute within DoD and industry an integrated "system of systems" to create, transmit, and use technical information in digital form to design, manufacture, and support weapon systems and equipment, and apply communication and computer technology to acquire and support major weapon systems and information systems. CALS focuses on integrating automation between the DoD contractor and the DoD. This is a Department of Defense program to acquire, manage, access, and distribute weapon systems information more efficiently. This includes all acquisition, design, manufacturing and logistics information. CALS focuses on an increase in reliability, maintainability, and availability through the integration of automation systems. In addition, CALS seeks improvement of the productivity, quality, and timeliness of logistics support while again reducing costs.

In-plant quality evaluation program

The In-Plant Quality Evaluation (IQUE) program changes the method by which in-plant government people evaluate contractor controls over product quality. The IQUE program changes some of the traditional methods of evaluation with a TQM approach. The IQUE approach focuses on measuring and continuously improving processes with the aim toward quality (customer satisfaction). It concentrates on the "what" versus the "how." The government provides the "what," and the contractor determines the "how." IQUE implements a cooperative team concept between government and contractors.

R&M 2000

The Reliability and Maintainability (R&M) 2000 approach is geared to increasing combat capability while reducing costs through R&M practices. It stresses improvements in R&M to increase combat availability and reduce logistics support requirements. The R&M 2000 principles build on the traditional Total Quality Management approach. R&M 2000 stresses the need for management involvement (leadership), requirements (vision/mission, involvement of everyone and everything focused on customer satisfaction), preservation (continuous improvement of processes and years of commitment and support), design and growth (training and ownership), and motivation (rewards and recognition).

Value engineering

Value Engineering (VE) is an organized effort directed at analyzing the function of systems, equipment, facilities, services, and supplies for the purpose of achieving essential functions at the lowest life cycle cost consistent with performance, reliability, maintainability, interchangeability, product quality, and safety. This specific DoD weapon system again stresses the need to improve the quality and productivity of the DoD and DoD contractors while reducing cost.

Measuring the Success of New Product Development Mainstreaming

Originators of successful new products and services often measure their success in terms of financial and marketing success. However, a major outcome of successful new product development is the *mainstreaming* of a product or service. Mainstreaming means that the product or service becomes an accepted part of a system to the point that it is no longer a new product. It is essentially assimilated into the sponsor or customer's way of doing business and meeting expectations, needs, and requirements. This is often seen as a negative outcome based on the view that when a sponsoring agent or client integrates a new product, system, or problem solving approach into their culture and their way of doing business, it is a sign of failure.

We would argue that a good measurement of new product success is in fact the extent to which a product or service is actually assimilated, the process of embedding the product into a business or consumer activity so that it is no longer seen as a new product. It becomes *part of the base*. While the originator may see this as loss of revenue and market share, the challenge is to build on that new product with follow-on ideas and concepts that build on the relationship already built on the success of the first product.

This is the continuous improvement concept that drives the search for new concepts and products, and which holds the key to *letting go* of new products in order to create new ones.

Measuring New Product Workmanship

Charts such as the one shown in Figure 8-1 are used to track new product quality and workmanship in test management. This particular chart shows quantities of defects in the following categories for a new product called "SAV" for this example: missing part, wrong part, misaligned part, lifted pads, damaged part, long leads, and other. Failure quantities, shown as dark gray vertical bars, are under FMEA standards (FMEA standard tests a product until it fails—to test product failure points); quantities shown in a light gray vertical bars are test results under customer requirement standards. In this example, failures attributed to *wrong part*, damaged *part*, and *misaligned* parts appear to be major failure categories under customer requirement testing and FMEA testing.

Product failure indicators like these must be tested under at least two conditions—those that test product performance against specifications under customer conditions, and those that test the limits of product performance under FMEA studies.

If verified through retesting, the source(s) of these product failures are investigated using root cause analysis. Root cause analysis uses the fish bone diagram shown in Figure 8-2 to get at the root of the failure. A small group of product team members is gathered to brainstorm root causes and to drill deeper until the *real* cause is uncovered. In the following case, it was determined that the root causes could be associated with misaligned parts:

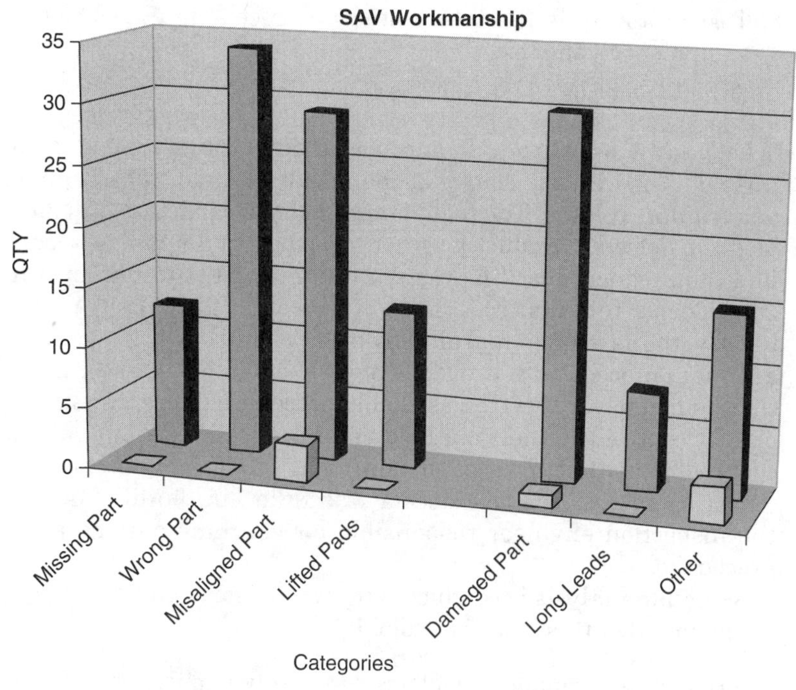

Figure 8-1 New product SAV test results.

1. Supplier quality—is a supplied part misaligned, and could the misalignment be due to a contractor error?
2. Design—is the design of the new product defective, and does it lead to misalignment of parts due to errors in tolerances?
3. Assembly—is the assembly process at the heart of misalignment, e.g., assemblers are not correctly assembling parts?
4. Training—is the lack of training of designers, developers, and/or assemblers the problem?

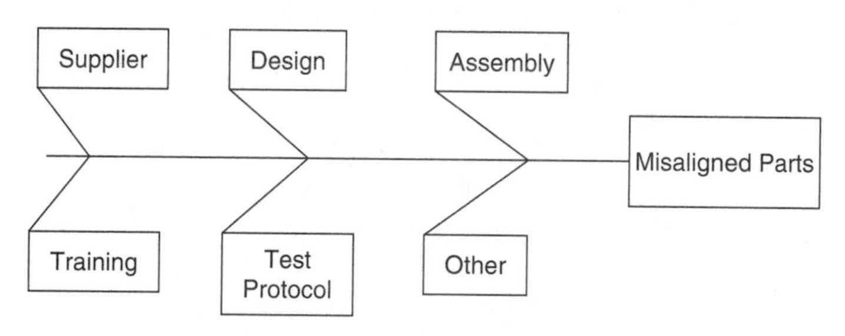

Figure 8-2 Fish bone diagram for root causes.

5. Test protocol—is the test protocol correct, and does the test process and equipment fit the test?
6. Other—new source, unspecified.

Let's say it is confirmed in another production and testing of a prototype after this analysis that the source of the misaligned part was found to be *assembly and training* related. The team found that the particular assembler who assembled the defected product was not aware of an alignment technique designed into the product, and this was not communicated to him in training. Drilling down deeper into that root cause, it was found that the training did not occur because the assembler had already been "checked out" by quality control on this assembly procedure; however, when asked about it, the assembler indicated that he had never seen the assembly procedure. More investigation found that quality control routinely checked veteran assemblers out on new product procedures without actually confirming that the training occurred.

In this case, corrective action was taken in quality control by replacing the inspection engineer responsible for checking out the assembler on this procedure.

A similar analysis is conducted for wrong parts and damaged parts.

The moral of this story is twofold:

1. Product performance problems are rarely simply technical or simply organizational or managerial. They often involve technical and organizational elements and are resolved through action on both fronts. Good measurement of new product performance can uncover technical and organizational/management issues in product quality. Unless project management insists on this kind of inquiry when these Six Sigma–type violations occur in test management, they will likely remain uncovered.
2. Unless the testing and measurement process is constrained by goals or standards, e.g., product specifications or FMEA standards, the process has no firm ground to declare the product fit for use.

Chapter

9

Project Management and Teamwork

Team Dynamics

Systems and processes don't produce and market new products successfully—people do. We would like to describe the typical new product development team as a *high-performance team*, but the reality is that new product development is a *messy* and often frustrating process. We believe that the common picture of a *high-performance team* is a construct of an unreal, abstract, and academic view of teamwork. Those of us with experience in program and project management in a new product development setting feel that this ideal vision of a team view simply does not describe what actually happens in producing a new product and rolling it out. It sets expectations too high, given human nature. The reality of guiding and managing a new product team involves the following realities of today's often irrational and *badly managed* new product company setting.

Small coalitions: There are many small coalitions of team members who work together to produce pieces of the new product process, with a lot of conflict involved. Communications are sporadic and vertical, often with no predictable patterns. Language and culture differences are increasing, especially with global outsourcing. Program managers *mix* with corporate managers and officers, customer and stakeholder representatives, and sometimes project sponsors and venture capitalists, but they often don't talk with each other enough.

Teamwork within a team: Teamwork within the team, e.g., work between the project manager and task leaders for major chunks of work at the high level, is again sporadic and unpredictable. Often we see individuals such as electrical engineers and software designers working relatively independently, and the team often splinters into small groupings.

Patience: A substantial amount of patience and tolerance is needed in the system so that people are able to work out differences for the good of the order.

Leadership: Sometimes leadership simply has to drive the process to work through brute organizational force.

Customer and User Diversity

Because new product development teams are ultimately focused on designing and producing products for the marketplace, it is important that these teams reflect in their composition and *texture* the wide diversity of customers and product users who use their products. Product users vary as demographics and marketing research suggest, but eventually products are sold to men and women of increasingly varied backgrounds, racial make-up, and national origin. While the new product development field is still the province of the more *male* model of management and still carried on largely by men in positions of power in marketing and project management, new product development teams are changing to reflect global economic realities. More women, minorities, and international team members are participating, generating more rich diversity of views but also more conflict and language issues.

Personal Growth in New Product Development

How does one survive and grow in the hurried and hassled environment of new product development? It begins with attitude. Because individual participants in new product development teams provide the energy and innovation of the team, it is important that team members have certain attributes. What are the characteristics of the optimal new product team member, and what guidance can we give to those who aspire to work in this kind of creative, innovative work setting? The process of building and sustaining a new product team through the product life cycle involves a wide variety of skills and experience. Product team members are tasked to perform particular technical tasks; the focus must be narrow and systematic. Team members are often disturbed by *interferences* in their work by what they see as irrational business and political factors in the organization.

Growth to New Products Program Manager Role

Because new product development is a messy and often muddled process, the management of new product projects requires a good deal of organizational savvy and insight. Progression from a narrow, technical view of new product development to a broader, more political view is necessary in building a good program manager.

Because leadership is key to establishing an effective framework for new products, and because that leadership comes from what we call the program management level, we now explore the progression from project manager to program manager. This growth process addresses the new skills and competencies required to make the transition from leading a single new product project to leading a longer-range, multiproject program in a real company setting. While there are many gray areas in the transition from project to program, it is clear that new competencies and skills are "called up" as the focus turns to long-term programs and the broader scope needed to monitor many projects at once.

Single Project Management

The management of single projects requires several analytic and leadership skills in addition to proficiency with project management software. Project managers must be able to work with a customer to develop the scope of work for the project, and to prepare a work breakdown structure that captures the deliverable in an organizational chart or outline. The project manager must be familiar with the technical field and changes in technology. The manager must be able to put a team together, assigning team members to activity and task areas in the work breakdown structure, and lead the team through the project life cycle. The manager must have the capacity to use project management software to prepare Gantt charts, assign resources, estimate costs, produce reports, and make presentations on the project.

There is an active debate in the field on the extent to which the project manager must be familiar with the technology of the project and technical aspects of design, development, testing, and product delivery. Some say the manager need only have a cursory sense of the deliverable and the technology involved, relying on team members and subject matter experts for technical assistance. Others say the manager must know enough "not to be snowed" by the customer, team members, or suppliers and subcontractors. They indicate that the manager will have to interpret technical progress reports of team members, and be able to communicate with technical counterparts in the customer organization. In any case, it is clear that the project manager must have a good grasp of the deliverable, and be comfortable in the field, if not an expert. If the field is changing rapidly, it is especially important for the project manager to grasp the implications of change for the current project.

Single project managers must be wholly focused on the day-to-day dynamics of the project because the project environment is always changing and shifting unexpectedly. The horizon is short in project management. This requires that the project manager manage through the 80-hour rule; that is, the focus in each weekly review is the current status of projects, indicated in earned value analysis, and anticipating the next 80 hours of work. This way the team is reminded of the next two weeks' work, challenges that can be anticipated, and key milestones in that 80-hour period. The single project manager is typically wrapped up in the project at hand.

Project managers must have several key skills:

- The ability to lead a team and resolve team problems
- The ability to communicate and report effectively to a wide variety of customers and stakeholders on technical and project issues

- The ability to manage a number of technical assignments all at once
- The capacity to deploy project management tools, such as Gantt charts and schedules
- A full understanding of the project life cycle
- Proficiency in project management software

In addition, the project manager must have judgment skills to make tradeoffs among cost, schedule, and quality during the progress of the project, and to make difficult decisions quickly to keep a project moving.

Finally, project managers are expected to be advocates for their projects and make effective arguments for resources and priorities based on their project needs and their project's "critical path." They are not necessarily expected to see the big picture or to make decisions on sharing resources with other project managers. They are expected to be narrowly focused on making their project goals, objectives, and deliverables, regardless of what else is happening in the company. If they start compromising their focus in the context of the needs of other projects, they do a disservice to their customers and to their project team.

Program (or Multiproject) Management

Program management is a different kettle of fish altogether. Program management is the process of managing a "portfolio" of projects, some of which are going on at the same time, and some of which are linked in a sequence of product enhancements over a longer time period. In either case, the program manager's span of control is wider and broader than the project manager's responsibility. Program managers can be responsible for a long line of products and product enhancements in one program area over time, say over a 10-year period, transitioning from one project to another based on customer and market feedback. Program managers can also be responsible for many projects and project managers across programs or product areas, thus complicating the management process. Program managers often have responsibility for broad corporate product lines and markets across wide technical boundaries.

Program managers typically hire and supervise other project managers. They are responsible for developing the company's project manager workforce, building and advancing project managers into higher levels of responsibility. They coach project managers. Program managers serve as the interface between projects and broader company strategies and business plans, thus they are called on to communicate a broad purpose to individual project managers and report individual project results to corporate executives in high-level program reviews.

The complex, multiproject environment that program managers face requires different skills in managing information. The program manager cannot get lost in the details of one project or one project schedule or report, but must have an "enterprisewide" perspective. Information on many projects must be managed so that the program manager can see the big picture and make the necessary tradeoffs between projects, if necessary, to resolve resource and priority problems. Program managers must be able to step away from project details, see the broader implications, and make decisions on the basis of corporatewide considerations and impacts.

Program managers mix in the milieu of vice presidents and CEOs of both their corporations and those of the customer. They tend to interface at high levels where decisions impact broad business performance, business-to-business relationships, workforce planning and management, regulatory issues, financial performance, stockholders, partnerships, and major supplier and contract issues. These factors often bring program managers into play with issues that single project managers may not fully understand, especially if program managers get in the way of achieving narrow project goals. This is the essence of program management—balancing between individual projects with broader performance issues and implications across projects.

There are many working definitions of a program manager. In Europe the program manager is seen differently. The focus of most UK program managers is likely to be change and change management. This view says that the program manager has the broad perspective to be able to manage several projects aimed at changing the company in fundamental ways, across a wide variety of company activities and divisions.

Gender and Minority Diversity in New Product Development

New product teams are becoming more *diverse and virtual*. The opportunities for women and minorities to assume project manager responsibilities are increasing as project firms recognize the need to reflect their increasingly diverse global customers *in the composition of their project teams*. One of the spillover impacts of globalization of project management is the need to generate a more diverse candidate pool for selection of project and task leaders. But what will more diversity bring to the project management field, especially in the area of project quality management? How will diversity affect the dual quality targets of conformance to specification and customer satisfaction?

Gender and cultural diversity in project management brings both benefits and potential conflicts. For instance, research suggests that in the traditional male-oriented project world, women are now bringing different dimensions to the project management workplace—perhaps more collaborative approaches to problem solving, more sensitivity to teamwork, more focus on new and innovative approaches, and maybe even more focus on scheduling and details. Experience suggests the sad truth that sometimes women have to work doubly hard to justify themselves, their ideas, and their management styles despite their qualifications and background, simply because they often find themselves in historically male project management work settings in new product development. This may be especially true in international project management settings in countries that have fundamentally different views of the role of women in society.

We deal here with women in traditional male new product roles, e.g., mechanical and electrical engineer, and their introduction to project management roles. Five key issues—none of them unique to women—will challenge women as they assume these new responsibilities in project management:

- Contingency management, being able to use a wide variety of management styles depending on the project situation
- Working to model leadership behaviors for other women and balance gender and cultural diversity in hiring to the project team
- Learning technical processes and business systems so they can see the "big" picture and perform with project peers who may have had more experience in broader areas of a given industry sector
- Adopting central themes that support project success, such as project quality management, and consistently "walking the talk" on those themes
- Developing their own career paths and development approach, starting with their own approach to key issues such as project quality management

Individual Responsibility as a New Product Team Member

This discussion explores some key personal and professional issues raised by the concepts of project quality and customer-driven new product project management. The purpose of this discussion is to concentrate on the key building block of people. Quality project management still comes back to individuals, growing and developing in an organization truly concerned with them and their growth as people. Despite the increasing tendency to assume that organizations have lost their allegiance to employees, companies will invest in their people, but their people—not the human resource office—have to do the groundwork and justify the expense.

Three subjects are discussed here to stimulate the thinking of women—and all candidates for project leadership positions:

- Self-assessment
- Using performance appraisal to get feedback
- Empowerment

The purpose of this discussion is to put some personal and professional meaning on the project quality management process. We start the process of change and empowerment *ourselves*, through our mental models, personal decisions, and choices that we make in the daily routine of the work place. While the employer and the organization can open up opportunities, the individual must take advantage of those opportunities for anything to happen. We must want to *be empowered to be empowered*, and we must seek out opportunities to grow and develop. We must draw the boundaries of our own jobs wide enough to allow creative uses of the new talents and competencies we acquire across larger and larger parts of the system.

But how should we "play" in such a world, one that, on the one hand, holds out so many opportunities for growth and empowerment in meeting customer needs, but on the other seems so determined to place accountability for failure at the level of the individual. What if the organization does not support empowerment and does not have the infrastructure to help widen the boundaries of a job?

Each individual member of the team is responsible for his or her specific change management strategy. Nothing will change unless there is awareness and understanding of one's potential in the context of quality and the curiosity and drive to listen to the customer. The organization and its leadership can help this happen through providing self-assessment tools and techniques, and through communicating with and empowering each employee, but the organization cannot totally make it happen. It requires the commitment of each working person, regardless of role or level in the organization. In effect, the focus on employee participation and involvement has never been clearer; those who take advantage of it will grow personally and professionally, and participate directly in making the organization more competitive both domestically and globally. Those who do not may find themselves regretful later—perhaps in different times with less stress on empowerment and self-development—that they did not take up the challenge when it was available.

Self-assessment starts with an attitude, one that welcomes feedback from several sources inside and outside the organization. The employee should be able to place that feedback into productive learning and growth. Because the whole process of looking at one's self through others' eyes raises questions of competency and comparison, it is not easy for team members and people in general to develop a healthy attitude about accepting feedback, especially if that feedback is negative. Women face the same challenge as men in this sense, but there will be a continuing tendency to attribute issues to women that derive from their being women instead of from their personalities per se. In other words, women need to be able to accept feedback and information in whatever form it comes, and to translate it to corrective action, if necessary, without taking it personally.

Self-assessment requires an attitude of openness. This openness evidences itself in many ways in the work place, and indicates a willingness to solicit and learn from others about your performance. Customers, peers, employees, and leaders will provide useful data if they feel the solicitation is genuine, and will welcome the opportunity to be candid. In most cases, they have harbored views of you long before you requested feedback, thus to open the conversation is often comforting for those who participate.

Here is the basic area of inquiry as one approaches the question, "How do I engage in a self-assessment process in the organization so that I can identify the best opportunities for my contribution to serving customer needs and continuous improvement?" Answers to this question can be solicited formally through questionnaires, but the richest feedback does not come from the written word. It comes from the daily interactions one has with customers and suppliers, and from the regular feedback that customers give about the "service."

Do it right

Do I do the right thing right the first time? Do I encourage quality in the way I do my job? Am I a good model for other team members? The process begins with the individual's sense of accountability for completing the job, adding quality and value to the workflow as it comes in and out of personal control, and assuring that one is not expecting someone else to fix a problem later. This challenge is at the heart of the enlightened approach to work: ensure that the inspection and appraisal functions are eliminated so that each point in the workflow can add value and not commit costly time and effort to doing work over. In project processes, this means that as the project moves from front-end analysis and planning in the first several phases, and as the project deliverable is designed, produced, and installed, each team member assumes that their individual component of the work breakdown structure is done right first, according to the needs of the next customer down the line in the project development process.

In literature about bureaucracy and project administration, the theory of the "tyranny of small decisions" is sometimes addressed. This theory indicates that the progress and productivity of an organization stems not from big investment decisions, but from the many small decisions that individuals make as the work progresses. In customer-driven project management, there are many decisions made that combine in effect to create the quality of the process itself. It starts with each member of the team and each party to the process feeling responsible and empowered to do the right thing right the first time, and to inquire of the next customer down the line if it is not clear what the right thing is.

Be a leader

As a leader, how do I come across in the organization with respect to serving internal customers, doing it right the first time, and continuously improving myself? This problem has to do with "the book" on an individual, those stories and anecdotal images that gather almost inevitably among peers, bosses, and colleagues about individuals. One can seek out "the book" on their own performance, but candid feedback is often difficult to get, especially if one of the issues in "the book" is one's inability to listen and seriously address opportunities to improve service to one's internal customers.

Serve the team

How do I contribute to gaining consensus in the team, and what is my role in moving the group from its orientation and dissatisfaction stages into resolution and production? This criterion has to do with whether one is a team player or an individual performer. Getting to this issue in a self-assessment involves understanding the cultural and social underpinnings of American workers and professionals. We have not been trained to work in teams; rather, we have been trained to do our own "thing" and pass it on, to hand off "my work" to the next work station. Part of the philosophy of project quality management has to do with the ability to rise above this attitude and enter into a full commitment to working every customer's agenda as we work our own. This means that in the analysis of opportunities, one is willing to contribute to the brainstorming process and educate the team. This means that in the development of corrective action and project development, one is willing to work within the context of meeting customers' needs down the line, not "handing off" the work or leaving it on the stump for someone else to pick up.

Develop yourself

What professional development opportunities do I seek in improving my grasp of my job? How about the tools and techniques of project quality management? This issue has to do with how willing you are to pursue training and education as the organization opens up possibilities to do so. The quality movement opens up new ways to learn and grow with the movement toward customer-driven work, but the energy to follow up on these opportunities must start with the individual, not the training or human resources office.

Dedicate yourself. What is my emotional investment in teamwork, and how much effort and energy am I able to bring to the process? This is a difficult issue to articulate, but it is characterized by the "effective" side of the workplace. Attitude drives behavior, and training and education guide the effectiveness of behavior in the workplace. Looking particularly at the issue of embracing the personal theme of quality management, the beginning of individual fulfillment in the organization for new women leaders may be in charting out the quality sphere as a key contribution to the project process.

The process of responding to the opportunities that project quality management brings to the organization begins with self-assessment and work planning—designing one's own career path based on feedback from others. This planning process does not look upward into the "home" organization, but rather looks outward into the customer's environment. Traditional upward work planning orients individuals to training and education that equip them for higher and higher levels of responsibility to manage resources and people. In contrast, outward work planning is oriented to learning internal and external customer needs and processes so that one can fine-tune the ability to work in a team process serving customer needs. Self-assessment is really "peer-assessment." Work plans for the former include training and education in supervising others and controlling programs; work plans for the latter include training and education in quality improvement, project management, customer service, understanding a market and a client base, and in team facilitation and management.

In a customer-driven organization, work planning does not start with discussions with the boss on program objectives, goals, and performance appraisal criteria, as in traditional performance management systems. It begins in a peer-assessment process that never ends—and that is driven by feedback from key colleagues and customers in many areas. These areas of personal inquiry include, for instance, your ability to work in teams and your customer orientation. Here are some of the questions that can be addressed in such a process, either conducted formally through questionnaires, or informally through open communication and dialogue. The idea here is to find your strengths and play to them.

Ability to work in teams

1. Do you spend time working on individual tasks or on team projects? Are you a team player?
2. Do you choose to be part of collective and shared effort, or are you a loner?
3. Do you participate effectively in team meetings?
4. Can you lead meetings effectively and control their agendas?

Customer orientation

1. Do you think about the "user" of your work; can you put yourself in the shoes of those downstream customers of your work and adjust what you are doing accordingly?
2. Do you concentrate on "doing it right the first time" so that others will not have to do it over again, or do you tend to assume that if your work is not satisfactory, someone will fix it?
3. Are you a good customer, in the sense that you are clear about your needs to suppliers upstream from you in the workflows?

Ability to see "workflows" and the "big picture"

1. Are you curious about the whole process in which you work, how workflows are designed and created and serve the ultimate customer?
2. Do you see the big picture issues in the organization and your role in the big picture?
3. Do you make suggestions on improvements in the work process?

Ability to do and lead quality work

1. Do you value doing work well and right the first time, and meeting your own expectations for the work?
2. Are you able to inspire others to develop their own personal themes of quality work?
3. Do you solicit customers' views of what high quality work is and adjust your standards accordingly?
4. Do you produce work that is thoroughly researched and accurate according to the customer?

Ability to do a high quantity of good work

1. Do you produce outputs and work products regularly?
2. How good is your turnaround, as customers see it?
3. Do you produce a high volume of work?

Use of resources and time

1. Are you an effective and efficient user of time and resources?
2. Are you aware of the passing of time and expenditure of money in the work that you do? Are you cost conscious?
3. Do you use the computer effectively to assist you in managing tasks, time, and resources?

Communications

1. Do you speak clearly and make effective presentations to groups, keeping to the point and "reading" the audience?
2. Do you prepare good written reports that are clear and concise?
3. Do you listen effectively, especially when your customers are talking and giving you feedback?

Interpersonal relationships

1. Do you interact successfully with a wide range of people?
2. Can you focus on the process and not the person in analyzing root causes and optional corrective actions?
3. Can you maintain control and composure in conflict situations?

Conceptual skills

1. Do you think critically about the issues, and look beyond superficial symptoms to discover underlying causes?
2. Do you have a good model in your mind about how things should go before you pursue a plan of action?
3. Do you have a good conceptual grasp of the customer's issues and needs so that you can meet them?

Problem-solving skills

1. How effective are you in identifying the problem and addressing it?
2. Are you able to simplify problems in order to manage them?
3. Are you comfortable with plans and programs to resolve problems, and are you able to stay with implementation to see the problems through to resolution?

Job knowledge

1. Do you keep current in your area of work?
2. Can you effectively translate your technical knowledge to guidelines for others?
3. Are you active in understanding the customer's job and technical expertise?

Organization of work

1. Are you good at setting objectives and sticking with them until completed?
2. Can you handle several tasks at once, or do you need to prioritize them and handle one at a time?
3. Do you use the computer to assist you in organizing the work?

Personal initiative

1. Do you take the initiative to change processes and procedures that do not work?
2. Are you willing to take on jobs and tasks that are not part of your job?
3. Do you communicate your ideas for continuous improvement and quality, even if they suggest more work for you?

Coaching and mentoring staff

1. Do your peers and staff seek out your advice on work-related problems?
2. Are you interested enough in the growth of those you manage to listen to their issues and give them guidance?
3. Do you act as a resource person, and are you accessible when needed?

Technical and professional competence

1. How professionally competent are you in your field, as evidenced by the number of peer inquiries you receive to help others do what you do?
2. What self- and professional-development activities do you engage in?
3. What measures do you implicitly use to gauge your own effectiveness as a technical member of a project team?

Doing Your Own Performance Appraisal

As you enter higher visibility positions such as project manager, performance appraisal comes differently. Less formal and more subtle, feedback from executives and high-ranking customers tends to be less clear than traditional performance feedback.

While most discussions of performance appraisal proceed from the manager's viewpoint, this discussion proceeds from the individual's viewpoint. How do you use the appraisal system to continuously improve? The assumption behind project quality management is that one seeks every possible avenue for feedback, even the often "dreaded" performance appraisal system, but often the more visible indicators are not there. For instance, as you assume more responsible project roles, the organization expects you to work individually, but think collectively. This means that in project reviews, you see the big picture and are able to think of the company and the market as a whole without getting buried in details, but that you are also able to represent the progress of a project in meaningful detail as well.

Customer-driven project management requires a high degree of alignment between individual members' goals, team goals, and organizational goals for meeting customer requirements. That alignment can be ensured through self-assessment, as described earlier in this chapter, but there is another important step in the customer-driven project management process. That is the process for gaining feedback on your individual and team performance.

Performance appraisals give everyone in an organization a unique chance to improve personally and professionally. But there is an important practical reason for focusing on improvement in the appraisal process. It is because many organizations consider a "meets standards" rating to equate with "job proficiency," the competent, high-quality performance they expect from all staff. In other words, you are expected to master your job requirements and be able to accomplish them simply to meet the standard. And while there are some exceptions to the rule, simply doing a good job does not equate with excellence and a high rating. What is increasingly valued in this process is the willingness to improve the job and the way things are done, and to improve personally and professionally at the same time. In general, this means that to achieve high performance ratings, one is expected to continuously improve the way things are done. Thus it pays to focus on improvement.

The following are seven steps that might help you use performance appraisal to achieve personal continuous improvement. These seven steps come from a model developed by the Logistics Management Institute.

The process first involves establishing a vision for your own improvement effort; then enabling that effort; then focusing your behavior and your expectations to achieve leadership skills and continuous improvement in your job and in the processes that you touch in your job; then helping others to improve; and, finally, evaluating your efforts to improve.

Step 1. Envision personal improvement

Before you can begin to improve and assume a project leader role, you have to decide that there is a need for improvement and then determine the general emphasis of your improvement effort. You therefore build you own self-awareness of the need and your ability to improve. Assessing your relationships within the organization and with your internal customers and your internal suppliers provides a fundamental understanding of where you are now. From this assessment, you can develop your expectations for your own behavior, and you can begin by creating a personal vision for your improvement. In other words, how do you see yourself improving, and what is your own personal vision for yourself?

This concept of envisioning your personal improvement and learning has been described by Peter Senge in *The Fifth Discipline: The Art and Practice of the Learning Organization* (Doubleday 1990). Senge says that if we have a clear vision of what state we want to reach and a clear picture of reality, e.g., where we are, we manage the resultant *creative tension* and work with the forces of change to get there.

Step 2. Enable personal improvement

Make your vision a reality by smoothing the road along which you will travel. This effort starts with educating yourself about which improvements are considered high priority. Seek training for yourself in the skills and principles you, your supervisor, your customers, and others see as essential to your effort. Enabling or empowering oneself is a process of learning—learning about your own potential, the "big picture," opportunities for training and education, participating in teams, and tools such as Total Quality Management that are available to achieve continuous improvement. You should also seek the support of others, not so much from the standpoint of gaining their approval, but more importantly to cultivate their help in removing the barriers to your effort. In effect, get yourself ready to improve!

Step 3. Focus on improvement

Focus your improvement effort through establishing goals for that effort and then ensuring that your improvement activities are aligned with those overall goals. You should develop a clear improvement strategy to guide your efforts and ultimately use that strategy to evaluate the success of those efforts. Making improvement a high personal priority and creating time in your schedule for improvement activities are vital to this effort. They are a clear demonstration to yourself and to others of your commitment to improvement. In other words, plan for improvement!

Step 4. Improve the job

Personify project quality management in the way you both do your job and perceive your job. Your job may be defined as the collection of the processes you own, and it may involve your leading others who process responsibility, such as a quality engineer. You should establish control over your job by defining your processes and understanding how those processes interrelate and relate to others, including your customers and your suppliers. By removing unnecessary complexity from your processes and pursuing small, incremental improvements, you will substantially increase the effectiveness of your own performance in your job, and you will greatly enhance your personal improvement effort. In essence, make the changes in your job that are necessary to make your life easier and more rewarding.

Step 5. Improve yourself

You can demonstrate leadership in the overall quality improvement effort through your commitment to personal improvement. This means that you establish and adhere to a structured, disciplined approach to improvement that clearly defines your goals and requires steady, consistent improvement in your personal performance. You should also facilitate communication between yourself and others. Remove the barriers you place in your own way, seek the assistance of others to remove the barriers you do not control, and work to eliminate your own fears of change and improvement. This is best done through education as well as communication with others. Depend on your vision as your guide for improvement, and use that vision to maintain your momentum. Start by looking at yourself!

Step 6. Help others improve

The traditional female attribute of nurturing and support has a place in project quality management. Through your improvement effort, you will help your unit as a whole improve. An essential part of your personal improvement effort should be to help others improve themselves and the organization. By training and coaching others, by creating more leaders, by working to create teams and eliminate barriers, and by encouraging others' improvement activities, you will spread your own example and enthusiasm throughout the organization. Personally, you can make a substantial contribution to the individual improvement efforts of others. Spread the word!

Step 7. Evaluate your improvement progress

Measure your success in your efforts to improve. By measuring your performance against your vision and your plan, and by documenting your improvement efforts so they may be shared with and used by others, including your supervisor, you will benefit the most from your efforts. Ensure through your evaluation that the improvement effort itself is rewarding and provides further incentive for the continuous improvement effort. Celebrate your success and the success of others!

Empowerment

Empowerment is first a personal strategy to improve. We cannot empower others until we are empowered. We empower ourselves to the extent that we take advantage of the process of ensuring that all employees, and particularly those close to the customer, have the flexibility and support to meet customer needs and expectations. This process is designed to open up the creative and innovative potential of the organization, and to put everyone to work thinking through their jobs in relation to customer needs. But what does it mean to each member of an organization when the organization communicates the message, "Since quality is the job of every employee, you are now empowered to carry it out." It sounds simple, and, of course, it assumes that each member of the organization seeks out and can handle the "handoff" of responsibility, authority, and resources that goes with empowerment. And it assumes the support is there to make it happen.

To the individual member of a customer-driven organization, the empowerment process can be a vexing problem. This is because it is not clear that the stated benefits of empowerment, improved quality of work life, professional and personal development, rewards and recognition, new opportunities to assume new jobs and roles, and increased latitude in decision making are really all so "beneficial" in practice. The following sections explore each and show how the individual can ensure an appropriate personal response in a project team environment.

Improved quality of work life

Empowerment is intended to allow employees to take more ownership for their jobs, and thereby improve the work experience. The assumption is that employees want more flexibility, and that change in the direction of more flexibility will make employees happier and more satisfied in their work. The underlying issue is whether the individual's quality of work life is indeed determined by having more flexibility. The personal issue here is: "If I accept more flexibility and empowerment in the project team, will I benefit in terms of the quality of my work life and work place?"

The answer is liable to be, "It depends." It depends on whether empowerment is granted to all colleagues and team members, so everyone can truly negotiate on relative roles. It does no good, and may do harm to the project team, if only some of its members are empowered. It also depends on ability and competence to perform; empowerment brings with it new challenges for creative and critical thinking, conceptual skills, communication skills, and a generally increased self-confidence in the workplace. Finally, will empowerment allow team members to really look at fundamental system and process design problems and correct them, or is the empowerment limited to narrow issues with little long-term consequence for the organization's performance? Thus with empowerment must come education and training.

Professional and personal development

Do you really want to grow and develop in the organization? How do you know when you really want to grow? How can you communicate your interest to the organization? The effectiveness of training and education is determined by the investment of the participants; therefore, there must be legitimate individual commitment before there is effective individual learning and consequent organizational development. Professional and personal development has many angles, but in a customer-driven project environment, professional development typically takes on the following four basic characteristics:

Planning and analytic skills. More and more, quality improvement teams are being charged to exercise analytic and critical thinking skills to fully understand the customer's business systems and workflows. This will require training in research design, data collection and presentation, interviewing, and the use of computer-assisted charts and graphs. It will not be enough to be able to collect and analyze data; meaning must be derived and articulated to the customer, placing a high premium on those who can quickly spot variation and nonconformance.

Team building skills. Increasingly, the workforce will be called on to exercise more effective teamwork skills and attitudes. This may require training and development in group and meeting management, communication and listening, and facilitation. Rather than taking responsibility for a group's work as the chairman always did, the facilitator takes responsibility for group development and support. The team is now accountable.

Presentation skills. Presentation skills include the ability to analyze data, extract meaning and essence, and make oral presentations using briefing material, computer projection systems, and multimedia systems. Computer conferencing will become common.

Scheduling skills. The capacity to schedule one's own time and to manage a scheduling and control process is at the heart of project management, but the connections between personal and project organization are not often made. Women can bring a high quality of organization and schedule control to the team, and they can ensure that quality is scheduled into the work through their own personal approach to their work day.

Rewards and recognition

While the emphasis seems now to be on giving recognition, the team must be able to receive recognition and rewards, and model behaviors for others who aspire to more recognition. This means that those who receive rewards have a responsibility to model for younger, more impressionable staff who look to those who are recognized for keys to success.

New job opportunities

Empowerment brings on new job opportunities because the process widens contacts and relationships in the working environment, thus opening up new career tracks. But with this newfound flexibility, employees need to see their responsibility to accept empowerment with some sense of longer-term commitment to the organization to live out the outcomes of their work. If they are granted new latitude, one assumes more responsibility for longer-term commitment to the organization, thus cutting expensive attrition.

Increased latitude in decision making

Remember that empowerment grants more latitude in decision making, but also assumes more trust that the decision will be the right one. Thus, along with the ability to decide comes the accountability to defend the decision in terms of facts and figures. This adds weight to the argument that empowerment and quality improvement leads to more data collection and documentation needs, and thus to more reliance on the computer in such areas as project management and financial management.

Preserving the *Wonder* in Project Management

Sometimes we simply don't know why one new product team performs and the next one does not. In the end, the factors that contribute to project success are sometimes difficult to trace. We wonder about what actually happened in a successful new product enterprise so we can replicate it, but we often cannot capture the invisible elixir. While we strive to reduce the project management process to disciplined procedures, systematic analysis, and structured control, the more we learn about the project management process, the more we see that in the end quality projects are produced by people working in the dynamic and sometimes mysterious environment of human enterprise. The human factor—as unpredictable as it is—remains the key determinant of project success. All the planning and scheduling in the world cannot motivate and inspire people to perform and work together with a customer. In the end, we continue to be amazed when collaborative human enterprise is able to produce complex quality outcomes together that could not be designed and assembled by any one person. In a sense, we respect the *wonder and mysteriousness* of the process while at the same time we try to plan, predict, and control it.

This is why successful projects are usually associated with leaders, people at every level in the organization who provide purpose and create the environment in which people find meaning in their learning and work. A leader can balance the use of tools and techniques with the wisdom of *wonder* and the appreciation that many of the factors of success lie in the hands of the people doing the work. This is the challenge to women and minorities—and any aspiring professional—who enter the project management field. Sometimes it makes sense to simply respect the wonder of team success, let people work out their issues, and "muddle through" without trying to understand why and how work gets done.

Integrated Product Development Teams

Integrated Product Development Teams (IPTs) are the organizational structure resulting from Integrated Product Development (IPD) implementation. IPT membership is made up of multifunctional stakeholders working together with a product-oriented focus. This team is empowered to make critical life cycle decisions for the new product. Because the product and system development activities change and evolve over its life, team membership and leadership will likewise evolve. While marketing personnel, acquisition planners, project managers, and design engineers may be the most prominent members early in the life cycle, task managers gain a bigger voice during engineering and manufacturing development. Equipment specialists and mechanics may be the lead members during the operations and maintenance phase, with the design engineers returning once again if a major modification is needed.

IPTs are what make IPDs work. They are created for the express purpose of delivering a product or managing a process for their customer(s). Implementation of IPD represents a transition from a functional stovepipe focus to a customer product focus. Teamwork within the framework of IPD drives the functional and product disciplines into a mutually reinforcing relationship that helps remove barriers to the IPT success.

IPTs are applied at various levels ranging from the overall structure of an organization to informal groups functioning across existing units. The purpose of an IPT is to bring together all the functions that have a stake in the performance of a product/process, and concurrently make integrated decisions affecting that product or process. The teams can be created, formed, and their talents applied at all levels of the organization ranging from the overall structure of the organization to ad hoc teams that address specific problems.

The key characteristics of IPTs include:

- Team is established to produce a specific product or service
- Multidisciplinary; that is, all team members/functions working together toward a common goal
- Members have mutual as well as individual accountability
- Integrated, concurrent decision making
- Empowered to make decisions within specific product or service goal
- Planned integration among teams towards system goal

Leading New Product Development

The direction and guidance given to the new product development process requires a unique combination of leadership skills and capacities. Because new product development is inherently uncertain and often open-ended, it is critical that the leadership of the process builds trust and integrity into the team and the process. Leaders of new product development must avoid micro-managing projects; they should steward rather than control the work. The key issue is to enable the team to proceed within a loosely administered schedule and with a full understanding that their job is to provide leadership with relevant information and data for the go or no-go decision in project review.

The team is trusted to proceed without getting co-opted by the product and to avoid feeling that their careers are dependent on the success of the product at all costs. Reporting and information exchange must be open, and bad news along with good news must be able to travel up the organization freely to support project review decisions. Team members are expected to document their work and ensure that the configuration of the final product is preserved in configuration management software for inventory and production transition.

Using the Critical Chain Concept in New Product Development Teams

The critical chain concept of project management stresses that fact that current projects often fail because of bad estimates of task duration that then become outside limits for performance. These bad estimates are often made with large fudge factors to protect the task manager from failure. This approach often leads to inflated durations and project life cycles.

Critical chain places more emphasis on removing key bottlenecks in the system, and on stressing task starting times rather than task durations. More flexibility is given to team members to get the job done right and as fast as they can, rather than within an arbitrary timeline. Project leaders trim down task durations to bare bones and then give team members buffers, or additional time, to complete work if not completed on time. Starting times are stressed. Meanwhile, the project leader manages the key bottleneck in the new product development process.

Project Team Charter

As discussed earlier in this chapter, top management charters teams to define the work but also to ensure the commitment of the whole team to success. Success is defined here as an outcome that helps to grow the business and reduce risk with limited payoff, thus the team is alerted that pushing new product beyond the point that a business case can be made is not consistent with business value. The charter can include:

- Scope of work
- Project deliverables
- Project objectives
- Reporting relationships
- Timeline and schedule
- Budget
- Constraints and assumptions

Team Training

New product development teams should be trained in project management systems and tools, as well as the technical process of new product development. This training should include:

- Project management principles and systems
- Applications to new product development
- Team functions and roles
- Function of project review as go or no-go

- Preparing a project plan using MS Project or equivalent software
- Scheduling and budgeting
- Monitoring progress using earned value and other evaluation tools

Cautionary Note on New Product Teams

While it is fashionable to promote teamwork and team development in new product development, in reality it is often a particular individual's effort at particular times that drives a new product development process to the next level. Sometimes the *creative juices* of a particular individual, often a person who is not even on the team, feed advancement of the process unexpectedly. Thus project management has to be attuned to empowering these individuals to proceed outside the traditional team process, even challenging it. Customers themselves often provide this kind of impetus by providing new insights on a product and its value and actually contributing to new product development in the process. Project management in new product development is always attentive to keeping the process open to this kind of input and ensuring that teams do not act to inhibit new ideas.

Chapter

10

Putting It All Together

Principles for Working in the Real World

So how do we integrate all the ideas addressed in this book on managing new product development in the real world of day-to-day company business? How do we handle situations when all the key processes described here are not in place? The purpose of this chapter is to get at the essence of these questions.

The overriding theme for putting it all together is strong leadership.

Strong leadership, including a vision for success and a clear product strategy, is key in every stage of the process. To successfully generate a customer base before their competitors, company people need to feel the inspiration and excitement of doing something new and different, but also sense that they are part of a team effort and a disciplined process. And the process needs to be *energized* by a product champion or leading sponsor who either (1) *lifts* the process when it inevitably loses momentum or (2) decides to terminate a project because it simply does not make business sense

Our earlier discussion of Sonoco's introduction of the plastic bag is illustrative of the need for this kind of leadership and energy. The plastic grocery sack was pushed into the marketplace through the day-to-day efforts of marketing. Salespeople worked with customers to discover and resolve their issues and provide feedback to headquarters for development or production changes. During that process, the sales force *lived for that product.* They were energized by the challenge, sustained by a champion who *never gave up,* and reinforced daily by customers and clients who changed their thinking and behaviors because of the new product.

Seven Principles of Project Success

Going beyond the leadership factor, the following sections examine seven key principles that underlie new product project success and help leadership and management put it all together:

Principle #1. Develop project management and new product development processes, and integrate the two

Principle #2. Open the company to new ideas and new partners

Principle #3. Define measures for choosing new product projects

Principle #4. Create a way through project reviews to stop bad products

Principle #5. Choose project managers who understand technology

Principle #6. Build cross-functional teamwork and accountability

Principle #7. *Ad hoc* it when necessary to succeed

Principle #1 Develop key processes

New product development is most successful when proactively managed as a disciplined project supported by a formal project management system and a defined new product development and marketing process.

This principle implies that new products will not succeed in the marketplace simply because they incorporate innovative ways to meet customer needs or because the new product team is enthusiastic and dedicated. The fact is that new product development must be *managed, integrated, and controlled* as a project within a key business process. Work with the whole system in mind. Define the generic process, and don't let new products bypass key steps in that process, especially project review. Cost, schedule, and quality are controlled through defined work breakdown structures and task assignments, structured schedules and budgets, and organized project reviews to assess product results. Marketing and business case issues are built into the process at every turn.

This does not mean that new product processes cannot be flexible, open to change, and supportive of collaborative teamwork. But it does mean that there must be controls on change in product requirements. Defined processes can be flexible and agile when managed effectively using change management tools.

If the organization does not have a disciplined project management system and a defined new product development process, then *getting started* requires a new product team to document the process so that management can begin to internalize the process into the company culture.

Principle #2 Open upto new ideas

To ensure that creative ideas and concepts are encouraged to jump-start new products, the new product development process requires management to *open the organization* to free thought and innovative ideas in a global process.

New products increasingly come from an accurate understanding of global market and technical developments in a given field; thus new product development must be open to new ideas from many sources. This requires the organization to not only look to company resources for new product concepts, but to use the Internet and other tools to tap sources wherever there is insight and data on given market opportunity. And the company should be open to new partnering arrangements on a global scale that provide opportunity for all partners to benefit from new product concepts.

Company management must *empower* the workforce to think about and discover new product concepts to fuel the engine of change. Constant change in the marketplace must be seen as inevitable, and the company must sense that it has two choices: to facilitate and guide that change with new products and services, or to fight it.

Opening up the organization requires a proactive use of the Internet. The company discovers new *growth engines*—key new product features that move the field and technology ahead in the marketplace and induce demand. For instance, progressive computer and software companies are experimenting with logic chips that will be able to *think* for the customer, e.g., in the sense of artificial intelligence. Some companies have started development centers within the company to house this discovery process and protect it from bottlenecks and traditional company processes.

The limit on this open-ended process of idea generation is the decision to commit funds to a given new product development and the need to *close up* the process to protect intellectual property and patentable ideas. That decision is typically made when a product concept is documented and made a part of the company's funded project portfolio.

Principle #3 Define measures to select

Three measures seem to hold the key to new product success:

- Predicting financial and business performance of a product in the marketplace by satisfying customers at a competitive price
- Ensuring alignment with company strategy and workforce competency
- Building a strong risk management system and way of thinking that anticipates risks and mitigates them through contingency planning.

Financial performance. Accurately predicting the financial performance of a product is key to success. Lack of good marketing and pricing data to make the *business case* for profitability is a major problem for new product development. The team must envision a cost and revenue scenario, discount it with the company's preferred discount or *hurdle* rate to calculate new present value, and then continuously update the picture as the product is developed.

Strategic thinking and company competency. The product should be closely aligned with the company's past history and future plans. New product designs must be used in strategic ways to build company competency and further the company's goals. And new products must be within the company's capacity to design and produce based on proven success in the past.

Risk management. The company must be attuned to risks and opportunities, e.g., what is likely to happen that will inhibit the success of the product, and what opportunities are created when the company overcomes those risks? And what will it cost to implement various contingencies in case of high-risk events?

These three measures—a strong case for product success and enhanced financial performance, alignment with the company's proven capacity and strategy, and a risk mitigation process that avoids surprises—appear to be the most salient ingredients to predicting new product success in selecting projects.

Principle #4 Use project reviews to stop bad products

Control of the new product process should occur at key points at the end of each phase in the process, in project reviews. Project and company management should have to sign off on advancing a project to the next phase.

Strategically scheduled project reviews are increasingly important to ensure that new products are aired at key points in design and development. This process of project review allows management to evaluate progress and make the business case for advancing a product to the next phase.

New product processes are thus designed to produce information for project review. In other words, the purpose of each phase is not simply to move the product to the next phase, but to inform management on whether the product *should* be moved to the next phase.

Principle #5 Choose technical project managers

New technology-based products such as electronic instrumentation and automated manufacturing tooling are complex and difficult to manage through to successful marketing without strong technical experience at the project management level and broad accountability from beginning to end.

Project managers need to have *technical and marketing sense* about the product, but they do not need to be technical experts in all aspects of the field. Technical and marketing sense means that the project manager has experience in the field, knows how to validate and verify technical data and technical sources, has direct access to the best marketing research available on the probabilities of product success in the target market, and has the respect of the product team.

We see the need for a new professional role in technology management, professional project and program managers who *know* technology, but who are *not captured by it* and who use products strategically in the global marketplace. This means they retain their objectivity in the midst of inevitable product advocacy, insist on and manage by the facts, and drive the process of developing new technologies into product design.

Principle #6 Build team accountability

New product development teams should be representative of the wide spectrum of competencies needed to pull a product through to the marketplace quickly, including a strong project manager, product designers and developers, engineers, technicians, configuration managers, production and inventory managers, training staff, and marketing and sales people.

The new product project team should be reflective of the broad factors in new product success from design to manufacturing to marketing and sales. This does not mean that all these factors need to be represented in every team meeting and decision, but rather that the team decision process in any given phase must be able to make the business case for continuing work.

Accountability is a major issue in the new product team. If a project or program manager is responsible for the whole process from new product design through to product production and marketing, decisions in the new product development process will reflect a broad look at the product as it advances. If, on the other hand, project accountability is limited narrowly to producing the product prototype within cost, time, and quality constraints, the company is likely to produce products that cannot be marketed *very efficiently*. This is the weakness of narrow project management guidelines that focus only on cost, schedule, and quality, and not on the entire company marketing process.

This guideline is important because a narrowly structured product team that reflects only product requirements and design—and that is focused solely on the traditional project management goal of producing a final, tested prototype within estimated cost, schedule, and quality limits—often fails in the longer term. This is because a product prototype *brought in* within cost, time, and quality constraints may have failed key business and financial tests simply to make narrow project goals. Sometimes a product that takes longer than estimated and costs more than hoped for becomes the best product in the longer term simply because of its market and customer value.

Principle #7 Ad hoc it when necessary

Sometimes it is better to *ask for forgiveness than to ask for approval* in the heat of a competitive, new product launch. Sometimes key people need to be empowered to do whatever they need to do to move a product to beat the competition, despite the lack of substantial supporting data. The six previous principles—and this book in general—support a disciplined, planned approach to managing new product development projects; sometimes you simply don't have time to plan in the field. Sometimes the forces close to the customer require seat-of-the-pants decisions that are not consistent with past practice, full product development, or management guidance. Sometimes products have to be launched that have not been fully aired in the development process, and for which there are no clear guidelines for marketing and sales. And sometimes marketing has to promise a product to a prospective customer or client when that product is in a preliminary, unproven state *back home*. Sometimes you have to ask for forgiveness after the fact.

We do not condemn decisions to proceed in an ad-hoc way in new product development, but we do feel that many such situations could be anticipated in a robust, risk management process. Contingencies can be drawn up for a wide variety of situations in new product launch simply based on worst-case scenarios. Worst-case scenarios anticipate situations where certain parts of the new product process must be bypassed simply to beat the competition or meet an urgent customer need. The point here is that simply depending on experienced people to make the right decisions when they need to does not always work, but asking those same people to anticipate situations that might require seat-of-the-pants decisions usually produces contingencies for about every risk event. In the end, you let good people in the field with an ear to the customer do what they have to do.

Appendix

A

Generic New Product Development Work Breakdown Structure

This appendix provides an example of a generic new product development process showing task descriptions and responsibilities. This model provides new product developers and project managers and teams with a prescribed procedure for the process. Companies and agencies are encouraged to develop this kind of generic process, to update it with lessons learned, to publish it on the company network, to use it in training project and product development managers, and to promote its use in all new product development projects as a guide.

Task/subtask/work package/ level of effort	Description	Responsibility
	Create a Culture of Ideas and Innovation	
Preparing the Business or Agency to Generate Innovation and Creativity	Ensure that all company or agency people understand that they are expected to participate and collaborate in new product development with ideas and insights from their perspective. Develop an ideation process to gather their ideas and demonstrate benefits.	VP for Product Development
Define the Ideation Process: Opportunity Identification, Evaluation, and Selection	Define ideation process, and gather new product and service ideas and concepts through process to solicit inputs from technical and marketing sources, evaluation systems for new products, portfolio analysis, problem analysis, lessons learned documents, modeling and simulations, marketing scenarios, computer/network-assisted creativity techniques, customers and stakeholders, gap analysis, qualitative and quantitative tools and processes, risk assessment, and individual suggestions.	VP for Product Development
Customer Focus: Understanding Customer Needs	Interviews and focus groups are conducted to identify customer needs and settings for possible product applications.	Project Team
Remove the Fear of Failure and Roadblocks	Be specific that the purpose of new product development teams is to provide phased information to management for go or no-go decisions in key project reviews.	VP for Product Development
Value of Time to Market Goals	Demonstrate that time to market or total project life cycle determines the success of new products.	VP for Product Development
Integrate New Product/ Service with Process Issues	Make sure new product development goes through normal business processes and does not separate itself from the company's mainstream activity.	VP for Product Development
Technology and Process Innovation	Encourage creativity not only for new products and services, but also in all business operations.	VP for Product Development
Identify New Processes and Process Reengineering	Consider process innovation, e.g., new processes for inventory, production, procurement, and so on, to be new product development activity.	VP for Product Development
Enable Alignment of Processes with New Product Concepts	Make sure that new products align with key business processes, including testing, design, and configuration management.	Project Manager
Identify Logistics Issues and Implications; Prepare Logistics Plan	Make sure product logistics are considered as part of the project plan, and costs and impacts of logistics are part of go and no-go decisions in project review	Project Manager
Build New Product Project Leader Pool	Train potential new product project leaders in both the technical process and project management tools.	VP for Product Development

(*Continued*)

Task/subtask/work package/ level of effort	Description	Responsibility
	Strategic Alignment, Business Value, and Portfolio Management	
Strategic Alignment and Weighted Scoring	Make sure each project is analyzed for consistency with business plans using the weighted scoring model. Train everyone in its use.	VP
Articulate Business Strategy and Strategic Objectives	Make sure the business strategies are clear to project management and team members, along with specific strategic objectives including those that relate to new products and services.	VP
Identify Program Areas	Identify and categorize new product projects of similar character and place them in a program category so that they can be managed.	VP
Develop Projects	Create new ideas in an ideation process and identify new projects in each program category for assessment.	VP and Product Managers
Analyze Alignment of Candidate Projects with Business Strategy	Make sure projects are scored with the weighted scoring model.	VP and PMO
Business Value and Advocacy	Business value is measured by financial performance using cash flow projections and net present value discounted at company hurdle rate.	VP
Define Project Portfolio Management Process	Define the way new products and new product projects are generated.	VP
Capture New Product and Service Ideas and Document	Make sure there is a way to capture all new ideas and concepts, and get them documented by marketing or engineering.	VP
	Set Up Project Management System for New Products and Services	
Project Management and Project Review Tools	Describe (as in this book) the new product development and project management process to be used in the company.	
Critical Chain Concept	Make sure schedules are based on resource availability; don't multitask new product work (make sure you do one project at a time with a team); and stress loosely, open-ended scheduling with emphasis on starting tasks.	Project Manager
Define New Product Life Cycle	Define product life cycle from concept through to production, marketing, and product support, and make sure someone is responsible for the whole process.	VP
Categorize New Products	To ensure a balanced portfolio, make sure new products are mixed in terms of risk, payoff, horizon for marketing, and cost.	VP
Develop Generic Work Breakdown Structure (WBS)	Develop this kind of generic WBS to apply to all new products.	VP
Decision and Data Issues Concept	Ensure that project options and alternatives are opened up for top management attention in project planning, and are driven by data and facts.	VP

(Continued)

Task/subtask/work package/ level of effort	Description	Responsibility
Project Planning Process	Define the project planning process and make sure project plans are fully prepared for new products.	VP and PMO
Product Platform Planning	Make sure functional managers are in charge of technical processes for a product, including platforms and supporting systems.	VP
Create New Product Team Concept	Identify new product teams and give them recognition.	VP
New Product Innovation Charter	Charter teams using charter team concepts.	Program Manager
Define Project Plan and Monitoring Process	Define how project progress will be measured, and especially how go and no-go decisions will be made in project review at the end of each stage.	VP
Define Resource Management Process: Cost Estimating and Funding	Make sure resource planning is integrated into project plans, so that key resources, both people and systems/equipment, are committed to implementing the project schedule when they are needed.	Program Manager and Project Manager
Ensure Interface Management	Interconnect various "silos" in the company so that there is integration and coordination between project, functional, and supporting staffs.	VP
Get Company Management Commitment	Make sure there is a sponsor for each new product project at the VP level whose job performance is tied to making good decisions on that product, particularly in view of business case.	CEO
Define Special Role of Project Review as Gate Decision	Define project review agenda and data requirements, and make sure key management review data and attend project review meetings to participate in go or no-go decision.	VP
Create Supply Chain and Contract Types	Review contractor partnerships and contract types to ensure that risk is shared appropriately with vendors.	Project Manager
Start Vendor Partnering	Establish long-term partnerships with key vendors to ensure consistency in supply function with company quality goals.	VP
Define Project Review Milestones	Use generic work breakdown structure to define key, standard milestones, e.g., design project review, that all new product projects must schedule.	PMO
Resources Review	Make sure there is a manpower planning function that tracks current and planned use of all new product resources to facilitate implementation of new projects.	VP
	Product Concept Generation and Evaluation	
Product Selection Based on Alignment, Financial Performance/Cash Flow Analysis, and Risk Assessment	Confirm company commitment to assessing each new product in terms of financial performance, alignment, and risk using the risk matrix approach.	VP

(Continued)

Task/subtask/work package/ level of effort	Description	Responsibility
Prepare Product/Service Concept Statement	Describe each new product idea or concept in a concept statement that is widely distributed throughout the company.	VP
Do Initial Concept Market Testing	Make sure initial market research and testing are accomplished before concept phase is completed to allow the marketing case to be made at project review.	Program Manager
Transition from Concept to Project: Get Approval	Get approval from project sponsor for transitioning product from concept to funded project.	Program Manager
Identify Project Sponsor	Project sponsor must be VP level committed to making the right decisions for the company.	VP
Identify Project as Implementation of Specific Business Strategic Objective	Link every new product to one or more business strategies.	Program Manager
Define Value to Business Growth	Make case for new product in terms of financial performance; update previous assessments.	Project Manager
Identify Risks and Contingencies	Convene small groups to identify risks in each new product project and complete risk matrix.	Project Manager
Define Value to Customer/ Client Growth and Benefit Segments	Marketing defines value of new concept or product to customer and identifies benefits.	Marketing
Concept Definition	Concept is defined in a concept statement that gives enough detail to make decision to proceed.	Project Manager
Identify Customer Requirements	Marketing defines customer requirements based on market research.	Marketing
Create Project Scope of Work	Scope of work is definition of the work to be done, timing and cost, and deliverables.	Project Manager
Develop Project WBS	Work breakdown structure is prepared in project team meeting, top down.	Project Manager
Define Project Budgeting and Funding	Projects are prioritized with estimated costs, then funded from earmarked funds allocated to each strategic objective, until funds run out.	VP and Program Manager
Project Review for Go or No-Go Decision	Project Review session is planned to make go or no-go decision.	VP and Functional Managers
	Set Up Product and Process Team and Resources	
Develop New Product Team Concept	New product teams are defined in terms of the unique function of new product development.	Program Manager
Compose New Product Team(s)	Ensure involvement of marketing, engineering, manufacturing, and support personnel from the beginning.	VP
Identify Team Member Roles	Write job description for each new product development team member based on their tasks and responsibilities.	PMO
Identify Project Leader	Project manager/leader is chosen from trained PM leader pool.	VP

(Continued)

Task/subtask/work package/ level of effort	Description	Responsibility
Team Training and Development	Train team on new product technical process and on project management tools, including reporting, performance tracking, and schedule adherence	VP
Identify Team Composition	Team members chosen based on task needs in defined WBS, matching task needs with personnel equipped to do assigned tasks.	Project Manager
Define Organizational Implications: Matrix Management	Make sure project teams are connected to functional home departments as well as assigned to project managers to ensure dual allegiance is clear.	Program Manager
	Full Product Design and Development	
Prepare Product Design and Architecture	Product drawings, graphics, and so on, of product architecture.	Project Engineer
Identify Platform Issues	Identify product platform issues and challenges, and make sure project includes platform design and development.	Project Manager
Design and Produce Prototype, CAD, and Design for Manufacturability	Produce technical prototype graphics.	Project Engineer
Design to Cost	Make sure design is evaluated in terms of cost of production as well as cost of design.	Project Manager
Test Prototype	Prototype is tested using accepted industry test protocols.	Test Technician
Ensure Quality Assurance and Control	Quality is built into design and ensured through process control.	Quality Assurance
Identify Regulatory, Safety, and Public Policy Issues and Opportunities	Political and public policy and regulatory issues, legal issues, morality, and ethics questions addressed; education.	Project Manager
Project Review 2		VP and Functional Managers
	Market, Commercialization, Launch, and Distribution	
Target Market Plan and Product Protocol	Alternative ways to segment market, micromarketing and mass customization, diffusion of innovation issues, speed to market goals.	Marketing
Platform Decisions	Permanence, aggressiveness, type of demand sought, competitive advantage, product line replacement, scope of market entry.	Marketing, Engineering
Branding	Trademarks and registration, brand name options, brand equity, brand name dilution, brand leadership.	Marketing
Packaging	Roles of packaging and options and impacts.	Marketing
Distribution Plan	Describe how product will be distributed to market channels.	Marketing

(Continued)

Task/subtask/work package/ level of effort	Description	Responsibility
Positioning	Make sure product is placed into market so that demand is clear and all conditions of the market described before field test.	Marketing
Competitive Analysis	Look at all real and potential competitors in terms of their decision to enter product market once product is public.	Marketing
Quality Function Deployment	Make a "house of quality" that matches every customer need or requirement with a product function or component.	Engineering
Product and Service Support Plan	Make sure product support and maintenance needs are identified and a plan created to address them.	Project Manager
Market Testing	Product use testing with user groups, methodology, documentation, and guidelines for interpretation, pseudo sale, controlled sale, full sale, speculative sale, simulations, other technologies.	Marketing
Business Decision to Launch	Top management makes decision to launch based on project review at end of development phase.	VP
Launch Management	Communications plan, alliances, strategy connections, sales, logistics, control events, product failure scenarios.	Marketing
Special Global Considerations	Look at economic, social, and political factors on a global scale to ascertain potential impacts on product success.	Marketing
Project Review 3		VP and Functional Managers
	Product Support, Maintenance, and Feedback	
Implement Product Support and Maintenance Plan	Make sure product support plan can actually be implemented.	Engineering and Marketing
Implement Logistics Plan	Identify logistics responsibility and capacity, and task support and logistics team to implement plan.	VP, Project Manager
Provide Product Performance Feedback to Production and New Product Development Process	Establish reporting system for product performance and customer feedback into new product development process.	VP
	Market Product In Field	
Logistics Plan Distribute Product	Execute logistics plan and provide feedback.	Marketing
Monitor Sales and Support	Track sales and provide feedback to new product development and top management.	Marketing

Appendix

B

Managing New Product Development Projects: Course Outline

This appendix contains a set of course objectives and a generic weekly schedule for an eight-week training or credit course in project management in new product development.

Course Objectives

A. Given information about an organization's industry, culture, performance, resources, objectives, and products, develop a new project management and new product development process and make recommendations to maximize the probability of success of new product introductions.

B. Given information on a new product development project, develop a project plan including work breakdown structure, schedule, budget, and risk management plan.

C. Given information about an organization's industry, culture, performance, resources, aims, and projects, design a new product strategy and marketing plan for the organization.

D. Given an organization's new product strategy, analyze the strengths and weaknesses of different methods of need and gap analysis and concept generation.

E. Given a new product concept, make recommendations to ensure that appropriate attention is paid to customer and consumer safety and other ethical concerns during the new product development process.

F. Given a new product development and project review process, determine what information management needs at each stage and recommend appropriate evaluation tools to make the go or no-go decision based on the business case.

G. Given a new product concept, assess the options for each step in the technical development phase and recommend decisions to maximize the success of the product's introduction.

H. Given a new product, design a marketing program for the product pricing, distribution, and new product launch.

I. Given a new product launch, design the appropriate market test, evaluation, and feedback to predict and measure the product's performance in the marketplace.

Generic Weekly Schedule

Session	Topics
Week 1	▪ Course Introduction and Overview ▪ Managing New Product Development Overview ▪ Project Management Process ▪ Product Development Process ▪ Strategic Planning for New Products
Week 2	▪ Project Setup ▪ Concept Generation ▪ Idea Creation
Week 3	▪ Project Management Systems ▪ Project Management Tools ▪ Project Management Teams
Week 4	▪ Concept Generation Concept Evaluation ▪ Mid-Term Exam
Week 5	▪ Technical Development/Design
Week 6	▪ Understanding the Marketplace ▪ Marketing Strategy/New Product Launch ▪ Market Testing
Week 7	▪ Market Testing ▪ Marketing Implementation ▪ Public Policy/Ethics/Safety

Appendix

C

Issues for Discussion

The following "issues for discussion" are provided as agenda items for company or agency new product project teams as they organize and meet to design and deliver new products.

Integrated Project Management in New Product Development

- Project management tools—schedule, cost, quality
- Requirements, features, scope of work
- Project life cycle controls key process gateways
- Work breakdown structure defines product deliverable and work to be performed
- Schedule defines work sequence/interdependency
- Resources, cost estimates, risk management
- Tracking, monitoring, and change management
- The project team and organization
- MS Project information support

Requirements and features

- Customer requirements captured in requirements document; performance versus design
- Freeze requirements in baseline project WBS; baseline schedule against frozen requirements
- Appreciate that change is expected dynamic in product development as learning occurs

- Accommodate change as development proceeds
- Exercise: identify user requirements

Scope of work

- Scope is general statement of the work to be performed, references product requirements
- Scope is written into contract or agreement
- Scope is frozen at baseline
- Monitor for scope creep; change management process; ensure orderly review and approval of change
- Exercise: write a scope of work

Project life cycle

- Concept: early customer requirements
- Project Planning: objectives, WBS, schedule, budget
- Design: preliminary product design
- Development: develop prototype
- Production: produce prototype
- Close-out: close out project
- Gateways at each entry point to next phase
- Exercise: define life cycle phases and illustrate with a real product

Work breakdown structure

- Defines all work to be performed in outline or organization chart form
- WBS controls process and deliverable
- Generic WBS ensures that schedules include all necessary work in the product development process
- Exercise: develop a WBS

Scheduling

- WBS
- Task list, duration estimates, calendar
- Key linkages between tasks
- Use linkages to generate concurrent work to accelerate process
- Use MS Project Gantt chart and resource usage table as baseline reference
- Exercise: enter tasks and durations into MS Project

Resources and costs

- Identify teams and resource needs
- Assign resources to tasks with percentage of total time
- Add resource costs
- Add fixed and variable costs
- Reports will produce budget and cash flows
- Resolve resource conflicts
- Exercise: enter resource assignments and costs

Risk planning and management

- Risk planning; preparing for risk management
- Risk identification; identify high risks
- Risk assessment; assess risk qualitatively and quantitatively
- Contingency management
- Risk management; manage and monitor risk
- Exercise: prepare risk matrix (risk description, probability, impact, severity, and contingency)

Monitoring and tracking

- Kick off project; use baseline as point of departure for monitoring variance
- Monitor for earned value, schedule, and cost variance
- Project reviews, schedule, cost, quality
- Engineering and design reviews
- Estimates of percent complete
- Capture actual costs
- Exercise: interpret earned value results and corrective action

Change management process

- Configuration management approach to product (bill of material)
- Engineering prototype defined early
- Change accommodated through CM process
- Exercise: develop change request and approval format

The project team and organizational structures

- Team is defined by work to be done
- Team is chartered

- Project manager appointed
- Team meetings and agendas
- Project reviews, data, information
- Alternative project organizations; matrix, pure project, functional
- Exercise: prepare team meeting agenda

Microsoft project

- Well suited to product development
- Gantt chart and optional tables useful in documenting schedules, resources, etc.
- Varied report formats
- Earned value calculations
- Risk-based schedule calculations
- Communication in team through MS Project
- Exercise: practice MS Project applications

Contract management issues

- Structure contract to share risk
- Procurement process
- Preparing the RFP
- Contractor bidding
- Contractor selection
- Contractor management
- Contract project reviews
- Exercise: write RFP

Nine Elements of Project Management in New Product Development

1. A uniform project life cycle. This consists of a definition of phases, deliverables, key milestones, and success criteria for each group involved in the project. This is also sometimes referred to as a methodology.
2. Project requirements, objectives, and scope must be documented. It is also essential that a system be in place to ensure that the project requirements and scope are stabilized as early in the life cycle as possible.
3. A work authorization and change control system. A frequent source of problems on projects involves the expansion of the scope of work without adding value to the overall project. Change control must include formal systems for reviewing, evaluating, and authorizing changes to scope after the project has begun.

4. Defined roles for project team members and functional supervisors must be identified and documented. Similarly, a system of communication between the project participants must be established.
5. A planning system must be in place that allows for the creation of plans based on organizational capability, not wishful thinking. The planning system allows for the creation of the scope (WBS), schedule, and budget plans.
6. Quality metrics and systems to ensure quality must be in place in the organization. These systems must include identified metrics for each element of the WBS as well as procedures for assessing quality.
7. Tracking and variance analyses are vital functions for controlling the project. Projects are managed by an exception process in which deviations from plans are reported and acted upon. An effective project management process requires regular reports and regular meetings of the project team to identify when things are off target. Schedule slips, cost overruns, open issues, new risks, and identified problems must be dealt with as early as possible.
8. An escalation process is a set of procedures that defines how problems, open issues, and risks will be addressed in a timely manner. Issues and problems are inevitable in projects. A good escalation procedure requires problems and issues be addressed by the lowest level of management first. If the lowest level cannot resolve the problem, it is escalated to higher levels until a resolution occurs.
9. Corrective action decisions are necessary when variations from the plan are detected. In some cases trade-offs must be made. Systems and procedures must be in place to address how corrective action decisions will be made.

Integrated Program Management Tools for New Product Development

- Organizationwide project management system
- Program/portfolio planning and development system
- Resource management system
- Program information technology system
- Product/service development process
- Interface management
- Portfolio management
- Program monitoring and control system
- Change management system
- Program evaluation system

Organizationwide project management system

- Project management culture
- Work breakdown structure

- Scheduling system
- Resource assignment
- Task linkages and interdependency
- Steering group
- Matrix team structure

Program management system

- Business planning system and strategic objectives
- Decision process
- Budgeting system
- Risk management system
- Program definition
  - Portfolio pipeline system
  - Criteria
  - Selection process

Resource management system

- Workforce planning
- Workforce utilization system
- Staffing
- Financial control
  - Earned value
  - Industry standards
  - Facilities and equipment management
  - Resource pool system

Program information technology system

- Network system
- Accessibility to key information
- Reliability
- Flexible formats
- Workforce training
- Web-based communication and reporting system

Product/service development process

- Key process definition for development of products and services
- Assigned functions in matrix organization structure

- Industry standards
- Technology support and testing
- Technically training workforce
- Uniform work breakdown structure

Interface management

- Matrix organization
- Program review meeting formats
- Assignment of support functions
- Control "gates"

Portfolio management

- Top management visibility of programs and projects
- Uniform project management system
- Pipeline management
  - Generation of projects
  - Evaluation of projects
  - Selection of projects

Program monitoring and control system

- Project management office
- Corrective action procedures
- Reporting system
- Escalation process

Change management system

- Change order system
- Change impact analysis
- Risk management assessment

Program evaluation and audit

- Document lessons learned
- Close out system
- Audit system
- Corrective action followup

Challenges in New Product Project Integration

- Scope creep
- Resource/budget/funding
- Client management
- Risk management
- Control reports and processes; clear reports
- Communication
- Vendor management
- Access to right tools
- Juggling too many projects, not enough resources; diversity
- Documentation and right software
- Changing requirements
- Schedule
- Contractors
- Office space
- Accountability without responsibility/authority
- Lack of functional support
- Lack of senior management support
- Lack of correct skills on team
- Changes in management
- Inadequate planning tools
- Scope, deliverables, etc., not well defined
- Scheduling team meetings
- No process to resolve and escalate issues
- No model for costing and budgeting
- No method for archiving project materials
- General workload pressures and time management
- Training
- Red tape
- Team member conflicts
- Communicating urgency
- Getting buy-in
- Coordinating cross functional group availability
- Managing remote contractor

- Unrealistic dates determined without team input
- Long projects; keeping everyone engaged
- Taking corrective action
- Unclear roles of participants

Earned Value in New Product Project Monitoring

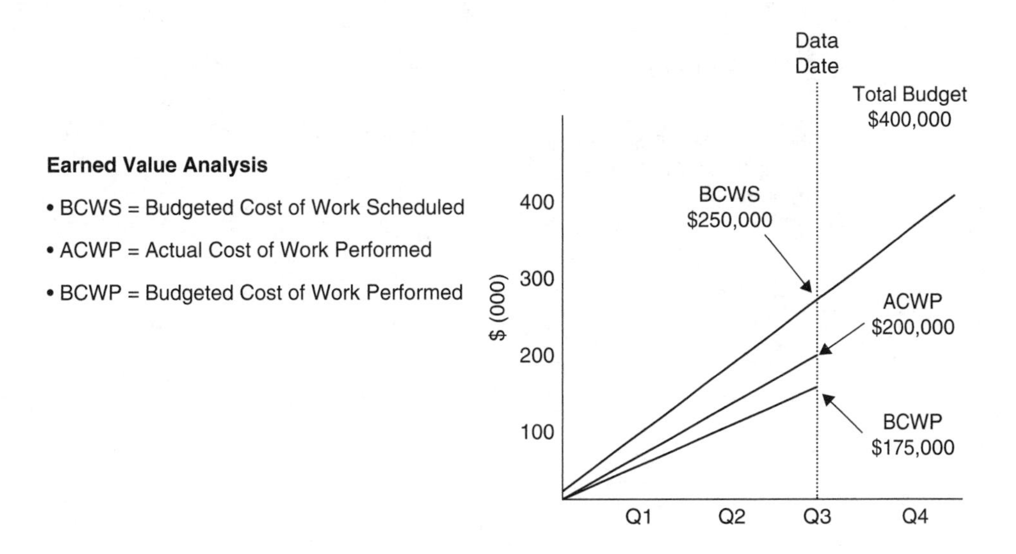

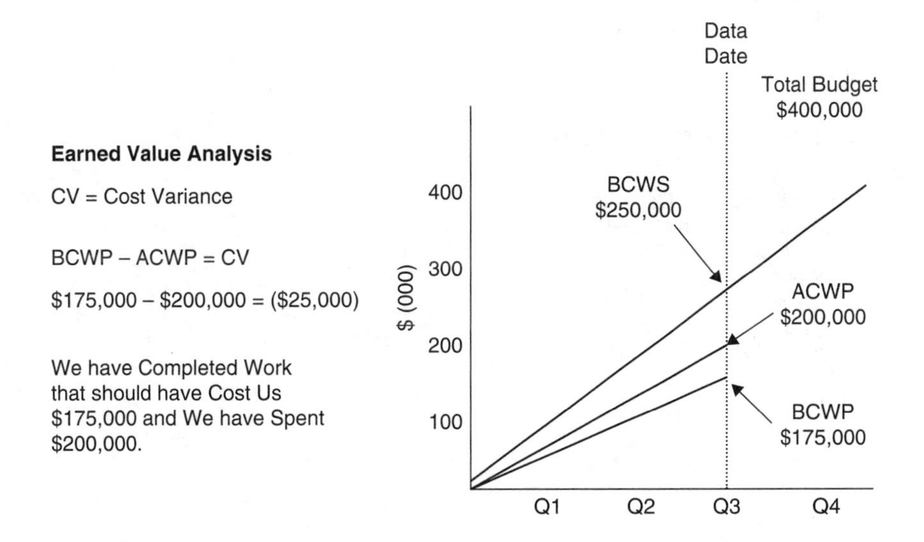

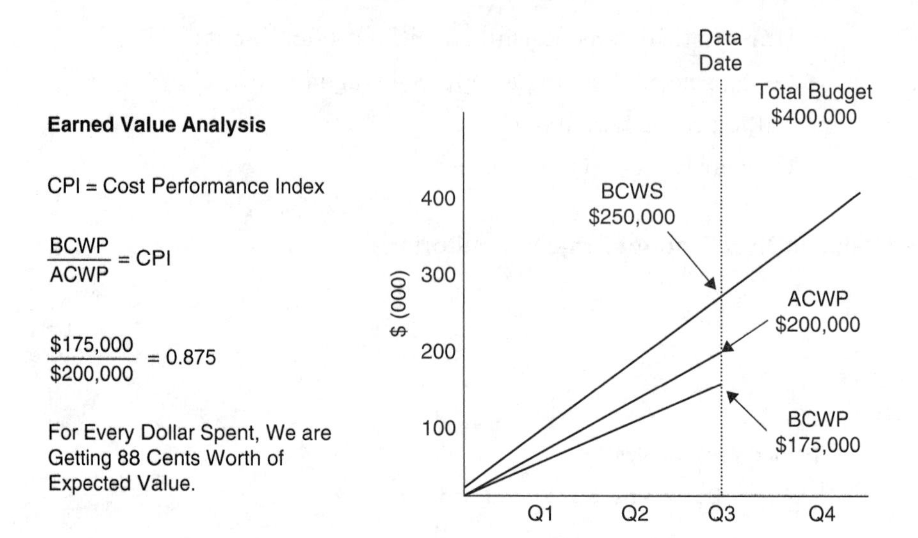

Earned Value Analysis

CPI = Cost Performance Index

$$\frac{\text{BCWP}}{\text{ACWP}} = \text{CPI}$$

$$\frac{\$175{,}000}{\$200{,}000} = 0.875$$

For Every Dollar Spent, We are Getting 88 Cents Worth of Expected Value.

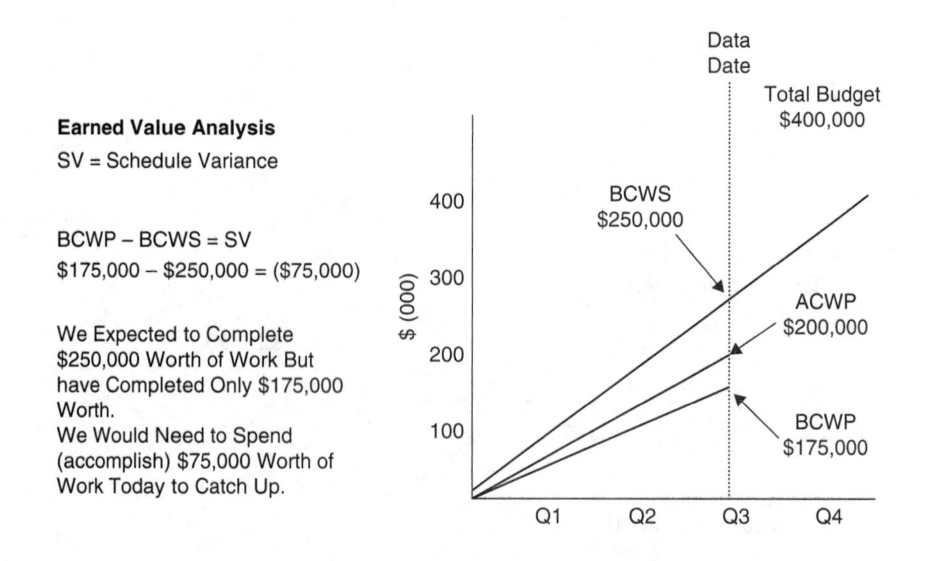

Earned Value Analysis

SV = Schedule Variance

BCWP – BCWS = SV

$175,000 – $250,000 = ($75,000)

We Expected to Complete $250,000 Worth of Work But have Completed Only $175,000 Worth.

We Would Need to Spend (accomplish) $75,000 Worth of Work Today to Catch Up.

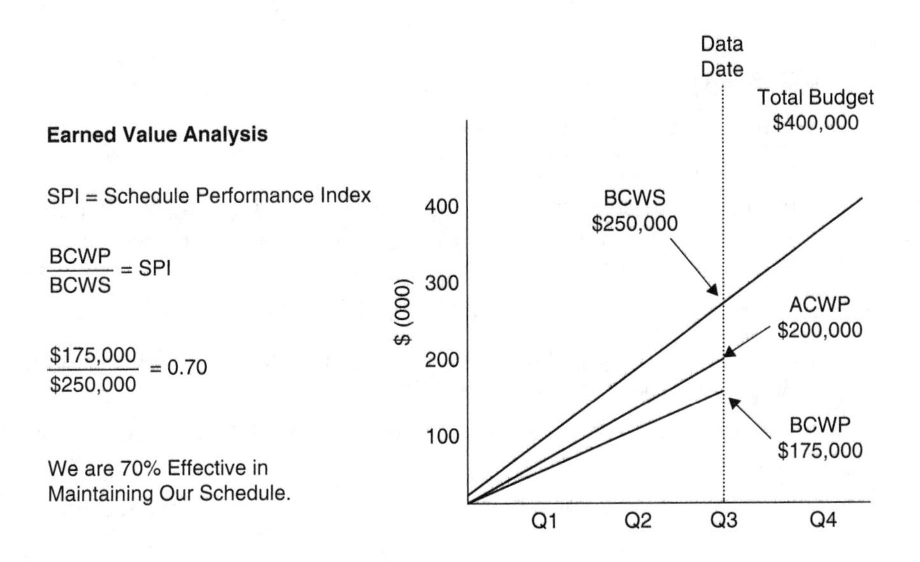

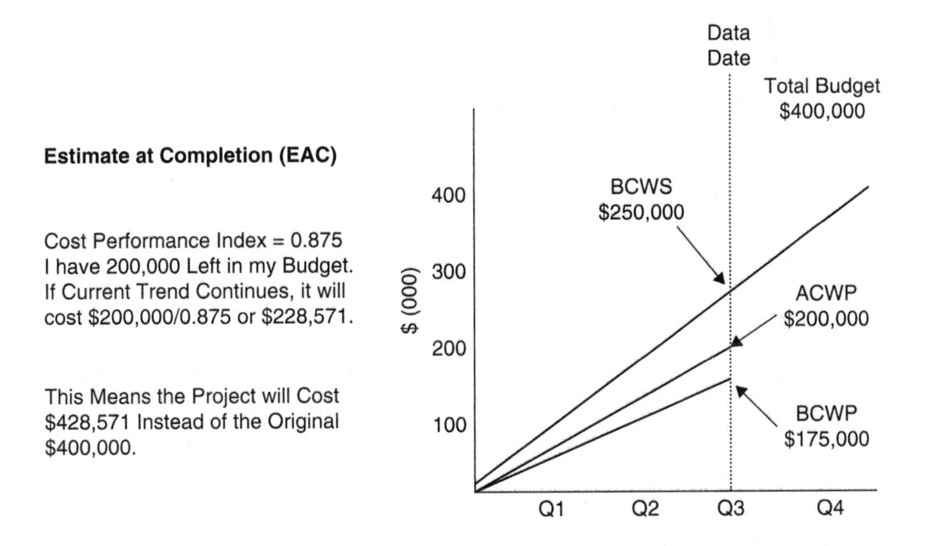

Critical Chain and Team Dynamics in NPD Project Management

- Goldratt; Theory of Constraints
- Focus on constraints
- Projects as a whole
- Measurements should induce parts to do what is good for the whole
- Managing projects as a portfolio

Discipline

- Putting flexibility where it counts
- Finding the ultimate constraint
- Critical chain schedule
  - Protects critical chain from noncritical task variation—feeding buffer
  - Protects due date from critical chain variation—project buffer

TOC dictionary

- Buffer: time or budget allowance used to protect scheduled throughput, delivery dates, or cost estimates on a project
- Drum: bottleneck work station; the most highly used resource; one that is not easy to elevate
- Rope: the information flow from the drum to the front of the line which controls project

Learning from projects

- Post mortem—a process and document
- Professional, factual, brief
- Postmortem meeting
  - Length
  - Room
  - Who should attend
  - Facilitator
  - Recorder

Learning

- Preparing
- Running
  - Opening
  - Timeline
  - What went poorly
  - What went well
  - Recommendations
  - Post mortem summary document
  - Next steps

The successful project

- Support environment
- Enterprise environment

- Business environment
- Success factors, user involvement, clear requirements, proper planning, competent staff, clear vision and objectives, hard working staff

Success

- Teams, balanced problem solving, decision making, conflict management, skills

Selecting and developing PMs

- Sense of ownership
- Political awareness
- Relationship development
- Strategic influence
- Interpersonal assessment
- Action orientation

Synergistic team relations

- Teams and productivity—related?
- Lessons
  - Communicate
  - Manage themselves
  - Facilitation

Criteria for competent PMs

- Enthusiasm
- Tolerance for ambiguity
- Team building skills
- Customer focus
- Business orientation

Leadership

- Strongman
- Transactor
- Visionary
- Superleader

Leadership impacts

- Overpowering
- Powerless
- Power building
- Empowerment

Developing core teams

- Continuous membership on team through project cycle
  - Reduce cycle time
  - Increase quality
  - Better plans
  - Overcome organizational problems
  - Encourage creativity
  - Technical expertise

Project team competence

- Team efficiency
- Matrix
- Disorganized
- Co-responsibility
- Customer focus
- Self-managed
- Balance team and self-recognition

Assessing team competence

- Goals
- Deliverables
- Skills
- Tools
- Discipline
- Cohesion
- Effective leadership
- Structure
- Integrate diversity
- Achieve desired results
- Work with customers
- Chutzpah

Motivation

- Theory X/Y
- Maslow
- Hertzberg Hygiene theory
- Locke: Goal Setting
- Understanding value and risk

Total manager

- True leadership: substance, humanity, morality
- Leadership
  - Show the way
  - Have a compass
  - Give credit
  - Take risks
  - Keep faith
- Act the part
- Delegate
- Be enthusiastic
- Be competent
- Thrive on change
- Don't ignore company culture

Communication

- Active listening
- Silence
- Perception checking
- Giving feedback
- Receiving feedback
- Nonverbal communication

Presentations

- Define objective
- Define audience
- Define approach
- Develop presentation
- Preparing and rehearsing

- Delivering
- Critique the presentation

Motivation and leadership

- Needs differ
- Management style affects motivation
- Satisfaction versus dissatisfaction factors
- Job satisfaction
- Leaders establish vision
- Challenge beliefs
- Take risks
- Leaders are honest
- Competency
- Align individual and project missions
- Get people involved
- Active
- Encourage contrary opinions
- Doable goals
- Recognize performers
- Make it fun

Project Management and Conflict

- Conflict is not a state, but a process
- Conflict is perceptual
- Sources
  - Reward
  - Scarce resources
  - Uncertainty
  - Differentiation

Steps in conflict

- Frustration
- Conceptualization
- Orientation
- Interaction
- Outcome

Getting unstuck

- Unclear goals
- Attitudes
- Missing skills
- Membership changes
- Lack of discipline
- Outsider intervention

Negotiating

- Negotiate with people
- Shape project quality
- Defeat the problem
- Deal with the problem
- Planning
- Goals, options
- Negotiation session

Conflicts

- Objectives
- Conflict environment
- Managing conflict
- Conflict intensity
  - Schedules
  - Priorities
  - Manpower
  - Technical issues
  - Administration
  - Personalities
  - Cost

Meeting skills

- Listening
- Meeting necessary?
- Who should attend?
- Where and when?
- How long?

- Best room
- Present ideas
- Visuals
- Stimulate, inspire, productivity

Personal effectiveness

- Set long-term goals
- Establish strategies and short-term goals
- Personal strategies and "to do" lists
- Expect the unexpected
- Confront procrastination
- Delegate
- Paperwork management
- Control interruptions

Managing Risk in New Product Development Projects

- Risk is inherent in business; competitive edge comes from overcoming risks better, faster, and cheaper than competition
- "Projects" imply risk; deliverable is new
- Project uniqueness implies risk and opportunity; integrates challenge and potential opportunity
- Advanced technology and complexity
- More project-based companies
- Good risk management skills make you more marketable as a project manager

Purpose of Project Risk Management

- Planning and control tool
- Understand project complexity and value
- Identify key success or failure factors that will impact project success—early
- Assess and rank
- Quantify probability
- Mitigate (control) risks through risk management plan
- Monitor change in the nature and impact of risks

Risk and Decision Making

- Why take risks?
- Difference between "taking risks" (implies lack of plan) and risk management
- Nature of business itself—serving customers *is* risk management
- What will be gained if you control risk–opportunities
- What could be lost—if risk occurs, could jeopardize project success
- Chances of success and failure?

Risk Elements

- Frequency of loss
- Information available
- Severity of loss
- Manageability of risk
- Potential for publicity
- Measure consequences
- Source of finances

Rewards of Taking Risks

- Achieving project success with minimum cost and time
- Cost of risk management increases the later it is undertaken
- Advance state of the art—make real contribution to field
- Enhanced profitability—competitive edge
- Improve market position—market share
- Ensuring customer satisfaction—customer shares risk in financing project

Potential Project Risk Factors

- Lack of top management commitment
- Failure to get user commitment
- Misunderstanding requirements
- Inadequate user involvement
- Changes in scope
- Lack of personal and professional skills
- New technology

- Staffing, conflicts
- Inadequate processes and procedures

PMI PMBOK: Project Management Body of Knowledge

- PMBOK is a standard for approaching project risk management
- Process approach: describes project risk management as "inputs/tools/outputs"
- Course organization is consistent with PMBOK
- Note that PMBOK was updated in 2001 to include a new risk management process—risk planning
- Following are more details on each process

PMBOK Section 11: Risk Management Processes

1. Risk management planning
2. Risk identification
3. Qualitative risk analysis
4. Quantitative risk analysis
5. Risk response planning
6. Risk monitoring and control

Risk management planning

Inputs

- Decide how to approach and plan the risk management activities for a project
- Project charter
- Organization's risk management policies
- Defined roles and responsibilities
- Stakeholder risk tolerances
- Template for risk management plan

Tools

- Planning meetings
- Agendas
- Review data
- Variances
- Special reports

Outputs

- Risk management plan
- Methodology
- Roles
- Budget
- Timing
- Scoring
- Thresholds
- Reporting and tracking

Risk identification

Inputs

- Risk management plan
- Project planning outputs
- Risk categories
- Historical information

Tools

- Documentation reviews
- Information gathering techniques
- Checklists
- Assumptions analysis
- Diagramming techniques

Outputs

- Risks
- Triggers
- Inputs to other processes

Qualitative risk analysis

Inputs

- Risk management plan
- Identified risks
- Project status

- Project type
- Data precision
- Scales of probability and impact
- Assumptions

Tools

- Risk probability and impact
- Probability/impact risk rating matrix
- Project assumptions testing
- Data precision ranking

Outputs

- Overall risk ranking for project
- List of prioritized risks
- List of risks for additional analysis and management
- Trends in qualitative risk analysis results

Quantitative risk analysis

Inputs

- Risk management plan
- Identified risks
- List of prioritized risks
- List risks for additional analysis and management
- Historical information
- Expert judgment
- Other planning outputs

Tools

- Interviewing
- Sensitivity analysis
- Decision tree analysis
- Simulation

Outputs

- Prioritized list of quantified risks
- Probabilistic analysis of project

- Probability of achieving the cost and time objectives
- Trends in quantitative risk analysis results

Risk response planning

Inputs

- Risk management plan
- List of prioritized risks
- Risk ranking of project
- Probabilistic analysis of project
- Probability of achieving the cost and time objectives
- List of potential responses
- Risk thresholds
- Risk owners
- Common risk causes
- Trends in qualitative and quantitative risk analysis results

Tools

- Avoidance
- Transference
- Mitigation
- Acceptance

Outputs

- Risk response plan
- Residual risks
- Secondary risks
- Contractual agreements
- Contingency reserve amounts needed
- Inputs to other processes
- Inputs to revised project plan

Risk monitoring and control

Inputs

- Risk management plan
- Risk response plan

- Project communication
- Additional risk identification and analysis
- Scope changes

Tools

- Project risk response audits
- Periodic project risk reviews
- Earned value analysis
- Technical performance measurement
- Additional risk response planning

Outputs

- Workaround plans
- Corrective action
- Project change requests
- Updates to the risk response plan
- Risk database
- Updates to risk identification checklists

Risk Planning in New Product Development Projects

- This process prepares for good project risk management
- Plans approach, support, and standards
- Establishes risk management policies
- Roles and responsibilities
- Provides template for risk management plan (used in the course project)
- Sets culture for risk management, defining it as integral to project management

Risk Planning

Inputs

- Decide how to approach and plan the risk management activities for a project
- Project charter
- Organization's risk management policies
- Defined roles and responsibilities

- Stakeholder risk tolerances
- Template for risk management plan

Tools

- Planning meetings
- Agendas
- Review data
- Variances
- Special reports

Outputs

- Risk management plan
- Methodology
- Roles
- Budget
- Timing
- Scoring
- Thresholds
- Reporting and tracking

Setting up company for risk management

- Policies and procedures set up risk management processes
- Company leadership establishes risk as integral part of planning and control
- Information databases
- Templates for risk planning documents
- Project histories and lessons learned

Issues in Risk Planning

- Scope of work risks
- Resource risks
- Quality risks
- Cost risks
- Time/schedule risks
- Technology risks
- Project information: what data is needed on risks

Scope risks

- Customer needs and expectations don't equate to requirements, leads to "scope creep"
- Scope does not adequately describe deliverable
- No process to manage scope creep
- No project process to trace from scope to requirements to deliverable

Resource risks

- Resources scheduled for work not the right ones; bad match of competence and task
- Resources not available when scheduled
- Resources leave company
- Key resource acts as bottleneck and cannot be managed
- Resources not trained adequately

Quality risks

- Quality processes not in place
- Quality assurance up front in process is not effective
- Quality control, inspections for conformance, etc. not effective
- Definitions of quality differ between customer and team

Cost risks

- Cost estimates inaccurate
- Unit cost information not up to date
- Cost controls not in place, leading to inaccurate cost capture systems
- Hidden costs not uncovered in project plan and budget
- Cost variance indicates lag in capturing and registering costs in earned value analysis

Time/schedule risks

- Schedule structure not adequate, leading to unscheduled tasks and costs
- Schedule durations wrong
- Schedule linkages not accurate
- Customer schedule and timing requirement for deliverable is not feasible
- No schedule review and update process

Technology risks

- Project equipment does not perform as planned
- Unproven tooling or project techniques
- Key contractor technology not available; no in-house competence
- Hidden technology issues do not surface early enough to address and respond

Risk information

- Information on past risks in similar projects not available
- Risk information templates, e.g., risk matrix formats, not available
- Lessons learned not captured
- Spreadsheet formulas for calculating probabilities not available

Risk Intensity in Project Life Cycle Phases

- Phase 1 (Concept) Low amount at stake; opportunity to discover and manage risk before impacts
- Phase 2 (Development): Higher amount at risk; impacts begin to occur
- Phase 3 (Implementation): Highest risk impacts; highest amount at stake; sunk costs
- Phase 4 (Termination): Too late!

Integrated Risk-Based Scheduling in New Product Development

- After risks are identified, categorized, and assessed, begin identifying optional scenarios
- Focuses on three scenarios—expected, pessimistic, and optimistic
- This week's focus is on impacts of three scenarios on schedule
- Microsoft Project (PERT tool) helps document options, assumptions, and calculates probable durations for tasks

Choose Risks for Three-Scenario Analysis

- Risk assessment has helped you identify and rank tasks with the highest risks considering impact, severity, and probability
- Choose the five highest-risk tasks for analysis
- Generate three scenarios for these tasks and review schedule impacts

Generating scenarios

- Generating scenarios involves thinking through the extent of the risk for each task
- For each task, identify the impact of the task risk on task duration
- Tailor the duration to the anticipated impact on schedule of that scenario
- Example: a two-week (expected) software review becomes five weeks in the pessimistic scenario

Scenario 1: Expected. Expected scenario is the option that, given all the risks and issues inherent in that task, will likely occur

- Expected scenario is generated by consensus, drawing on team members and assigned resource, to determine *normal* delays
- Expected scenario would be the baseline schedule under normal circumstances

Scenario 2: Pessimistic

- Pessimistic scenario is the task duration that results from the worst case
- Worst case implies that all risks inherent in the task *all go wrong*
- Worst case (pessimistic) implies Murphy's Law—if something *can* go wrong, it *will go* wrong

Scenario 3: Optimistic

- Optimistic scenario results when all risks are controlled—everything goes right
- All task risks are managed effectively so that there is no delay in any task
- No unanticipated risk impacts
- Can imply some tasks are finished early, allowing some float

Microsoft Project PERT Tool

- See PERT analysis toolbar, provides buttons to perform PERT analysis
- PERT Entry box allows entry of duration for three scenarios
- PERT Weights box allows entry of weight to be given to that scenario—judge weight on best estimate of *severity* of impact
- Pert Entry Sheet schedule calculates and presents the three schedules that result from entries—in one Gantt chart

Pert is "what if" analysis

- PERT analysis gives project manager a way of establishing the outer bounds of "what if" scenarios
- Working with team, project manager identifies highest risk tasks and brainstorms possible "what if" discussions

- Task managers participate to ensure that their best estimates of outcomes is reflected
- Process itself helps task managers plan for the unexpected

New Product Project Risk Assessment: Two Approaches

- Qualitative Approach
  - Describing and defining risks
  - Rank ordering risks
  - Prioritize
- Quantitative Approach
  - Probabilities
  - Sensitivity
  - Decision tree

Risk Assessment Inputs

Qualitative

- Risk management plan
- Identified risks
- Project status
- Project type
- Data precision
- Scales of probability and impact
- Assumptions

Risk Assessment Tools

Qualitative

- Risk probability and impact
- Probability/impact risk rating matrix
- Project assumptions testing
- Data precision ranking

Risk Assessment Outputs

Qualitative

- Overall risk ranking for project
- List of prioritized risks

- List of risks for additional analysis and management
- Trends in qualitative risk analysis results

Risk Assessment Inputs

Quantitative

- Risk management plan
- Identified risks
- List of prioritized risks
- List risks for additional analysis and management
- Historical information
- Expert judgment
- Other planning outputs

Risk Assessment Tools

Quantitative

- Interviewing
- Sensitivity analysis
- Decision tree analysis
- Simulation

Risk Assessment Outputs

Quantitative

- Prioritized list of quantified risks
- Probabilistic analysis of project
- Probability of achieving the cost and time objectives
- Trends in quantitative risk analysis results

Assessment Methodology

- Step 1: Select risk event
- Step 2: Assess probability
- Step 3: Assess consequences, severity, impact
- Step 4: Plan response and mitigation
- Step 5: Document

Risk Response/Monitoring/Communication in New Product Development Projects

- Risk response involves whole spectrum from analysis to action
- The key action to control risk based on results of risk assessment
- Implement risk management plan
- Develop strategies, options, and contingency plans
- Monitor risk through project reviews

Risk response: Inputs

- Risk management plan
- List of prioritized risks
- Risk ranking of project
- Probabilistic analysis of project
- Probability of achieving the cost and time objectives
- List of potential responses
- Risk thresholds
- Risk owners
- Common risk causes
- Trends in qualitative and quantitative risk analysis results

Risk response: Tools

- Avoidance—Eliminate risk by change in scope, process, or deliverable
- Transference—Shift consequence of risk to third party, or find way to share risk in contract relationship
- Mitigation—Take early action to reduce impact of risk
- Acceptance—No targeted action on risk except preparation to address—contingency

Risk response: Outputs

- Risk response plan
- Residual risks
- Secondary risks
- Contractual agreements
- Contingency reserve amounts needed
- Inputs to other processes
- Inputs to revised project plan

Elements of Risk Response (or Risk Management) Plan

- Identified risks
- Risk owners; who is accountable to address
- Results from risk assessments and analyses
- Residual risks remaining after risk plan is implemented
- Secondary risks that arise as a result of implementing risk response plan

Risk Monitoring and Control

Inputs

- Risk management plan
- Risk response plan
- Project communication
- Additional risk identification and analysis
- Scope changes

Tools

- Project risk response audits
- Periodic project risk reviews
- Earned value analysis
- Technical performance measurement
- Additional risk response planning

Outputs

- Workaround plans
- Corrective action
- Project change requests
- Updates to the risk response plan
- Risk database
- Updates to risk identification checklists

Risk Monitoring Indicators: Some Examples

- Computer purchase—contractor reluctant to sign up to "expected" schedule delivery dates
- Equipment performance—preliminary equipment testing results in failures; possible worst case

- Resource constraint—key resource not available when scheduled; past history not promising
- Cost—major variance between cost estimates
- Scope and requirements—repeated changes in scope; indications that scope and requirements are not clear

Risk Communication

- Communication of risks is key to enabling task managers and stakeholders/sponsors to anticipate and address risks, and to avoid surprises
- Identify stakeholders, customers, sponsors, clients
- Schedule regular communications and reports to each, based on risk monitoring
- Identify key decision points given actual performance
- Facilitate sharing of risks and responsibility for risk response

Procurement Planning and Contract Risk in New Product Development Projects

- Decision to "make or buy" has special risk implications for project success
- "Buy" (contract) decision allows use of contract vehicle to control risk in selected tasks
- Type of contract and nature of relationship with contractor determines how you will share risk with contractor
- Contacting allows more control based on incentives

Procurement Planning

Inputs

- Scope statement
- Product description
- Procurement resources
- Market conditions
- Other planning outputs
- Constraints
- Assumptions

Tools

- Make-or-buy analysis
- Expert judgment
- Contract type decision

Outputs

- Procurement management plan
- Statement of work

Contract Types and Risk Impacts

- **Fixed**—Places major risk responsibility on good cost estimates and assumes contractor produces
- **Cost plus fixed fee**—Covers eligible costs; contractor can charge risk mitigation costs
- **Unit**—Pays on unit basis, incentive to contractor to cut corners and perhaps take unwarranted risks
- **Target**—Negotiated price but provides relief for contractor if risks are costly
- **Time and material**—Combines cost and fixed types

Contract Issues

- Financial objectives
- Contractor involvement
- Contractor capacity to assess and schedule risk
- Client involvement
- Claims resolution
- Final cost forecast
- Payments for cost of risk events

Contractor Design Reviews

- Regular contractor reviews
- Address risk issues in review of designs, drawings, prototypes, etc.
- Require contractor to communicate risk issues as they surface
- Renegotiate contract if worst case, pessimistic risk scenario occurs

Bibliography

Advanced Product Quality Planning and Control Plan APQP, Ford, Chrysler, GM joint publications, Detroit, 1994. [Technical guide for product engineers, designers and developers.]

Advanced Project Portfolio Management and the PMO, Gerald Kendall and Steven Rollins, J. Ross Publishing, Boca Raton, Florida, 2003. [Great source on how to develop a new product portfolio and manage it using a Project Management Office.]

Creating an Environment for Successful Projects, Robert Graham and Randall Englund, Jossey-Bass, San Francisco, 1997. [Perhaps the best source on executive roles in project management; introduces the concept of project sponsor.]

Customer Driven Project Management: Building Quality into Project Proceses, 2d ed., Bruce T. Barkley and James Saylor, McGraw-Hill, New York, 2001. [Good source on quality, customer requirements, and SPC tools. Available in Chinese.]

Fast Cycle Time: How to Align Purpose, Strategy, and Structure for Speed, Christopher Myer, Free Press, New York, 1993. [Early treatment of new product development process improvement; set the stage for later work on new product development.]

Implementing Concurrent Project Management, Quenten T. Turtle, Prentice Hall, Englewood Cliffs, New Jersey, 1994. [Good source on integrated project teams and new product development in engineering.]

Integrated Project Management, Bruce T. Barkley, McGraw-Hill, New York, 2006. [Good source on PMBOK integrated project requirement standards.]

Intelligence Reframed: Multiple Intelligences for the 21st Century, Howard Gardner, Basic Books, New York 1999. [Excellent treatment of various intelligences including creativity and leadership intelligence.]

Leading Change, John Kotter, Harvard Business School Press, Boston 1996. [Excellent source on building coalitions in an organization going through change and providing change leadership.]

Managing Performance Improvement Projects, Jim Fuller, Pfeiffer, San Francisco 1997. [A good source for "soft" projects focused on organizational improvement.]

Marketing Management 12th ed., Philip Kotler and Kevin Keller, Pearson, Prentice Hall, Upper Saddle River, New Jersey, 2006. [Good market launch text]

Maximizing Project Value, Jeff Berman, AMACOM, New York, 2007. [Good reference on seeing costs and benefits of projects and business case for projects.]

New Generation Product Development, Michael McGrath, McGraw-Hill, New York, 2004. [Best source on resources and resource management in new product development.]

New Products Management, Merle Crawford and Anthony Di Benedetto, McGraw-Hill, New York, 2006. [Good text in the field for graduate programs in new product development and marketing.]

On Time Within Budget: Software Project Management Practices and Techniques, 3d ed., John Wiley and Sons, Inc., New York, 2000. [Excellent discussion of software development project issues and challenges.]

Portfolio Management for New Products, Robert Cooper, Scott Edgett, and Elko Kleinschmidt, Basic Books. New York 2001. [Excellent, original work on new product portfolio development using the Cooper approach.]

Process Innovation: Reengineering Work Through Information Technology, Thomas Davenport, Harvard Business School Press, Boston 1993. [Excellent treatment of process improvements and information technology issues.]

Production and Operations Management: Planning, Analysis, and Control, Richard Hopeman, Merrill Publishing, Merrill, Visconsin 1980. [Good basic text in production planning and control principles.]

Project Risk Management, Bruce T. Barkley, McGraw-Hill, New York, 2004. [Excellent source on project risk analysis, PMBOK standards, and cases.]

Project Management: Strategic Design and Implementation, David Cleland, McGraw-Hill, New York, 1999. [Excellent resource for project management process, with some information on senior management strategic roles and functions.]

The Business of Innovation, Roger Bean and Russell Radford, AMACOM, New York 2002. [Good source on the culture and behavior of innovation in organizations.]

The Structure of Scientific Revolutions, 3d ed., Thomas Kuhn, University of Chicago, 1996. [The seminal book on paradigms and how paradigmatic change occurs.]

The Wisdom of Teams, Jon Katzenbach and Douglas Smith, Harper Business, New York 1994. [The standard in the field on teams and teamwork.]

The World is Flat, Thomas L. Friedman, Farrar, Straus, and Giroux, New York, 2005. [Excellent treatment of internet impact on global economy.]

Winning at New Products, Robert Cooper, Basic Books, New York, 2001. [The seminal work in stage gate approach to new product development.]

Index